Keys to Behavior-Based Safety

from Safety Performance Solutions
featuring E. Scott Geller

Edited by
E. Scott Geller, Ph.D.
Joshua H. Williams, Ph.D.

ABS Consulting
Government Institutes
Rockville, MD

ABS Consulting

Government Institutes
4 Research Place, Rockville, Maryland 20850, USA
Phone: (301) 921-2300
Fax: (301) 921-0373
Email: giinfo@govinst.com
Internet: http://www.govinst.com

Published 2001.

04 03 02 01 5 4 3 2 1

Library of Congress Cataloging-in-Publication Data

Keys to behavior-based safety / from Safety Performance Solutions, featuring E. Scott Geller / E. Scott Geller and Joshua H. Williams editors.

p. cm.

Includes index.

ISBN: 978-0-86587-888-4

1. Industrial safety--Psychological aspects. 2. Industrial safety--Management. I. Geller, E. Scott, 1942- II. Safety Performance Solutions (Firm).

T55.3.B43 K49 2001
658.3'82--dc21

2001040502

Printed in the United States of America

Dedication

To Our Clients:

※ For all you do to bring these principles and procedures to life.

※ For going beyond the call of duty to improve the culture and safety of your colleagues.

※ For your sustained efforts to achieve a Total Safety Culture.

※ For spreading the *actively caring* philosophy throughout your organizations and communities.

Summary of Contents

Contents

List of Figures and Tables

Foreword

It was at one of the annual National Safety Congress meetings back in the 1980s that I first came across Dr. E. Scott Geller. All that separated us was a standing-room-only crowd of six- or seven-hundred safety pros at a session where Dr. Geller paced up and down aisles and across the speaker's platform. Actually, I couldn't even get into the room at first; the audience spilled out into the hallway.

So I listened from outside and soon realized I'd never heard anyone talk about safety like this guy. His voice boomed like a football coach as he told jokes and stories and ripped through a stack of overhead transparencies. The man was seriously pumped and passionate, and his vitality was contagious. Laughs rolled through the large room and echoed out into the hall, where a group of us stood on our toes peeking through the door, trying to figure out this wiry professor with the energy of a rock-and-roll drummer.

But this was not another motivational speaker. What separated Scott was his message, his substance. He's always had this knack for making ivory tower research somehow interesting. Scott's research is all about why we do the things we do—behaviors and attitudes. He never scares you with OSHA and rarely mentions the word "compliance," itself a novel concept in a field long dominated by regulations. What matters to Scott is the human side of safety. Why we take short cuts and ignore our training. How to get employees to "actively care" about one another on the job.

"I've got to get him to write articles for Industrial Safety & Hygiene News," I thought after listening for maybe five minutes. It didn't take long to decide, because Scott's approach to safety is so distinct—upbeat and natural, yet grounded in science. "Command-and-control rules and enforcement are turn-offs," he'd proclaim. "Stop using that word 'accident'—safety is within our control." "Enough with all the investigations and fault-finding. Say something positive when you see someone working safely."

Scott has written more than 100 articles for ISHN at this point. Many have focused on the topics you'll read about in the following pages: culture, systems, actively caring, self-management, observation and feedback, coaching, ergonomics, conversations, recognition, humor, performance measures, leadership, involvement, signs and slogans, myths and barriers, and critical success factors.

They say a preacher has maybe ten good sermons in him—after that it's all recycling and repackaging. With a microphone in his hand, Scott works a room with a missionary's zeal, but his mission is to always give his audience, and readers, something new. New research, insights and ideas to reach his goal of a "Total Safety Culture."

The fruits of Scott's drive to make a difference and the breadth of his knowledge are here for you to enjoy and put to use. Not only will your safety program benefit, but you'll have answers to the question that tugs at teachers, parents, friends, neighbors and safety pros everywhere: Why do people act like that?

Among the lessons I've learned from Scott over the years is this: The principles behind working safely—empathy, recognition, feedback and the like—are universal. They're the basis for relationships, leadership, teamwork, even communities and organizations. Scott gives us a map for getting along. Use it to find safety success, and much more!

Dave Johnson
Editor & Publisher
Industrial Safety & Hygiene News

About Dr. E. Scott Geller

E. Scott Geller, Ph.D. is a Senior Partner at Safety Performance Solutions, Inc.—a leading-edge organization specializing in behavior-based safety training and consulting. Dr. Geller and his partners at Safety Performance Solutions (SPS) have helped companies across the country and around the world address the human dynamics of occupational safety through flexible, research-founded principles and industry-proven tools. In addition, for more than three decades, Professor Geller has taught and conducted research as a faculty member in the Department of Psychology at Virginia Polytechnic Institute and State University, better known as Virginia Tech. In this capacity, he has authored more than 300 research articles and over 60 books or chapters addressing the development and evaluation of behavior-change interventions to improve quality of life.

He has written many books on occupational health and safety, including: *The Psychology of Safety*; *Working Safe*; *Understanding Behavior-Based Safety*; *Building Successful Safety Teams*; *Beyond Safety Accountability: How to Increase Personal Responsibility*; *The Psychology of Safety Handbook*; and the primer: *What Can Behavior-Based Safety Do For Me?* His latest book, entitled *The Participation Factor*, will be published next year by the American Society of Safety Engineers.

Dr. Geller is a Fellow of the American Psychological Association, the American Psychological Society, and the World Academy of Productivity and Quality Sciences. He is past Editor of the *Journal of Applied Behavior Analysis* (1989-1992), current Associate Editor of *Environment and Behavior* (since 1982), and current Consulting Editor for *Behavior and Social Issues*, *Behavior Analyst Digest*, and *Journal of Organizational Behavior Management*.

Scott Geller's caring, dedication, talent, and energy have helped him earn every university teaching award offered at Virginia Tech and, in

1982, a teaching award from the American Psychological Association. In 1983 he received the Virginia Tech Alumni Teaching Award and was elected to the Virginia Tech Academy of Teaching Excellence; in 1990 he was honored with the all-university Sporn Award for distinguished teaching of freshman-level courses; and in 1999 he was awarded the prestigious W. E. Wine Award for Teaching Excellence. In 2001, Virginia Tech awarded Dr. Geller the University Alumni Award for Excellence in Research.

Dr. Geller has written almost 100 articles for *Industrial Safety and Hygiene News*, a trade magazine disseminated to more than 75,000 companies. He has collaborated with Tel-A-Train Inc. and J. J. Keller and Associates, Inc. in the development of various training series, which include videotapes, workbooks, and facilitators' guides. For these innovative and creative works, beginning with a 1995 nationally renowned seminar series—Actively Caring for Safety—the Society for the Advancement of Behavior Analysis honored Dr. Geller with an award for Effective Presentation of Behavior Analysis in the Mass Media.

Dr. Geller and his colleagues at SPS have worked with numerous organizations, both public and private, to tailor training and implementation efforts related to the psychology of safety. Past customers include both small and large organizations representing manufacturing, chemical processing, transportation, and service industries, including: Allied Signal, American Standard, ARCO Chemical, AT&T, Bayer, Bechtel, BF Goodrich, BHP Cooper, Cargill, Coca-Cola, Champion International, Chevron, Corning, ExxonMobil, Eli Lilly, Florida Power & Light, Ford, General Motors, Georgia-Pacific, Hercules, Hewlett-Packard, Hoechst Celanese, Koch Refining, Leprino Foods, Lockheed, Lucent Technologies, Monsanto, Pacificorp, Power Bar, Rohm and Haas, Solutia, Textron, 3M, Toyota, TrusJoist MacMillan, Union Pacific Railroad, Wal-Mart, Westinghouse, Westvaco, Weyerhaeuser, and Wisconsin Electric.

Dr. Geller has been the Principal Investigator for more than 75 research grants that involved the application of behavioral science for the benefit of corporations, institutions, government agencies, or communities in general. Dr. Geller's research has been funded by both government agencies and corporations, including: the National Science Foundation; the National Institute for Occupational Safety and Health; the U.S. Department of Health, Education, and Welfare; the U.S. Department of

Energy; the U.S. Department of Transportation; the National Highway Traffic Safety Administration; the National Institute on Alcohol Abuse and Alcoholism; the Centers for Disease Control and Prevention; General Motors Research Laboratories; the Alcoholic Beverage Medical Research Foundation; and the Virginia Departments of Energy, Transportation, Litter Control, Agriculture and Commerce, and Welfare and Institutions.

About Safety Performance Solutions

Scott Geller Ph.D.	Sherry Perdue, Ph.D.	Mike Gilmore, M.S.	Steve Roberts, Ph.D.	Anne French, Ph.D.	Josh Williams, Ph.D.	Chuck Pettinger, Ph.D.

In the mid 1980s Dr. Geller and several colleagues began assisting organizations in the application of behavior-based and person-based psychological principles to industrial safety problems. In 1990, they teamed with Virginia Tech's Center for Organizational Performance Improvement located in the Department of Industrial and Systems Engineering. In 1995, the group collectively formed Safety Performance Solutions, Inc., a company dedicated to helping organizations understand and manage human dynamics to reduce workplace injuries.

Our team includes professionals from the fields of organizational behavior management, industrial/organizational psychology, industrial engineering/ ergonomics, and organizational design and development. Working with senior management, union leadership, and hourly work teams, we have significant experience designing, delivering, and evaluating safety improvement activities in hundreds of organizations.

Our mission is to help organizations cultivate a Total Safety Culture (TSC). In a TSC, everyone not only feels self-accountable for their own safety, but they also feel responsible for their peers' safety. When people act on that responsibility, individuals in a TSC have the necessary tools and methods as well as the appropriate person states—self-esteem, empowerment, and a sense of belonging—to actively care for the safety of co-workers. The organization's formal management systems and leaders' personal interactions contribute to developing an injury-free workplace by encouraging, recognizing, and reinforcing safe behaviors as well as removing barriers to achieving a TSC.

To learn more about Safety Performance Solutions, visit our website at www.safetyperformance.com.

Acknowledgments

I am honored to co-edit this book with E. Scott Geller. This book represents a large body of work from Dr. Geller and other members of Safety Performance Solutions (SPS). We hope this book will inspire and educate safety professionals to build more positive, healthy, and effective safety cultures in their organizations.

For my part, I would like to thank SPS for providing such a warm and supportive work environment. From my early days with Mike and Sherry in California, the SPS Partners have been patient and helpful in allowing me to develop, mature, and expand my skills. Co-editing this book with Scott Geller is an example of such an opportunity.

SPS is a small group of hard working (as in "road weary"), talented, and caring individuals—"good people" as my grandfather used to say. Our entire office is made up of good people and I am lucky to be a part of it. I hope to be co-editing the "Nth" edition of this book 10 years from now. So I'd like to say "Thank You" to Mike, Sherry, Steve, Bix, Anne, Chuck, and Scott for your support and for making SPS such a nice place to work.

In particular, I want to show my appreciation to my mentor, Scott Geller. To me, Scott is a great friend, loyal advisor, and the exemplar of a Winner.

Scott was my doctoral advisor at Virginia Tech and we now work together as colleagues on the SPS team. In graduate school, Dr. Geller led the Center for Applied Behavior Systems (CABS; which is still going strong today) doing large-scale community projects ranging from unobtrusively videotaping drivers in high tech "smart cars," to attempting to reduce alcohol abuse at fraternity parties and developing and implementing behavior-based safety for mining operations (to name a few).

Our mission at CABS was to research and then apply behavior analytic principles to improve the health and safety of our communities. We funded all of our graduate students and many of our undergraduates through numerous public and private research grants. Not bad for a handful of graduate students running a Center, managing 60-75 undergraduate students and trying to cross the doctoral finish line. Scott's vision and passion fueled CABS. His ability to get the most out of his students has made CABS prosper.

Joining Scott's team allowed me to flourish as a student and avoid some of the trappings of the "traditional" graduate school experience—ivory tower, non-applied, statistics-oriented, fear-driven education (i.e., survival). There was pain and suffering at CABS too, but it was shared and worked through together, as a team—actually, an interdependent "family." This taught me that working hard together as a team for a mutual goal in a positive and supportive environment is infinitely superior and more enjoyable than going it alone. It's a shame more graduate programs (and companies) don't understand this.

With the help of Scott and SPS, I now get to do what I love best — teach. And I'm able to do it with Fortune 500 companies all over the world, sometimes in Spanish. It is a privilege, and it all started the day I walked into Scott's office in the summer of 1996 and asked, "How can I get involved in your applied psychology research?" That request led to a summer research "job" that changed my life forever, and I am extremely appreciative. Thank you Scott!

As to this book, the first chapter is titled "General Principles for Achieving a Total Safety Culture," and includes Dr. Geller's seminal articles, "The Psychology of Safety" and "The Social Dynamics of Safety." The second chapter, "Actively Caring and Self-Management," focuses on the "person" side of safety and provides guidelines for increasing helping behavior in ourselves and others. The third chapter, "Behavior-Based Observation and Feedback," deals with practical issues related to interpersonal observation and feedback, a tool and method to help achieve a Total Safety Culture.

Chapter 4, "Communication Skills for Safety," provides specific guidelines for improving the frequency, openness, and effectiveness of safety discussions. The fifth chapter, "Measuring Safety Performance," is dedicated to defining proactive, upstream measures of safety performance. Chapter 6, "Safety Leadership," is designed to educate safety leaders about key psychological principles to help people feel more self-accountable or personally responsible for their safety and the safety of others.

Chapter 7 is entitled "Maximizing Behavior-Based Safety Efforts" and provides principles and procedures for sustaining a behavior-based safety effort, as well as pitfalls to avoid. Finally, Dr. Geller closes the book with a discussion of his recent concerns regarding the future of behavior-based safety. He ends by giving all readers an important assignment.

I hope you enjoy this book!

Joshua H. Williams, Ph.D.
July, 2001.

Keys to Behavior-Based Safety

I

General Principles for Achieving a Total Safety Culture

Achieving safety excellence requires going beyond the traditional safety focus of engineering and regulation. Because human behavior is a contributing cause to most incidents and injuries, safety excellence can only be achieved by addressing the human dimensions of safety. Reducing at-risk behavior and increasing safe behavior requires understanding how such behaviors are motivated by system and cultural influences.

Developing and sustaining a supportive safety culture hinges on understanding and applying the principles of behavioral science and person-based psychology to build tools and methods that encourage personal responsibility and interpersonal interaction about safety. We call this "actively caring," and it is an integral part of a Total Safety Culture. Each of the papers in this book explains an important aspect of the human dynamics involved in an organization's safety performance. Plus, practical guidelines are offered for addressing the psychology of safety and thereby improving safety performance. In other words, if the guidelines presented in this book are followed, workplace injuries and property damage incidents will surely decrease.

This first chapter defines the human dynamics of safety by describing those principles from psychology that are most directly related to occupational safety and health. The first paper reviews 50 principles of research-based psychology that serve as the foundation for the practical strategies described throughout the rest of this book. The role of social psychology in workplace safety is elucidated in the second paper, along with techniques for applying social influence principles to the human aspects of safety improvement. The competence theory explained in the third paper is relevant for understanding lack of participation in safety-related activities and for increasing such participation. The last paper in this chapter describes exemplary safety participation through two industry-based case studies.

The Psychology of Safety[1]

E. Scott Geller

This paper presents an overview of the psychology of safety by reviewing 50 principles relevant for understanding the human dynamics of safety. When people understand and believe these principles and use them to guide the design, execution, evaluation, and continuous improvement of interventions to change safety-related behaviors and attitudes, they are well on their way to achieving a Total Safety Culture. The principles are based on rigorous research in behavioral science, not on common sense. Interested readers can learn more about the theoretical rationale, research support, and procedural ramifications of each principle from the author's recent book *The Psychology of Safety Handbook* (Geller 2001).

At each of the Deming workshops I attended, one or more persons asked a practical question like "How should we translate these concepts into real-world application?" or "Would you please put your theory into procedures or practices we could follow in our plant?" Seemingly disappointed with such questions, Dr. Deming (1991, 1992) asserted that the purpose of the seminar was not to teach specific procedures. Rather, the seminar participants should have expected to learn theory and principles. The participants must return to their own organizations and develop specific methods and procedures that fit their culture. Dr. Deming stressed the importance of starting with theory and then customizing practices for a particular system and culture.

Similarly, there is no one-size-fits-all approach to managing the human dynamics of safety. Yet some safety consultants would like us to think they have such a thing. Consulting agencies, professing to have a quick-fix solution, send videotapes and workbooks to organizations for program implementation. Many companies have been willing to pay the high price for such outside solutions to their safety problems. The approach certainly sounds good. The consulting firm has passed a network check among a few sister plants, and the lean and mean cutbacks prevent the use of company personnel.

This "outside" approach often works over the short term. In fact, regardless of the consulting firm or procedures implemented, improvements in safety may be realized just because the new effort draws people's attention to safety. Management is spending money to promote safety, so working safely and not having injuries is especially important these days. The worse a company's safety record, the easier it is to show notable improvement. But, this "flavor-of-the-month" approach can only improve the worst organizations for a limited period. For organizations with a good safety record, this approach will not work even for the short term.

As Dr. Deming well knew, specific procedures for long-term improvement need to fit the culture. Outside consultants can be invaluable as teachers of the appropriate principles and as facilitators of customization and implementation.

If a majority of the people, however, do not understand and believe in the principles to begin with, well-intentioned efforts will have limited and short-lived impact.

Given the need to start with the right theory, this paper reviews 50 important principles relevant to understanding the human dynamics of safety. This list is clearly not exhaustive, but it serves to summarize the psychology of safety. Each principle can serve as a guideline for developing, changing, or refining a safety-related process. When most of a workforce understands and buys into these principles, procedures can be developed for improving safety over the long term.

The 50 principles are not explained in great detail; complete information is available elsewhere (Geller 2001). Some principles will prompt the design and implementation of programs or processes. Other principles will enable a better understanding of failures to reach safety standards. Most principles can be used as guidelines for checking the potential long-term benefit of a specific safety-improvement process. All principles in this list are relevant for appreciating the complex human dynamics of safety and health promotion.

The order of this list does not reflect priority. I have attempted to combine the principles in meaningful categories and then to provide series within and between categories that allow for coherent transition. Readers with some background in psychology or human relations can use this list as a guide for teaching others the psychology of safety and helping to promote safety excellence. Teach these basic principles to others and you will facilitate the achievement of a Total Safety Culture.

Principle 1: Safety should be driven by the local culture and not by an outside agency.

It is common to hear employees at all levels of an organization talk about their safety rules and procedures as government mandates or regulations (for example, from OSHA, the Occupational Safety and Health Administration). Their language implies that the things they do for safety are more for an outside regulatory agency than for themselves. This translates into a perception of top-down control and motivation to perform in order to avoid failure (like a citation for noncompliance) rather than to achieve success (an environment in which fewer people are getting hurt).

Ownership, commitment, and proactive behaviors are more likely to occur among people working to achieve outcomes for themselves than among people working to avoid missing goals set by someone else. This statement is intuitive to anyone who compares his or her own motivation when working for personal gain versus someone else's gain and when working to earn a reward versus to avoid a penalty.

The language used to define programs and activities can influence attitudes toward personal participation in such efforts. Thus, it makes sense to talk about corporate safety as a company mission owned and achieved by the very people it benefits. A safety process is not accomplished to comply with government mandates or standards. It is achieved by the people and for the people whose safety and health it protects.

Principle 2: Culture change requires that people understand the principles and know what to do about them.

Although the terms "education" and "training" are used interchangeably, they do not refer to the same process. Education refers to teaching and learning the theory (principles) and rationale behind a certain set of procedures, whereas training occurs when people learn how to carry out the procedures of a particular process. Lectures, videotape presentations, demonstrations, and group discussions are useful for educating people. However, training requires an additional format for assuring the appropriate practices or behaviors involved in a procedure are learned. Role playing or one-to-one interaction is needed so participants can receive specific feedback about their behaviors relevant to executing the procedure or process appropriately. Obviously, long-term culture change requires both education and training for all participants.

Principle 3: Champions of a Total Safety Culture will emanate from those who teach the principles and procedures.

When people teach, they develop optimal understanding, commitment, and ownership for the principles and procedures they teach. They "walk the talk" and become leaders of change relevant to what they teach. In other words, teaching leads to the development of internal mental scripts that support the material taught, and this translates to behaviors that set the right example.

For more than 30 years, I have assisted organizations in the education, training, and implementation of principles and procedures from the field of psychology. Although such efforts never made a company's safety record worse, the quantity and quality of benefits from using these principles and procedures have varied dramatically across cultures. One factor stands out as a critical determinant of success—the presence of champions to lead the effort. I have seen no better way to develop champions of a campaign than to teach them relevant theory and method, teach them how to teach the information to others, and then allow them opportunities to teach their colleagues and co-workers.

Principle 4: Leadership can be developed by teaching and demonstrating the characteristics of effective leaders.

Becoming the champion of a cause (perhaps as the result of teaching) does not guarantee the person will lead the change effort effectively. In other words, knowing what needs to be done to advance safety and being motivated to get it done does not necessarily result in successful leadership. Leaders have certain characteristics, including enthusiasm, flexibility, confidence, integrity, expertise, motivation, and critical thinking skills

(Baron 1995; Yukl 1989). These attributes can be learned and cultivated in others.

Thus, it is important to conduct both education and training programs on effective leadership. People need to understand the principles behind good leadership (education) and know what specific behaviors reflect good leadership qualities (training). Also, it is important to look for effective leadership behaviors in others. Noting leadership skills can provide personal observational learning and possibly provide an opportunity to offer rewarding feedback to support exemplary behavior.

Principle 5: Focus recognition, education, and training on people reluctant but willing rather than on those resisting.

People resist change for many reasons. Some feel insecure leaving their comfort zones, some mistrust any change in policy or practice that was not their idea, and others resist change for the special attention they get for resisting. Attempting to persuade (or force) change among these individuals is usually a waste of time. In fact, the more defensive reactions you elicit from an individual (as a result of your attempts to get them to join your effort), the more difficult it will be for that person to admit they were wrong and accept your position. People often resist beneficial change to "save face" or avoid admitting they were wrong. So, the more you force a person to defend a position you would like them to relinquish, the tougher it will be for that person to give up their position.

Given the difficulty in changing resisters, it makes much more sense to attend to those who "came to the dance." Those who believe in the change effort and are willing to get involved need to learn leadership skills. These people need opportunities to work with those who are reluctant to get involved but are willing to consider a change proposal. Enabling such peer instruction can cultivate change champions (Principle 3), as well as increase participation. When a critical mass of individuals gets involved and demonstrates beneficial results from the change, many initial resisters will join in—not because they have ***got*** to but because they ***get*** to.

Principle 6: Increase commitment, ownership, and involvement by giving people opportunities for choice.

This principle evolves naturally from Principle 5. A basic reason for preferring the use of positive consequences over negative consequences to motivate behavior is that people feel more free (perceive more choice) when working to achieve rewards than when working to avoid punishment (Skinner 1971). And, as illustrated in numerous laboratory experiments and field applications, increasing perceptions of choice leads to more motivation and involvement in the process.

Personal choice also implies personal control, which enhances empowerment and willingness to actively care for the safety of others. It is important to realize, of course, that eliminating the perception of choice (by imposing, for example, a top-down mandate that adds restrictions, discomforts, or constraints to work behavior) can sap feelings of ownership, commitment, and empowerment, and can inhibit involvement (Byham 1988).

Principle 7: A Total Safety Culture requires continuous attention to factors in three domains: environment, behavior, and person.

"The Safety Triad" introduced by Geller (1989) reflects the need to consider factors in each of these domains with two areas (behavior and person) representing the psychology of safety. Factors in these domains are dynamic and interactive; changes in one domain eventually have an impact on the other two. For example, behaviors that reduce the probability of injury often involve environmental change and lead to attitudes consistent with the safe behaviors, especially if the behaviors are viewed as voluntary. When people choose to act safely, they act themselves into safe thinking; such behaviors often result in some environmental change. The environment includes the general atmosphere or ambiance regarding safety management (for example, safety can be considered a top-down condition of employment or a bottom-up opportunity to get involved), and this certainly influences people's attitudes and behaviors toward occupational safety.

Principle 8: Do not count on common sense for safety improvement.

Most common sense is not common but is biased by an individual's subjective interpretation of unique experiences. As a researcher of psychological principles for more than 35 years, I have become quite committed to this basic principle. Indeed, I have dedicated most of my career to discovering principles of human behavior through systematic application of the scientific method. Thus, I get quite disturbed when I read or hear pop psychology based on unfounded intuition or "common sense."

Many statements I have read or heard related to the psychology of safety sound good but are incorrect. Profound knowledge comes from rigorous research and theory development and often runs counter to people's common sense that is derived from everyday experience. Dr. Deming (1991, 1992) supported this principle by frequently stating at his quality enhancement seminars, "There's no substitute for knowledge," and "Experience teaches us nothing; that's why American business is in such a mess."

Principle 9: Safety incentive programs should focus on the process rather than outcomes.

One of the most frequent common-sense mistakes in safety management has been the implementation of outcome-based incentive programs. Offering people rewards for not having an injury seems reasonable and logical, and such an incentive program is certainly easy to administer. In many instances, this kind of program has worked to bring down the numbers of minor injuries—those that can be covered up and not reported. Research-supported principles from behavioral science lead to the realization that safety incentives need to focus on process activities (or safety-related behaviors).

Outcome-based incentive programs often lead to a detrimental attitude of apathy or helplessness regarding safety achievement. In other words, employees develop the perspective that they cannot really control their injury record but must cheat or beat the system to celebrate the "achievement" of an injury reduction goal. I once heard of an incident whereby an employee was disqualified from an outcome-based lottery after he was burned on the leg by a welding torch. The employee whose unsafe use of the torch had caused the burn remained eligible for the raffle prizes.

Principle 10: Safety should not be considered a priority but a value with no compromise.

The ultimate vision should be for safety to become a value linked to every priority of a particular work assignment or recreational activity. Our priorities often change according to the situation or circumstances, but values are constant. Values are deep-seated personal beliefs that never get compromised.

Obviously, making safety a value is a tough assignment—much easier said than done. This idea is certainly not achievable overnight, and there is no quick fix to make this happen. However, the principles reviewed here can serve as the foundation for developing procedures and practices that enable people to give safety the value status it deserves.

Principle 11: Safety is a continuous fight with human nature.

If you wear earplugs, safety glasses, and steel-toed shoes when mowing the lawn in your back yard, you undoubtedly hold safety as a value. Indeed, you have made such personal protective equipment (PPE) available for every member of your family and make sure such PPE is used appropriately whenever it is called for. In addition, you have offered this PPE to neighbors when seeing them in situations where certain PPE provides protection from possible injury.

I have known people who meet the behavioral criteria described above for holding safety as a value, but the numbers are few. Why? Because human nature (or natural motivating consequences) typically encourages at-risk behavior, making it difficult and awkward to impose inconvenient safety precautions on others. The soon, certain, positive, and natural consequences of comfort, convenience, and expediency for at-risk behavior are difficult to overcome. This is especially the case when the safe alternative is naturally followed by the soon, certain, and *negative* consequences of discomfort, inconvenience, and inefficiency. Thus, when you must compete with natural supportive consequences in order to teach, motivate, or change behavior, you are fighting human nature.

Principle 12: Behavior is learned from three basic procedures: classical conditioning, operant conditioning, and observational learning.

Through naturally-occurring events (or human nature) and through planned instructional activities, we learn what to do each day in various circumstances, and we develop attitudes and emotional reactions to people, events, and environmental stimuli. The mechanisms for such learning of voluntary and involuntary behavior and emotions are covered in introductory psychology courses (e.g., Baron 1995) and will not be repeated here. (See Geller 2001 for details relevant to safety.)

The critical aspect of this principle to understand and teach others is that we do what we do (and feel what we feel) because of what we have learned from experience (both planned and unplanned). In addition, basic learning principles can be applied to change what people do and feel. Often, however, the learning of new practices and attitudes must overcome a history of past learning and practice that developed habits difficult to break. Moreover, it is possible that natural contingencies and social support systems support the bad habit or negative attitude. So, learning new behaviors and attitudes often requires another fight with human nature.

Principle 13: People view behavior as correct and appropriate to the degree they see others doing it.

This principle follows directly from the observational learning component of Principle 12. However, I feel observational learning warrants a separate principle because it should serve as the rationale (or internal script) for going out of our way to substitute safe behavior for at-risk behavior. Since personal experience often convinces us that "it" is not going to happen to us, we need a powerful reason to perform safely when personal injury is improbable. This principle provides that rationale.

Consider that everyone (including children) who sees you performing safe or at-risk behavior either learns a new behavior from your demonstration or receives verification that a particular behavior is appropriate. Now consider the vast number of people who observe your behavior every day (whether safe or at risk). Thus, our large-scale and long-term influence as a social model (or focus of observational learning) gives us a special social responsibility to go out of our way for safety.

Principle 14: People will blindly follow authority, even when the mandate runs counter to good judgment and social responsibility.

This principle is termed obedience in social psychology (Milgram 1974) and is a potential barrier to safe work practices. The fact that people will often follow top-down rules without regard to potential risk is alarming. This principle puts special responsibility on managers and supervisors who give daily direction for work practices. These first-line leaders could signal (even subtly) the approval of at-risk behavior, perhaps to reach additional production goals. People could read unintentional cues to get the product out on time even if safety must be compromised. They are apt to follow even implicit demands from their supervisor to whom they readily delegate responsibility for injury that could result from at-risk behavior.

Principle 15: Social loafing can be prevented by increasing personal responsibility, individual accountability, group cohesion, and interdependence.

Understanding this principle, as researched by social psychologists (Latané, Williams, and Harkins 1979), is relevant to finding ways to increase group productivity and synergy. Giving up personal responsibility for safety to another person (as referred to in the review of Principle 14) could be due to the social mechanisms presumed to influence social loafing. It is possible to decrease blind compliance to rules that implicate at-risk behavior by manipulating factors found to decrease social loafing. Thus, workplace interventions and action plans need to be implemented with the aim of increasing employees' perception of personal responsibility and accountability, as well as their sense of group cohesion and interdependence. Increasing these perceptions will also facilitate the occurrence of actively caring behavior (see Principle 34 below).

Principle 16: On-the-job observation and interpersonal feedback is key to achieving a Total Safety Culture.

This principle is critical to the development of procedures for reducing injuries. It is reflected in a most effective methodology for reducing at-risk behaviors (Geller 1995; Krause 1995; McSween 1995). More specifically, critical behavior checklists are developed and the results of checklist observations are communicated in order to put this principle to work. This and the next five principles provide the foundation and basic direction for a methodology that can dramatically improve the safety of an organization.

Principle 17: Behavior-based safety is a continuous DO IT process with D = Define target behaviors, O = Observe target behaviors, I = Intervene to improve behaviors, and T = Test impact of intervention.

This four-step process enables continuous improvement through an objective behavior-focused approach (See Figure 1.1.1). First, people need to decide on critical target behaviors to observe. After baseline observations are taken, an intervention is developed and implemented. By continuing to observe the target behaviors, the impact of the intervention program can be objectively evaluated. This evaluation reflects the scientific method and allows for continuous improvement. The results might suggest a need to refine the intervention, carry out another intervention, or define another set of behaviors to work on. The next four principles provide guidance for designing behavior-change interventions.

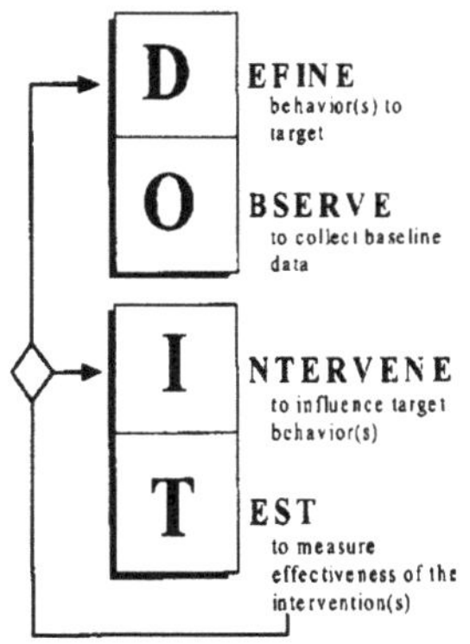

Figure 1.1.1 The "Do It" Process

Principle 18: Behavior is directed by activators and motivated by consequences.

At first, this principle might seem contrary to common sense, partly because pop psychologists have taught us that our intentions, attitudes, or "positive affirmations" motivate action. Pop psychologists have also told us that people cannot be motivated by others; we can only motivate ourselves. This leads us to believe that we behave in certain ways to fulfill a wish, intention, or goal preceding our behavior.

Behavioral research has shown that events (external or internal) antecedent to an event (referred to as activators) only motivate behavior to the extent that they signal or specify consequences (Skinner 1953, 1969). Thus, our intentions and goals can motivate behavior if they stipulate positive consequences to gain or negative consequences to avoid by performing a designated target behavior. Understanding this principle is critical to developing effective behavior-change techniques.

Principle 19: Intervention impact is influenced by amount of response information, participation, social support, and external consequences.

Interventions that give specific instructions (response information) and get participants actively involved are likely to influence behavior and attitude change. And, if the intervention facilitates support from others (such as co-workers or family members), the intervention can have lasting effects.

The role of external consequences is a bit tricky. The larger the positive consequence achieved or negative consequence avoided by performing the target behavior, the greater the *immediate* intervention impact. However, if participants perform a behavior only for the consequence (as is often the case when the consequence is large), the behavior is likely to stop if the consequence is no longer available to support the behavior. On the other hand, if the consequence is small and only sufficient to get the response started, people will develop inner responsibility for their performance. They will add an internal consequence to the external consequence in order to justify the behavior. Then, when the external consequence is removed or when people are in situations where the external supports (or pressures) are not available, they will continue to perform the target behavior.

Pioneering research by Freedman (1965) demonstrated the need to limit external consequences to motivate behavior if we want people to develop internal motivation. Dr. Freedman used mild or severe threats to prevent seven- to nine-year-old boys from playing with an attractive toy (an expensive battery-controlled robot). In the Mild Threat condition, the boys were merely told "It is wrong to play with the robot." Alternatively, the boys in the Severe Threat condition were told "It is wrong to play with the robot. If you play with the robot, I'll be very angry and will have to do something about it." Four other toys were available for the boys to play with when the experimenter left the room. From a one-way mirror, the researchers observed that only one of 22 boys in each condition touched the robot.

About six weeks later, a young woman returned to the boys' school and took them out of class one at a time to perform in a different experiment. She made no reference to the earlier study but instructed each boy to take a drawing test. While she scored the text, she told the boy he could play with any toys in the room. The same five toys from the previous study, including the robot, were available. Of the boys from the Severe Threat condition, 17 (or 77 percent) played with the robot, whereas only 7 (33 percent) from the previous Mild Threat condition played with the robot. Presumably, more boys in the Mild Threat condition developed an internal rationale for avoiding the robot and as a result avoided this toy when the external pressure was not available.

Other researchers have followed up this study and demonstrated that people are more apt to develop internal motivation when the external consequences (rewards or threats) are relatively small and insufficient to completely justify the target behavior (Riess and Schlenker 1977). This phenomenon is referred to as the "less-leads-to-more effect" (Baron 1995) and is most likely to occur when people feel personally responsible for their choice of action and the resultant consequences (Cooper and Scher 1990; Goethals, Cooper, and Naficy 1979; Lepper and Green 1978). These findings support the next principle.

Principle 20: Extra and external consequences should not over-justify the target behavior.

For reasons discussed above, rewards should not provide complete justification for the desired behavior. Rewards that over-justify the target behavior can motivate people to perform safely for the wrong reasons. We do not want people complying with safety rules only to gain a reward or avoid a penalty. Under these circumstances they are not likely to develop internal justification for the safe behavior. For example, when severe threats are used to force compliance with safe work practices, employees are not inclined to develop internal motivation for the behavior. The external pressure is more than sufficient to justify compliance. As a result, the same safe behavior should not be expected when the individual is performing similar tasks without the severe threat. This is one reason people rarely wear proper protective equipment when they mow the lawn in their back yard.

Principle 21: People are motivated to maximize positive consequences (rewards) and minimize negative consequences (costs).

This principle also explains why people are not likely to follow safe operating procedures when performing in situations without external controls (or behavior-consequence contingencies). As reflected in Principle 11, the natural external consequences usually (if not always) support at-risk behavior over the safe alternative (which is usually more inconvenient, uncomfortable, or time consuming).

Of course, this principle relates to behaviors other than those related to safety. If helping another person is perceived as having more costs than benefits, actively caring behavior is probably unlikely. Given this principle, a prime strategy for increasing safety and actively caring behavior is to overcome costs (negative consequences) with benefits (positive consequences). Various kinds of consequences are defined by the next principle.

Principle 22: Consequences have three dichotomous characteristics: positive vs. negative, natural vs. extra, and internal vs. external.

The three dichotomous characteristics of consequences defined in this principle define eight different consequences. Understanding these characteristics can enable significant insights into the motivation behind observed behavior. Appreciating these various consequences can also suggest whether external intervention is called for to change behavior, and what kind of intervention to implement. It is possible the natural external consequences supporting ongoing at-risk behavior cannot be overcome with extra external consequences. In this case, long-term behavior change requires the modification of the natural consequences or the application of techniques to alter internal consequences (details are given in Geller 2001).

Principle 23: Negative consequences have four undesirable side effects: escape, aggression, apathy, and counter-control.

To appreciate this principle, you need only to reflect on how you felt the last time you received punishment (perhaps a parking fine, a speeding ticket, or reprimand from a supervisor). And, consider how you feel about an individual who frequently threatens negative consequences in an attempt to manage others. Some groups and entire organizational cultures are managed predominately with threats (disincentives) and negative consequences. Without adequate benefits (such as a salary) the participants would avoid these situations completely. As such, these situations often promote apathy and sometimes counter-control (including sabotage tactics).

Consider the basic strategy used to manage traffic safety. Many people, including me, would not drive unless they had to. Driving is only a means to an end for many of us who feel controlled by the threat of negative consequences when we drive. Some of us are constantly on the lookout for a police car and look for ways to "beat the system" by purchasing radar detectors or following closely behind large speeding trucks (to escape radar detection). Could some of the violence on the highways these days (including the firing of guns into other vehicles) be due to the excessive use of negative consequences to control driving behavior?

Principle 24: Natural variation in behavior can lead to a belief that negative consequences have more impact than positive consequences.

Besides the fact that undesirable behavior sticks out, disrupts the flow, and seemingly demands immediate corrective attention, this principle explains the overuse of negative consequences. Behavior fluctuates from good to bad for many reasons within a system of influences. Usually person states and a variety of natural contingencies influence peak performance to get worse and poor performance to improve. When people intervene to recognize exceptional behavior or to correct undesirable behavior, they are apt to attribute natural variation to the success or failure of their intervention. They could mistakenly presume their recognition did not work (because behavior naturally regressed to the mean), or believe their reprimand worked (because of natural improvement in noticeably poor performance).

Principle 25: Long-term behavior change requires people to change "inside" as well as "outside."

The psychology of safety requires a consideration of both behavior factors (outside) and person factors (inside). For example, certain person states determine people's propensity to actively care for another person's safety and health as well as their own (See Principle 38). Various outside factors can be manipulated to influence person states, thereby increasing the probability of actively caring behavior.

This need to consider both "inside" and "outside" factors is also reflected in Principle 20. If the outside controls are too large, people will not feel obligated to change their inside (internal) justification for the performance. And, without inside influence such as giving oneself internal activators to direct behavior and internal consequences to motivate behavior, desired behavior will continue only as long as the external controls are in place. Thus, a Total Safety Culture requires an integration of both behavior-based and person-based psychology. The next several principles focus on inside factors (person-based) that influence the safety and health of a culture.

Principle 26: All perception is biased and dependent upon personal history, prejudices, motives, and expectations.

Appreciating this principle is key to understanding people and realizing the importance of actively listening to others before intervening for change. This principle also supports the need to depend upon objective, systematic observation of knowledge rather than common sense (Principle 8). Of course, our common sense improves with research-based knowledge. So, when people learn and believe in the right principles, they acquire knowledge that is, in fact, common and useful for intervention development, implementation, and evaluation.

It is also important to realize the reciprocal relationship between perception and behavior. Our perceptions influence our actions, and in turn our actions influence our perceptions. Thus, if we perceive risk, we will act to reduce that risk; if we act to reduce risk, we will become more aware of other risks to reduce. The current popularity of safety perception surveys in industry reflects increased awareness of perception as a determinant of the safety of an organization. Perception surveys can help pinpoint issues that need attention and activities that need an intervention. They can also be used to assess the person (or internal) factors influenced by a particular intervention program.

Principle 27: Perceived risk is lowered when a hazard is perceived as familiar, understood, controllable, and preventable.

When people perceive risk they adjust their behavior to reduce it. Hence, factors that increase risk perception also reduce at-risk behavior. The reverse is also true and explains why it is difficult to get workers to work defensively. Research has shown that hazards perceived as familiar, understood, controllable, and preventable are viewed as less risky (Sandman 1991; Slovic 1991). Thus, the risk of workplace hazards are underestimated to the extent these hazards are perceived as familiar, understood, controllable, and preventable.

Principle 28: The slogan "all injuries are preventable" is false and reduces perceived risk.

Given that "prevent" means keeping something from happening (*The American Heritage Dictionary* 1985), "preventable" implies that all of the factors that cause something to happen can be anticipated and controlled. It is difficult enough to anticipate and control all environmental and behavioral factors contributing to injuries, but controlling the person factors (or internal characteristics and states) of people is clearly impossible.

Frankly, I believe that telling people all injuries are preventable insults their intelligence. They know better. Such a slogan can make it embarrassing to report an injury and could influence a cover-up. An employee may believe that if employers think all injuries are preventable, then the employer might also think the employee was really stupid to have this injury; thus, the employee does not report the injury. But, the most critical problem with this popular slogan is that it can reduce the perception of risk, as discussed above for Principle 27. Hazards considered controllable and preventable are perceived as relatively risk free.

Principle 29: People compensate for increases in perceived safety by taking more risks.

Engineering devices that provide protection from possible injury (from machine guards and lockout devices to personal protective equipment) are certainly invaluable interventions for safety. They prevent numerous injuries and save countless lives daily. However, some of these interventions reduce perceived risk, and as a result, their prevention impact is not optimal. Researchers have shown that some people will compensate for a decrease in perceived risk by performing more risky behavior. In other words, some people increase their tolerance for risk and perform more at-risk behavior when feeling protected with a safety device (Wilde 1994). This is not the case for people who hold safety as a value (Principle 10).

Principle 30: When people evaluate others, they focus on internal factors; when evaluating personal performance, they focus on external factors.

This principle is termed "The Fundamental Attribution Error" and contributes to systematic bias whenever we attempt to evaluate others, from completing performance appraisals to conducting an injury investigation. Because we are quick to attribute internal (person-based) factors to other people's behavior, we tend to presume consistency in others because of permanent personality characteristics or personal dispositions.

To explain other people's injuries, we use expressions like "He's just careless," "She had the wrong attitude," and "They were not thinking like a team." This may be one reason safety professionals seem so impressed with the ineffective personnel selection approach to injury control. On the other hand, when we evaluate our own behavior we point to external factors first, thus realizing the dynamic influence of variables that can be measured and often changed to improve behavior and prevent injuries.

Principle 31: When succeeding, people over-attribute internal factors, but when failing, people over-attribute external factors.

This research-based principle is referred to as the "self-serving bias" (Miller and Ross 1975) and is sure to bias an inquiry regarding a person's injury. Placing the blame of a mistake on outside variables is just a basic defense tactic to protect one's self-esteem. In most organizations, even a minor injury is perceived as a failure rather than an opportunity to fact-find and learn how to prevent a more serious injury in the future. As a result, the victim of an injury is sure to avoid discussing inside, person-based factors contributing to the mishap.

Statements like "I was fatigued," "I didn't know the proper procedure," or "My mind was on other things" are far less probable than "The work demands were too severe," "The trainer didn't show me the correct procedure," or "The excessive noise and heat distracted me from my job." My advice is to accept the self-serving bias and allow people their ego-protecting excuses. Then search for measurable external factors (including behaviors) that can be changed to reduce the probability of another injury.

Principle 32: People feel more personal control when working to achieve than when working to avoid failure.

The sense of having personal control over life events is one of the most important person states contributing to our successes and failures. When we feel in control we are more motivated and work harder to succeed. We are also more likely to attribute less than optimal results to our personal efforts and therefore try harder to make improvements. Thus, the value of increasing people's sense of personal control over safety is obvious. One way to do this is to develop scoring procedures for safety achievements rather than focusing on the number of reported injuries as a measure of success. This requires an emphasis on measuring the process activities that can lead to loss control or injury prevention.

Principle 33: Stressors lead to positive stress or negative distress depending on appraisal of personal control.

Perceptions of personal control can determine whether stressors put us at increased risk for personal injury. When we believe we can do things to reduce our stressors (such as work demands, interpersonal conflict, boredom), we experience increased motivation (or stress) to take control. This person state is not nearly as detrimental to safety as distress. We feel distress when we believe there is little we can do about the stressors in our life. This state can lead to frustration, exhaustion, burnout, and at-risk behavior.

It is important to recognize states of distress in others and attempt to help them. After actively listening to another person's concerns, you might be able to offer constructive suggestions. Sometimes it is useful to help people distinguish between their domain of influence (the stressors they can control) and their domain of concern (the stressors they are concerned about but cannot influence). Helping people focus on the stressors they can reduce builds their sense of self-efficacy, personal control, and optimism. These are the person states that imply empowerment ("I can make a difference") and increase one's willingness to actively care (Geller 1994).

Principle 34: In a Total Safety Culture everyone goes beyond the call of duty for the safety of themselves and others—they actively care.

While behavior-based psychology provides methods and techniques to improve the human dynamics of safety, principles from person-based psychology need to be considered to ensure that the behavior-based tools are used. The ultimate purpose from such integration of behavior-based and person-based psychology is that everyone participates in efforts to achieve a Total Safety Culture. In other words, in the ideal culture everyone actively cares for the safety and health of others (Geller 1994, 1995). The next principle defines the purpose of actively caring more specifically.

Principle 35: Active caring should be planned, purposed, and focused on environment, person, or behavior.

This principle is key to achieving a Total Safety Culture. We need to plan ways to enable and facilitate as much actively caring behavior as possible, rather than expecting "random acts of kindness" (*Random acts of kindness* 1993). By considering the three domains of an actively caring focus, we can sometimes get more benefit from an effort to actively care. Including behavior-focused actively caring in an effort to help another person can often result in the most long-term gain. For example, it is often possible to include specific behavioral advice, direction, or motivation with crisis intervention and active listening (person-focused caring), and with a donation (environment-focused caring).

Principle 36: Direct, behavior-focused active caring is proactive and most challenging and requires effective communication skills.

The suggestion to add behavioral counseling to environment-focused and person-focused helping behavior is much easier said than done. It is relatively painless and effortless to contribute to a charity, send a get-well card, or actively listen to another person's problems. However, offering specific recommendations for behavior change can be confrontational and challenging, especially when it is direct. In other words, it is relatively easy to offer behavioral advice indirectly through another person or with written expression.

Thus, even though direct communication to an individual who needs behavioral advice can have the most beneficial impact, we are most apt to avoid this type of actively caring behavior. This is partly due to lack of confidence in our communication skills. Proper training and practice as a safety coach increases our perceived empowerment to actively care for safety in the most beneficial way.

Principle 37: Safety coaching that starts with Caring and involves Observing, Analyzing, and Communicating leads to Helping.

Each letter of COACH signifies a label for the sequence of events in an effective safety coaching process (Geller 1995). Safety coaching should start with an atmosphere of interpersonal **C**aring and an agreement that the coach can **O**bserve an individual's performance, preferably with a behavioral checklist. Then the coach **A**nalyzes the observations from a fact-finding, system-level perspective. Subsequently, the results are **C**ommunicated in one-to-one conversation with the sole purpose to **H**elp another individual reduce the possibility of personal injury. Effective safety coaching is clearly an optimal way to actively care and cultivate a Total Safety Culture. How to increase such actively caring behavior is addressed in the next several principles.

Principle 38: Actively caring behavior can be increased indirectly with procedures that enhance self-esteem, belonging, and empowerment.

Substantial research is available to support each component of this principle, but prior to my journal editorial (Geller 1991), no one had combined these components (or person states) into one actively caring model. Procedures that enhance a person's sense of self-esteem ("I am valuable"), belonging ("I belong to a team"), and empowerment ("I can make a difference") make it more likely that person will actively care for the safety or health of another person. Nourishing each of these person states in people leads to the belief that "we can make valuable differences."

Principle 39: Empowerment is facilitated with increases in self-efficacy, personal control, and optimism.

This principle was mentioned earlier when reviewing the distinction between stress and distress (Principle 33). When people's sense of self-efficacy (I can do it), personal control (I am in control), or optimism (I expect the best) is increased, they are less apt to experience distress and more likely to feel empowered (I can make a difference). In addition, this person state increases one's inclination to perform actively caring behaviors.

Notice that empowerment does not necessarily result from receiving more responsibility (as in "I empower you to do more"). In order to truly feel more responsible (empowered), people need to perceive they have the skills, resources, and opportunity to take on the added responsibility (self-efficacy), believe they have personal impact over their new domain of responsibility (personal control), and expect the best from their efforts to be more responsible (optimism).

Principle 40: When people feel empowered, their safe behavior spreads to other situations and behaviors.

In a Total Safety Culture, people go beyond the call of duty for safety. This means they perform a variety of safe behaviors in a variety of situations. More specifically, they show both stimulus generalization (performing a particular safe behavior in various settings) and response generalization (performing a variety of safe behaviors related to a particular target behavior). Both types of generalization naturally occur when safety becomes a value rather than a priority (Principle 10). Obviously we need to intervene in special ways to promote safety as a value. Our applied research has shown that facilitating empowerment is one special way to increase generalization and cultivate safety as a value (Ludwig and Geller 2000).

Principle 41: Actively caring behavior can be increased directly by educating people about factors contributing to bystander apathy.

This principle expresses the most basic procedure for directly promoting actively caring behavior. Specifically, research has shown that educating people regarding the barriers to helping behavior can reduce some obstacles and increase the probability of actively caring behavior (Beaman, Barnes, Klentz, and McQuirk 1978). Similarly, I have found that discussing the barriers to safe behavior can influence people to improve safety provided they also learn specific techniques for doing this. Actively caring behavior can be increased throughout a culture by first educating people about the factors holding them back from helping others and then teaching them how to actively care in specific ways.

Principle 42: As the number of observers of a crisis increases, the probability of helping decreases.

This principle, supported with substantial behavioral research (Latané and Nida 1981), is probably the first barrier to actively caring behavior that should be taught. It runs counter to common sense and means that people cannot assume someone else will intervene to help a person in a crisis or situation calling for proactive intervention.

It is human nature to distribute the responsibility of intervention to other observers, especially when the number of potential helpers is large. In fact, the most common excuse for non-occurrence of actively caring behavior is probably something like, "I thought someone else would do it," or "I didn't know it was my responsibility." This principle reflects the need to cultivate a norm that it is everyone's responsibility to actively care for safety, and one should never assume that someone else will correct an at-risk behavior or condition.

Principle 43: Actively caring behavior is facilitated when appreciated, but it is inhibited when unappreciated.

This principle combines the basic activator-behavior-consequence contingency of behavior change (Principle 18) with the need to increase actively caring behavior throughout our culture (Principle 34). When we realize the direct impact of our reactions to caring behavior, we become obligated to accept and appreciate the attempts of others to actively care, even when the helping behavior is done incorrectly.

Making an effort to actively care directly for someone else's safety is a big step for many people and deserves genuine recognition. Then, if advice is called for to make the caring behavior more effective, corrective feedback should be given appropriately. Be sure to make your deposits first. All actively caring behavior is well-intentioned, but it is frequently not practiced with the kind of feedback that shapes improvement. A negative reaction to caring behavior can be quite punishing and can severely discourage a person from trying again. Consequently, much of the future of actively caring behavior is in the hands of those who receive people's attempts to actively care.

Principle 44: A positive reaction to actively caring behavior can increase self-esteem, empowerment, and belonging.

This is a follow-up to Principle 43 and supports the need to sincerely support observations of actively caring behavior. While consequences to behavior have direct influence on how often that behavior will reoccur (Principle 43), one's reaction to caring behavior can have an indirect effect on an individual's person states. Although research in this domain is lacking, it is intuitive that feeling successful at actively caring should benefit the very person states that facilitated the actively caring behavior in the first place. Thus, we have the potential for a continuous mutually supporting cycle of actively caring influence, provided the reactions to the caring behaviors are positive. Positive and supportive reaction to people's actively caring behavior should enhance their self-esteem, empowerment, and belonging, and thereby indirectly increase the probability of more actively caring behavior from these individuals.

Principle 45: The universal norms of consistency and reciprocity motivate everyday behaviors, including actively caring behavior.

These social influence norms have a powerful impact on human behavior. Sometimes people apply these norms intentionally to influence others. At other times, these norms are activated without our awareness. Regardless of intention or awareness, behavior-change techniques derived from these norms can be very effective. The next three principles offer more direction in this regard.

Principle 46: Once people make a commitment, they encounter internal and external pressures to think and act consistently with their position.

This principle reflects the consistency norm (Principle 45) and reflects a basic premise of attitude and behavior change: you can act people into thinking differently or think people into acting differently. If people act in a certain way on the "outside," they will adjust their "inside" behaviors (including perceptions, beliefs, and attitudes) to be consistent with their actions. The reverse is also true, but I recommend targeting behavior first because behavior is easier to change on a large scale. We know much more about changing behavior than changing perceptions, beliefs, and attitudes because behavior is easier to measure objectively and reliably.

Principle 47: The consistency principle is responsible for the impact of "foot-in-the-door" and "low-balling."

The "foot-in-the-door" and "low-balling" techniques of social influence are successful because of the consistency norm (Principle 46). These influence techniques can be used to increase safety-related activities, including actively caring behavior. When an individual agrees with a relatively small request (for example, to serve on a safety committee), you have your foot in the door. To be consistent, the person is more likely to agree later with a larger request (for example, to give a safety presentation at a plant-wide safety and health meeting). Similarly, when people sign a petition or promise card that commits them to act in a certain way (for example, to actively care for the safety of others), they experience pressure from the consistency norm to follow through with actual caring behavior (Freedman and Fraser 1966).

The technique of "throwing a low ball" (or low-balling) occurs when a person is persuaded to make a particular decision (for example, to serve on the safety steering committee) because of the relatively low stakes associated with the decision (the monthly safety meetings are not too inconvenient or time consuming). Then, when the individual gets committed to the decision (after attending the first two safety meetings), the stakes are raised (more meetings are requested because of a special safety campaign). Because of the consistency principle (Cialdini, Cacioppo, Bassett, and Miller 1978), the individual will likely stick with the original decision (remain an active member of the safety committee).

Principle 48: The reciprocity norm is responsible for the impact of the door-in-the-face technique.

The reciprocity norm is a powerful determinant of human behavior (Cialdini 2001). Its influence is reflected in the popular expression "one good turn deserves another" and the well-known Golden Rule "Do unto others as you would have them do unto you." When you do someone a good turn, they typically feel obligated to return the favor. Likewise, when someone does us wrong, we usually feel a strong urge to get even. This norm should keep us on our toes to actively care for the safety of another person, because one good turn for safety is likely to motivate another actively caring behavior.

The success of the "door-in-the-face" technique depends on the reciprocity norm. If you receive a refusal (or a "door-in-the-face") for a certain request, the probability of receiving acceptance for a lesser request is increased. That is, if, after the refusal, you retreat to a request that is less costly, imposing, or inconvenient, your chances of receiving an acceptance is greater than if you started with the minor request (Cialdini et al. 1975). Your willingness to withdraw the larger request sets up an obligation for the other person to reciprocate and accept the smaller request.

Principle 49: Numbers from program evaluations should be meaningful to all program participants and should provide direction and motivation for intervention improvement.

The last two principles relate to the critical issue of program evaluation. Given that the primary purpose of evaluation is continuous improvement, this principle is basic and intuitive. However, program evaluators often forget this principle and transform their results to numbers that are either difficult to understand or do not relate readily to the ongoing process activities which influence program outcome. In other words, program evaluations frequently do not provide feedback which program participants can use to improve the process.

In safety, the total recordable injury rate (TRIR) is the most popular evaluation number used to rank companies for safety rewards. It is calculated by multiplying the number of workplace injuries by 200,000 and dividing the product by the total person-hours worked in that time period (U.S. Department of Labor 1994). This outcome calculation is an obvious example of an evaluation number that gives little direction or motivation for improving the process that will eventually affect the bottom line—injuries. The most direct measure of process comes from behavioral observations, and this is facilitated with a critical behavior checklist (Principle 16).

Principle 50: Statistical analysis often adds confusion and misunderstanding to evaluation results, thereby reducing social validity.

This principle follows directly from Principle 49 and suggests that statistical analysis may not be the most appropriate evaluation index if social validity (or real-world application) is desired. Of course, complex statistics are appropriate (and often necessary) for the research journals. But if the purpose of a program evaluation is to improve a safety process, we need to provide numbers that provide the most meaningful feedback to those who can improve the intervention program—the program participants.

Sandman (1991) and Slovic (1991) have shown that group statistics have minimal impact on risk perception. If your goal is to increase people's awareness of risk and their motivation to contribute to a safety campaign, the most influential evaluation result you can provide is a case study (which cannot be statistically analyzed). In other words, the most motivational evaluation usually comes from the personal report of a victim of an injury.

The work culture needs to support the reporting of personal injuries as well as the discussion of ways to prevent future incidents. When people are willing to give personal testimony, the presentation will be more useful than a statistical analysis in raising awareness and motivating safe behavior. We should probably spend less time calculating summary injury statistics and spend more time eliminating the barriers to the personal reporting and investigating of safety-related incidents (from "near-misses" and first-aid cases to lost-time injuries).

Summary

This paper reviewed principles of human dynamics relevant to improving organizational safety. The principles were founded on research published in scientific journals and enable profound understanding of the psychology of safety. If you use these principles as guidelines for developing, implementing, evaluating, and refining safety-improvement programs, you will make beneficial differences in the safety of your organization, community, or culture. In other words, safety interventions consistent with the relevant principles reviewed in this paper will prevent injuries.

Champions are needed to lead a safety-improvement process. Some champions for safety are already available in work settings; others will evolve when the principles reviewed here are taught. When potential champions are identified, they should be given opportunities to teach these principles and help develop interventions based on these principles. Such active participation will increase both belief in the principles and empowerment to apply the principles for the achievement of a Total Safety Culture.

There is no quick fix to culture change. The journey will not be without bumpy roads, forced detours, and missed turns. Use these principles as your map to reach an enviable destination, but be prepared to blaze new paths, traverse difficult terrain, and construct connecting bridges. And, please do not forget to take a break now and then to appreciate journey milestones with recognition of process behaviors that contribute to a successful journey.

At the end of the second Deming workshop I attended, a participant raised his hand to ask one final question. When acknowledged, he stood, walked to the nearest microphone, and stated, "Dr. Deming, you have taught us many important principles to consider when designing procedures to transform a culture. But frankly, the challenge seems overwhelming. Can we really expect to make a difference in our lifetime?"

W. Edwards Deming, at age 92, replied, "That's all you've got."

References

Baron, R. A. 1995. *Psychology.* 3rd ed. Boston, MA: Allyn and Bacon.

Beaman, A. L., P. J. Barnes, B. Klentz, and B. McQuirk. 1975. Increasing helping rates through information dissemination: Teaching pays. *Personality and Social Psychology Bulletin,* 4:406-411.

Byham, W. C. 1988. *Zapp! The lightening of empowerment.* New York: Fawcett Columbine.

Cialdini, R. B. 2001. *Influence: Science and practice.* 4th ed. New York: Harper Collins College Publishers.

Cialdini, R. B., J. T. Cacioppo, R. Bassett, and J. A. Miller. 1978. Low-ball procedures for producing compliance: Commitment then cost. *Journal of Applied Social Psychology*, 15:92–500.

Cialdini, R. B., J. E. Vincent, S. K. Lewis, J. Catalan, D. Wheeler, and B. L. Darby. 1975. Reciprocal concessions procedure for inducing compliance: The door-in-the-face technique. *Journal of Personality and Social Psychology*, 1:206–215.

Cooper, J., and S. J. Scher. 1990. Actions and attitudes: The role of responsibility and aversive consequences in persuasion. In *The psychology of persuasion*, eds. T. Brock and S. Shavitt. San Francisco, CA: Freedman.

Deming, W. E. May 1991. *Quality, productivity, and competitive position*. Four-day workshop presented in Cincinnati, Ohio, by Quality Enhancement Seminars, Inc.

———. January 1992. *Instituting Dr. Deming's methods for management of productivity and quality.* Two-day workshop presented in Washington, D.C., by Quality Enhancement Seminars, Inc.

Freedman, J. L. 1965. Long-term behavioral effects of cognitive dissonance. *Journal of Experimental Social Psychology*, 1:145–155.

Freedman, J. L., and S. C. Fraser. 1966. Compliance without pressure: The foot-in-the-door technique. *Journal of Personality and Social Psychology*, 4:195–203.

Geller, E. S. 1989. Managing occupational safety in the auto industry. *Journal of Organizational Behavior Management,* 10(1):181–185.

———. 1991. If only more would actively care. *Journal of Applied Behavior Analysis*, 24:601–612.

———. 1994. Ten principles for achieving a Total Safety Culture. *Professional Safety*, 40(7):16–22.

———. 1995. Safety coaching: Key to achieving a Total Safety Culture. *Professional Safety,* 39(9):18–24.

———. 2001. *The psychology of safety* handbook. Boca Raton, FL: CRC Press.

Goethals, G. R., J. Cooper, and A. Naficy. 1979. Role of foreseen, foreseeable, and unforeseeable behavioral consequences in the arousal of cognitive dissonance. *Journal of Personality and Social Psychology*, 37:1179–1185.

Krause, T. R. 1995. *Employee-driven systems for safe behavior: Integrating behavioral and statistical methodologies.* New York: Van Nostrand Reinhold.

Latané, B., and S. Nida. 1981. Ten years of research on group size and helping. *Psychological Bulletin,* 89:308–324.

Latané, B, K. Williams, and S. Harkins. 1979. Many heads make light the work: The causes and consequences of social loafing. *Journal of Personality and Social Psychology*. 37:823–832.

Lepper, M., and D. Green, eds. 1978. *The hidden cost of reward.* Hillsdale, NJ: Erlbaum.

Ludwig, T. D., and E. S. Geller. 2000. An organizational behavior management approach to safe driving intervention for pizza deliverers. *Journal of Organizational Behavior Management*, 19(4).

McSween, T. E. 1995. *The values-based safety process: Improving your safety culture with a behavioral approach.* New York: Van Nostrand Reinhold.

Milgram, S. 1974. *Obedience to authority.* New York: Harper Collins.

Miller, D. T., and M. Ross. 1975. Self-serving biases in the attribution of causality: Fact or fiction? *Psychological Bulletin*, 82:213–225.

Random acts of kindness. 1993. Emeryville, CA: Conari Press.

Riess, M., and B. R. Schlenker. 1977. Attitude changes and responsibility avoidance as modes of dilemma resolution in forced-compliance situations. *Journal of Personality and Social Psychology,* 35:21–30.

Sandman, P. M. 1991. *Risk = Hazard + Outrage: A formula for effective risk communication*. Videotaped presentation for the American

Hygiene Association. Environmental Communication Research Program, Rutgers University, New Brunswick, NJ.

Skinner, B. F. 1953. *Science and human behavior.* New York: Macmillan.

———. 1969. *Contingencies of reinforcement: A theoretical analysis.* New York: Appleton-Century-Crofts.

———. 1971. *Beyond freedom and dignity.* New York: Alfred A. Knopf.

Slovic, P. 1991. Beyond numbers: A broader perspective on risk perception and risk communication. In *Deceptable evidence: Science and values in risk management,* ed. D. G. Mayo and R. D. Hollander, 48–65. New York: Oxford University Press.

The American Heritage Dictionary. 1985. 2nd College ed. Boston, MA: Houghton Mifflin Company.

U.S. Department of Labor. September 1994. *Recordkeeping guidelines for occupational injuries and illnesses.* Washington, D.C.: Bureau of Labor Statistics.

Yukl, G. 1989. *Leadership in organizations.* 2nd ed. Englewood Cliffs, NJ: Prentice Hall.

Wilde, G. J. S. 1994. *Target Risk.* Toronto, Ontario, Canada: PDE Publications.

Social Dynamics of Safety[2]

E. Scott Geller

The gurus of total quality management have told us to think win/win, interdependency, teamwork, synergy, and systems (e.g., Deming 1986, 1993; Senge 1990). To optimize systems we need to foster synergy, and this is facilitated when win/win consequences are achieved through teamwork. Teamwork is promoted with shifts from independent to interdependent thinking and acting. This total quality management perspective needs to be applied to occupational health and safety. Research in social psychology offers insight in understanding some of the special challenges in developing a sense of interdependency among people. Such research also suggests ways to increase interdependency throughout a work culture.

This paper addresses the need to cultivate interdependency and teamwork by discussing certain social dynamics of industrial safety and health. I explain six social influence principles derived from social psychology research that affect our behavior and attitudes every day. Each principle implicates certain procedures we should use or avoid using in our quest for a Total Safety Culture—a culture in which people actively care on a regular basis for the safety and health of others.

From Dependency to Interdependency

Years ago workers typically relied on safety engineers to keep them safe. In fact, the traditional three Es of industrial safety (Engineering, Education, and Enforcement) reflect this dependency perspective. That is, the organization is expected to provide a safe environment (including equipment and protective devices) and then to educate all workers regarding the use of the engineering interventions. If people do not comply with the rules and regulations of safe operating procedures, enforcement techniques are applied to motivate compliance and support the slogan "safety is a condition of employment." This approach has dramatically reduced injury frequency and severity in the workplace, at home, and on the road.

Over the years safety professionals have learned that people cannot be completely dependent on their work organization to keep them safe. The engineers cannot make equipment fail-safe, and safety professionals cannot anticipate every potential injury when defining a safe operating procedure (SOP). Even when everyone complies with all SOPs, workplace injuries are possible. People cannot expect the engineers and safety professionals to think of everything. They need to look out for themselves.

In many organizations, workers are encouraged to develop a personal commitment to keep themselves free from injury. They are urged to take a self-management perspective toward industrial safety and health. Do not count on the company to keep you safe; accept personal responsibility for injury prevention. In these work cultures, workers are encouraged to be on continual lookout for environmental hazards, and to always practice safe work behaviors.

This perspective reflects an independent approach and has enabled organizations to improve their safety records beyond plateaus established with a dependency perspective. The safety records of many companies improved dramatically with the involvement of line workers in implementing safety programs (from near-miss reporting and incident investigations to the administration of environmental audits and incentive programs). However, many of these companies report another plateau in their safety performance. The independence perspective is clearly an improvement over dependency, but a Total Safety Culture requires an interdependent approach.

It is not enough for people to look out for their own safety and health. They need to contribute to the protection of others. This approach naturally implicates social dynamics—the theme of this paper. Certain principles of social influence help us understand social barriers that must be overcome for interdependency to become the norm. They also offer direction for the design of interventions to help cultivate interdependent actively caring behavior throughout an organization. The consistency principle, for example, not only explains many of our interpersonal behaviors and attitudes, but it can also be used to improve safety-related behaviors and attitudes.

We Try to Be Consistent in Thought and Deed

Many psychologists consider the consistency principle a weapon of influence lying deep within us and directing our everyday actions. It reflects our motivation to be (and to appear) consistent. Simply put, when we make a choice or take a stand, we encounter personal and social pressures to perform consistently with our commitment. We obtain this pressure to be consistent from three basic sources: 1) society values consistency within people, 2) consistent conduct is beneficial to daily existence, and 3) a consistent orientation allows for shortcuts in information processing and decision making (Cialdini 2001). Instead of considering all relevant information in a certain situation, people need only remember their commitment or decision and then to respond consistently. This principle explains people's resistance to change while also suggesting particular ways to motivate lasting improvement in both behavior and attitude.

Public and Voluntary Commitment

When people sign their name to a petition or pledge card, they are making a commitment to behave in a certain way. Later, they behave in this way to be consistent with their commitment. Safety professionals can use this variation of the consistency principle to increase various safety-related behaviors (Geller and Lehman 1991). After a discussion about a particular work procedure, for example, the audience could be asked to make a commitment to perform the desired behavior. What kind of commitment should be requested?

Commitments are most effective (or influential) when they are public, effortful, and perceived as voluntary or not coerced. Thus, it would be more beneficial to have employees make a public rather than private commitment to perform a certain safe behavior. And, it would be better to have them sign their name to a card or public declaration display than to merely raise their hands. In addition, it is very important for those pledging to believe they made the commitment to follow a certain work practice voluntarily. The reality might be that decisions to make a public commitment are dramatically influenced by external factors like peer pressure. Nevertheless, if people write an internal script that they made a personal choice, consistency is most likely to follow the commitment. Thus, the promoter of a commitment strategy needs to realize the influence of personal choice and make statements that allow participants to believe the commitment is not coerced but is their own choice.

Start Small and Build

Sometimes referred to as the "foot-in-the-door" technique (Freedman and Fraser 1966), this influence strategy also follows directly from the consistency principle. To be consistent, a person who follows a small request is likely to comply with a larger request later. Thus, after agreeing to serve on a safety steering committee, an individual is more willing to give a safety presentation at a plant-wide safety and health meeting. Research has found this commitment strategy to be successful in boosting product sales, blood donations, and monetary contributions to charities (Geller 2001).

The pledge-card commitment technique referred to above uses this principle. After people sign a pledge card that commits them to perform a certain behavior for a specified period of time (such as "Buckle vehicle safety belts for one month," or "Use particular personal protective equipment for two months"), they are more likely to actually do the safe behavior.

The "start small and build" technique only works to increase safe behaviors when people comply with the initial small request. In fact, if a person says "No" to the first request, this individual might find it even easier to refuse a subsequent, more important request. Thus, if the circumstances suggest a "No" to your request, you did not start small enough. In this case, you should be prepared to retreat to a less demanding request. This technique is described later under the reciprocity principle.

Raise the Stakes Later

This technique occurs when a person is persuaded to make a decision or commitment because of the relatively low stakes associated with the decision. For example, a person may decide to serve on the plant safety steering committee because the monthly meetings will not require too much time and effort. Then, when the individual gets committed to the decision (in the example, attends the first two safety meetings) the stakes are raised—more meetings are requested for a special safety effort. Because of the consistency principle, the individual in the example will likely stick with the original decision and remain an active member of the committee.

More than 20 years ago Robert Cialdini and colleagues (1978) first demonstrated the powerful influence of this technique when attempting to get college students to sign-up for an early 7:00 a.m. experiment on "thinking processes." During the solicitation phone calls, the 7:00 a.m. start time was mentioned up-front for half the subjects. Only 24 percent of these individuals agreed to participate. For the other subjects, the caller first asked if they wanted to participate in the study. Then, after 56 percent agreed, the caller raised the stakes and said the experiment started at 7:00 a.m. The caller gave subjects a chance to change their minds, but none did. Furthermore, 95 percent of these individuals actually showed up at the 7:00 a.m. appointment time. After making an initial commitment to participate, practically all of the subjects showed consistency and kept their commitment—in spite of the higher stakes.

This procedure is similar to the "foot-in-the-door" technique in that a larger request occurs after the target person agrees with a smaller request. A key difference, however, is that there is only one basic decision in this procedure, with the costs or stakes raised after initial commitment. This compliance tactic is common among car dealers. Once a customer has agreed to purchase a car at a special price, the price is raised for a number of reasons. The salesperson's boss might have refused to approve the deal, certain options may not have been included in the special price, or the dealership manager may have decreased the value of the customer's trade-in. Customers who have agreed to the special price will usually not change their minds following a price increase because reneging on a purchase decision may suggest a lack of consistency or indicate a failure to fulfill an obligation, even though the obligation is only imaginary. Often customers will develop a set of new reasons to justify their initial choice and the additional costs.

Don't Stifle Trust

The stakes-raising strategy raises a critical issue with regard to using certain techniques to increase safe behavior. Even though this influence strategy is used rather frequently to increase compliance, how would you feel about change agents (people who use the technique) if you knew they used the procedure intentionally to get more money, commitment, or safe behavior from you? For example, do you trust the waiter who brings you an expensive wine list only after you have been seated and made selections from the food menu? Your answer probably depends on whether you believe this sequence of events was done intentionally to get you to buy more.

Similarly, you might not dislike or mistrust the car salesman who adds cost to a vehicle's advertised purchase price unless you are suspicious that the price differential was fabricated deliberately to increase revenue. In other words, our trust, appreciation, or respect for people might decrease considerably if we believe they intentionally used a particular influence technique to trick or deceive us into modifying our attitude or behavior. Of course, there may be no harm done if the result is clearly for our own good (i.e., our health or safety) *and* we realize this.

Which First—Attitude or Behavior?

Because of the consistency principle, it does not matter whether attitude or behavior changes first. The issue is whether a technique is available to influence one or the other. The three influence techniques discussed under the consistency principle were introduced as targeting behavior. It could be argued, however, that internal (attitudinal) dimensions were intertwined throughout each technique. For maximum influence, for example, the pledge-card procedure requires the person to believe (internally) that the commitment was voluntary. Following successive compliance with escalating demands, internal commitment is developed until an "attitude" results. The stakes-raising technique depends on the target individual developing an internal justification for the initial decision, which then strengthens commitment and leads to more behavior consistent with a particular attitude. Consequently, the key lesson is that people attempt to keep their internal dimensions (attitude) and external actions (behavior) consistent. Thus, whether attitude or behavior is influenced first, if the person does not feel coerced, the other will likely follow.

We Reciprocate to Return the Favor

Some psychologists, sociologists, and anthropologists consider reciprocity a universal norm that motivates a good deal of interpersonal behavior. It can be used to increase people's involvement in a safety-improvement process and to cultivate interdependence. Simply put, the reciprocity principle is reflected in the slogan "Do for me, and I'll do for you." In other words, if you are nice to someone, they will feel obligated to return the favor. Furthermore, research has shown that the favor might be returned to someone other than the original source (Berkowitz and Daniels 1964).

I witnessed the reciprocity norm rather dramatically when I worked in the Virginia prison system in the mid-1970s. Several inmates used gifts and other forms of personal assistance to dominate other inmates. I knew an inmate who went to great lengths to place a gift in the cell of a new inmate he wanted to dominate. Accepting such a gift meant the target inmate (or victim) was now obligated to return a favor in order to "save face."

Have you ever felt uncomfortable after someone did a favor for you? I certainly have felt this way, and I interpret my discomfort as the reciprocity principle in action. Another person's favor makes me feel obligated to reciprocate. What does this mean for safety management? I think it means we should look for opportunities to go out of our way for another person's safety. When we actively care for someone else's safety, we increase the likelihood they will actively care for the safety of someone else.

It is important to realize that how we react to people after doing them a favor can either stifle or mobilize reciprocity. When a person thanks you for actively caring, you should not demean the favor by saying things like,

"No problem" or, "It was really nothing." Anything that makes the actively caring behavior seem insignificant or trivial will reduce the impetus for reciprocity. On the other hand, adding words to make the actively caring behavior appear more significant or meaningful can be quite awkward and create an uncomfortable or embarrassing situation.

To maintain a comfortable verbal exchange that does not belittle actively caring behavior and stifle reciprocity, you could react to someone's "Thank you for actively caring" with something like "Thank you for appreciating my effort to promote health and safety, but I know you'd do the same for me." This reaction shows admiration for the thank you and thus increases the likelihood more thanks will be given. Plus, it activates the reciprocity principle in a way that will be perceived as genuine and valid.

Gifts Aren't Free

Has someone ever influenced you to listen to a sales pitch after giving you a free gift? Have you ever felt obligated to contribute to a charity after receiving gummed individualized address labels and a stamped envelope for your check? Have you ever purchased a certain food in a supermarket after eating a free sample? Did you ever feel obliged to purchase a commodity after using it for a 10-day "free" trial period? Did you ever contribute to the Hare Krishna Society after a Krishna member gave you a flower at an airport? If you answer "yes" to any of these questions, it is likely that you have been influenced by the reciprocity principle. Many marketing or sales-promotion efforts count on this "free sample" approach to influence purchasing behavior.

In one experiment, Isen and Levin (1972) observed that 84 percent of those individuals who found a dime placed by researchers in the coin-return slot of a public phone helped an accomplice pick up papers he dropped in the subject's vicinity. In contrast, only four percent of those who did not find a dime helped the man pick up his papers. Similarly, students given a cookie while studying at a university library were more likely than those not given a cookie to agree to help another person by participating in a psychology experiment (Isen and Levin 1972).

Does this justify the distribution of free safety gifts such as pens, tee shirts, caps, cups, and other trinkets? Yes, to some extent, but the amount of reciprocity activated depends on the recipient's perceptions. How special is the gift? Was the gift given to a select group of people, or was it distributed to everyone? Does the gift or its delivery represent significant sacrifice in money, time, or effort? Can the gift be purchased elsewhere, or does the safety slogan on it make it special?

The bottom line here is that the more "special" the safety gift—as perceived by the recipient—the more reciprocity activated. Remember also, the way a safety gift is presented can make all the difference in the world. The labels and slogans linked with the gift can influence the amount and kind of reciprocity implicated. If the gift is to represent the actively caring behavior that safety leadership expected from a "special" group, a certain type of reciprocity is activated. People will tell themselves they are considered safety leaders and that they need to justify this label by actively caring for the safety of others.

Door-in-the-Face: Start Big and Retreat

Suppose the plant safety director pulls you aside and asks you to chair the safety steering committee for the next two years. Let's assume you perceive this request as outrageous, given your other commitments and the fact you have never even served on a safety steering committee before—let alone chaired one. You say, "Thanks for asking, but no." The safety director says he understands and then asks whether you would be willing to serve on the committee. According to social dynamics research, because the safety director "backed down" from his initial request, you will feel subtle pressure to make a similar concession—to reciprocate—and agree to the second, less demanding assignment.

Dr. Robert Cialdini and his associates (1975) were among the first to demonstrate the influence of this "door-in-the-face" technique. Posing as representatives of the "County Youth Counseling Program," they approached college students walking on campus and requested their volunteer service to chaperone a group of juvenile delinquents on a day trip to the zoo. When this was the first and only request, only 17 percent of those approached volunteered to help. However, three times more students volunteered to chaperone the day trip to the zoo when the researchers first asked whether the student would be willing to counsel juvenile delinquents for two hours a week for two-years. All subjects refused this much larger request, but then half of them agreed to serve as unpaid chaperones for a day at the zoo. Apparently the researchers' willingness to retreat from their larger initial request influenced several college students to reciprocate and comply with a smaller favor.

Have you ever wondered why lawyers ask for such outlandish amounts of money at the start of a civil trial? Or, why labor negotiators start with extreme demands? It is likely that these influence agents do not expect to receive their initial request. But, perhaps they have learned they are more apt to gain a second request after retreating from the first. They have learned the power of this reciprocity-based technique through real-world negotiation.

We Follow the Crowd

In research conducted by Solomon Asch and his associates in the mid-1950s, more than 33 percent of intelligent and well-intentioned college students were willing to publicly deny reality in order to be consistent with the obviously inaccurate judgments of their peers. Asch's classic studies of conformity (1951) involved six to nine individuals sitting around a table and verbalizing their judgment of which of three comparison lines was the same length as the standard. All but the last individual to voice a decision in sequence were research associates posing as subjects. In some trials, these confederates uniformly gave obviously incorrect judgments. The last decision in the sequence was from the real subject of the experiment. The correct answer was always obvious, and on several trials everyone gave the correct answer. On some judgment trials, however, an obviously incorrect comparison line was selected by each research associate according to plan, and the critical question was whether the subject would deny the obvious truth in order to conform with the group.

This and similar procedures were used in numerous social psychology experiments to study the social dynamics that influence amount of conformity. For example, a subject's willingness to deny reality in order to go along with the group consensus was facilitated by increasing group size (Asch 1955), apparent competence or status of group members (Crutchfield 1955), and anonymity (i.e., by having group members wear disguises). On the other hand, the presence of one dissenter (or nonconformist) in the group was enough to significantly decrease conformity by increasing a subject's willingness to choose the correct line even when the previous decisions of only one of 15 individuals reflected this correct choice (Nemeth 1986).

The phenomenon of social conformity is certainly not new to any reader. We see examples of conformity every day, from the types of clothes people wear to their particular styles of communication in both written correspondence and verbal presentation. The role of conformity as an outside factor influencing at-risk behavior cannot be overlooked. Furthermore, we need to realize that group pressure to conform, even to at-risk behavior, is greater when the group is larger and the group members are perceived as relatively competent or experienced. Also important is the fact that one dissenter (i.e., a leader willing to ignore group pressure and do the right thing) is enough to prevent another person from succumbing to group pressure (or social conformity) to work at risk. And when the actively caring dissenter has authority in the situation, many others will likely follow suit. This is explained by the next basic principle of social influence.

The Power of Authority

In research conducted by Stanley Milgram and his associates in the 1960s, 65 percent of intelligent and well-meaning college students followed orders to administer 450-volt electric shocks to a screaming peer. Imagine you are one of nearly 1,000 participants in one of Milgram's 20 obedience studies at Yale University. You and another individual are led to a laboratory to participate in a human learning experiment. First, you draw slips of paper out of a hat to determine randomly who will be the "teacher" and who will be the "learner." You get to be the teacher. The learner is taken to an adjacent room and strapped to a chair wired through the wall to an electric shock machine containing 30 switches with labels ranging from "15 volts – light shock" to "450 volts – severe shock." You are seated behind this shock generator and instructed to punish the learner for errors in the learning task by delivering brief electric shocks, starting with the 15-volt switch and moving up to the next higher voltage with each of the learner's errors.

If you comply with the experimenter's instructions, you hear the learner moan when you flick the third, fourth, and fifth switches. When flicking the eighth switch (labeled "120 volts"), the learner screams, "These shocks are painful," and when the tenth switch is activated the learner shouts, "Experimenter, get me out of here!" At this point, you may show some concern, but the experimenter prompts you to continue with words like, "Please continue—the experiment requires that you continue." If you continue to increase shock intensity with each of the learner's errors, you reach the 330-volt level. At this point you hear shrieks of pain from the learner. At the 330-volt level, the learner pounds on the wall and then becomes silent. Nevertheless, the experimenter urges you to flick the 450-volt switch when the learner fails to respond to the next question.

At what point do you think you would refuse to obey the experimenter's instructions? If you believe you would stop playing this sadistic game soon after the learner indicated the shock was painful, your prediction would be the same as people who Milgram surveyed before conducting the experiment, including 40 psychiatrists. Thus, Milgram and his associates (1963) were surprised that 65 percent of his subjects, ranging in age from 20 to 50, complied fully with the experimenter's requests—right up to the last 450-volt switch.

Did the subjects figure out the learner was a confederate of the experimenter and was not really receiving shocks? Did they realize they were being deceived in order to test their obedience to authority? No, all of the subjects displayed genuine concern and stress when giving the shocks. They sweated, trembled, bit their lips, and some laughed nervously. Some openly questioned the experimenter's instructions, but with an additional directive from the experimenter, most of the subjects obeyed.

As with the conformity research, Milgram and associates studied the influence of various situational factors on amount of obedience. Full obedience exceeded 65 percent with as many as 93 percent of subjects flicking the highest shock switch when 1) the authority figure was in the room with the subject, 2) the authority figure was supported by a prestigious institution, 3) the shocks were given by a group of "teachers" in disguise, 4) no other subject was observed deviating from the experimenter's instructions, and 5) the victim was depersonalized or distanced from the subject who was in another room.

According to Stanley Milgram, an important lesson from this research is that, "Ordinary people, simply doing their jobs, and without any particular hostility on their part, can become agents in a terrible destructive process" (Milgram 1974). In a similar vein, people might perform unsafe acts or overlook obvious safety hazards and put themselves and others at risk as a result of social obedience or social conformity. The statement "I was just following orders" reflects this obedience phenomenon, and "everyone else does it" implies conformity or peer pressure.

As safety professionals, we need to realize the powerful impact of both conformity and authority and plan interventions to overcome their potential negative influence. Remember that a leader who deviates from the norm and sets a safe example can decrease destructive conformity and obedience. And, when a critical mass of individuals actively care for the safety and health of others, constructive conformity and obedience that supports a Total Safety Culture results.

We Actively Care for People We Like

For whom would you most likely actively care—someone you dislike or someone you like? The answer is obvious, of course, and reflects the value of increasing our personal appeal to others when cultivating interdependency and a Total Safety Culture. Social psychology research has verified a number of intuitive techniques to facilitate ingratiation: 1) demonstrating agreement with target persons on other matters before making a request, 2) offering genuine praise, recognition, or rewarding feedback, 3) using "name dropping" to show association with people respected by target persons, 4) radiating positive nonverbal cues that show appreciation and interest in the target person, and 5) modifying personal appearance to be more acceptable and appealing to the target person.

In previous publications (Geller 1996, 1997a, b) I have discussed the influence power of interpersonal recognition and offered guidelines for making the recognition process most effective. I will not repeat the principles of effective recognition here except to point out the special value of recognition in increasing ingratiation and spreading a sense of belonging and interdependency throughout a work culture. When you recognize a person appropriately, you do more than support that person's desirable behavior and build his or her self-efficacy and self-esteem. You influence that person to like and appreciate you more. This increases your ability to exert positive social influence on that person.

There are other benefits from an effective recognition process. Not only does liking and appreciating increase from the person being recognized to the person doing the recognizing, but the reverse also happens. After all, when we find behavior to recognize, we learn to appreciate the person performing the desirable behavior. This mutual ingratiation increases feelings of belonging and interdependency. Through the power of reciprocity, one appropriate recognition interaction will likely lead to more interpersonal recognition. This, in turn, affects more ingratiation and interdependency throughout a work culture.

In industrial safety, we frequently experience the opposite of ingratiation. Safety management is too often a confrontation between an authoritative individual and a subordinate who is not following a rule. Although the principle of ingratiation is known and understood through personal experience, it is often not followed when making a safety request. In fact, safety is too often presented as a top-down condition of employment, which leads to the perception that we follow safety rules because we "have to" not because we "want to." This can limit individual perceptions of freedom and choice when it comes to safety. As discussed under the final social influence principle, this reduced sense of personal freedom or control can lead to counterproductive behaviors and attitudes.

Value Increases with Scarcity

I wrote a significant portion of this paper while on a five-day speaking tour in California for CareAmerica. It was ironic that when I got to this final part of this paper, I was in the midst of experiencing the very essence of this principle. To make a long story short, the airline misplaced my luggage on my Sunday trip from Pittsburgh to San Francisco. As a result, I spent the first two days and nights with only the clothes on my back.

It was amazing how valuable my luggage became in the context of such deprivation. I literally spent hours on the telephone attempting to locate my luggage. After my Monday presentation in Sacramento, I traveled back to the San Francisco airport to check on the whereabouts of my "valuable" luggage. How excited I got when learning it was found (in New Bern, NC) and would be delivered to my room at the Embassy Suites that evening. That night I anxiously awaited a phone call to notify me that my luggage was at the front desk. I declined an opportunity to go out for dinner so I could be available when my luggage arrived. My luggage became so valuable in this context that I wanted to get it as soon as possible.

My luggage did not arrive that night, and I slept restlessly in anticipation of a critical phone call. Could it be I must wear the same "traveling" clothes for a third day and a second presentation? Just prior to leaving my room the next morning for breakfast in my only available clothes, I got a phone call from the front desk. My luggage was there, and I was thrilled. I skipped breakfast in order to have enough time to retrieve my luggage and change my clothes before my speech. Sometimes Maslow's hierarchy of needs does not work (Maslow 1954). We don't always fulfill our physiological needs before seeking higher-level needs (like the superficial social need to alter our attire).

Readers who have not lost luggage on a trip and subsequently felt its value increase dramatically have certainly experienced the scarcity principle in other situations. Have you ever gone out of your way to purchase front-row tickets to a sports event or music concert? Have you ever jammed into a department store in order to take advantage of special, limited-time bargains? Or, perhaps you are a collector of rare coins, stamps, baseball cards, or antique furniture. Have you ever participated at an auction where one-of-a-kind items are displayed and sold to the highest bidder? Scarce items become even more valuable when other people want them. Have you ever paid more than a fair price at an auction in order to win a bidding competition?

The Forbidden Fruit Phenomenon

Why do teenagers consume drugs, including alcohol, cigarettes, and marijuana? The principle of social conformity, or peer pressure, discussed above is certainly a factor, but why is the purchase and consumption of these drugs considered valuable? They do not taste good, at least at first, and everyone knows they are not good for you. For years these teens have heard the slogan "Just say no to drugs." Why is it difficult for some teenagers (and adults) to say "No"?

Consider that the message "Say no" can make the drug seem desirable. It is the "forbidden fruit" phenomena, as told in the Biblical story of Adam and Eve in the Garden of Eden. Forbidding something makes that something seem valuable, and if we can beat the system to partake of the "forbidden fruit," we might experience an extra rush of pleasure because we asserted our freedom in a top-down situation perceived as stifling our individuality and creativity.

Items or opportunities that appear scarce in the eyes of the beholder seem more valuable. Thus, drugs seem desirable to some people because they are scarce, and consuming drugs can feel more pleasurable because the behavior itself represents a scarce freedom. I am not advocating the legalization of drugs or the lowering of the legal age for alcohol consumption, but understanding the scarcity principle can help us appreciate why certain illegal acts occur and why increased enforcement of laws does not help to develop individual values.

Reacting with Countercontrol

When individuality or perceived personal control is made scarce with top-down control, some people will exert contrary behavior in an attempt to assert their freedom. This is referred to as psychological reactance by social psychologists (Brehm 1966) and countercontrol by behavior analysts (Skinner 1971). Whatever you call it, the results can be devastating to industrial safety and health.

I once met a worker who wore safety glasses without lenses. That's right, not safety glasses, only safety frames. He had removed the lenses. When a supervisor walked up the aisle, the employee would look right at him and wave. His co-workers gave him a lot of peer approval for thumbing his nose at the system. Safety was perceived as a top-down mandate that restricted individual freedom. This worker increased his status in that culture by pushing against the system and demonstrating his independence.

Situations that appear to take away personal freedom or control can activate behaviors opposite to what we want. When people perceive that the system is restrictive, they might attempt to beat it. However, when the threat of punitive consequences is severe and the probability of getting caught high, most people will comply, but compliance will typically be limited to situations in which the rule is reinforced. Do not expect people to internalize the rule and follow it when they do not have to. The more control you exert on the outside of a person, the less control the person will develop on the inside.

External Control and Values

When we perform in certain ways because of external controls or threats, we say we are doing it because we *have* to, not because we *want* to. Under such circumstances we feel no obligation to adjust our inner self, our beliefs and values, to conform to our outer self, our behavior. We can live with the inconsistency between what we do (follow the rules) and our belief (the rule is silly and unreasonable). When we are in situations where enforcement of a rule is difficult, we are apt to break the rule. Hence, most workers use their personal protective equipment (PPE) when it is called for in the workplace, but many do not use appropriate PPE when doing even riskier jobs at home. What percentage of your neighbors wear earplugs, safety glasses, and steel-toed shoes while mowing their lawns?

Our outside self (behavior) influences our inside self (attitudes, beliefs, values) when we believe our behavior was our idea (at least to some extent). That is why it is so important for people to volunteer for safety promotion efforts. When people choose to get involved in a safety process, their behaviors—from developing components of the process to teaching it to others—help to develop or support an internal value or belief system that drives the appropriate behavior in the absence of external controls. This is the consistency principle in action, as discussed earlier.

Consequently, values-driven safety will not come from increasing external controls over behavior. Rather, it is cultivated in situations that provide a clear purpose or mission (the value), promote a variety of straightforward methods for accomplishing the mission, and allow people to choose and customize procedures for their particular situation. Of course, this is easier said than done, and it is often tempting to exert external controls when things do not go our way.

An external threat is often most convenient and expedient, but it has clear disadvantages when attempting to make safety a value. In some cases, severe threats can actually make the undesirable behavior seem desirable. In situations with exclusive top-down controls and minimal bottom-up empowerment, getting away with noncompliance is reinforced by outside peer support and inside feelings of individual freedom.

We need to understand and believe that values-driven safety cannot be dictated. Values can be developed from outside sources that encourage voluntary participation in activities representing or supporting a particular value. We internalize the principles and lessons we choose to experience, but we are apt to resist those principles and lessons we feel are forced upon us. Obviously, it is a difficult but important challenge to provide just enough outside influence to make a program or process appear worthwhile and inviting without inhibiting feelings of personal choice and control. In this regard, please remember another psychological principle: we perceive more choice and personal control when working to achieve positive consequences (rewards) than when working to avoid negative consequences (penalties).

Conclusion

Safety improvement requires a transformation from dependence and independence to interdependence. In other words, it is not enough for people to rely on the organization to keep them safe through engineering intervention, nor is it sufficient for people to count on only their own individual effort to keep them free from injury. Rather, people need others to remove environmental hazards they do not notice and to provide corrective feedback for at-risk behavior they might not realize they are performing. Such interdependency requires interpersonal interaction. The social dynamics of the situation determine whether such exchanges are likely to occur and whether the impact of such interaction is beneficial.

The social dynamics of an organization reflect and influence the culture at the same time. That is, certain aspects of a work culture affect social dynamics and these social dynamics in turn influence the work culture.

The six social influence principles described in this paper reflect basic social dynamics that can either inhibit or facilitate the cultivation of a Total Safety Culture.

Our desire to be consistent influences resistance to change. When people choose to change even a little bit, the consistency principle can facilitate more commitment and more beneficial change. With this principle in effect, small changes in behavior can result in supportive attitude change followed by more behavior change and more desired attitude change. This spiraling of behavior feeding attitude, feeding more behaviors, feeding more attitudes (and so on) can lead to employees becoming totally committed to achieving a Total Safety Culture.

The reciprocity principle can also facilitate the achievement of a Total Safety Culture by increasing actively caring behaviors and interdependency throughout a work culture. Specifically, when people go beyond the call of duty for the safety or health of another person they induce a sense of obligation in that person to return the favor. They also set an example that spreads more actively caring behavior through the principle of consensus or social conformity. Thus, when people actively care they activate two principles of social influence that lead to more actively caring behavior.

The principle of authority reminds us that people might follow orders blindly. We have learned from childhood to follow authority—from mother knows best to the boss knows best. This becomes a barrier to safety when a manager or supervisor asks a worker to perform a behavior that could endanger the worker, other co-workers, or the environment. On the other hand, people in positions of authority can have the opposite effect. In fact, their role-modeling and support of interdependency and the actively caring model is key to cultivating a Total Safety Culture.

Safety leaders should also remember the principle of ingratiation and should use quality interpersonal recognition to support other people's actively caring behaviors, increase the appreciation and respect that other people have for the safety leaders, and increase their own admiration for those persons they recognize.

Finally, the social influence principle of scarcity explains why some employees actively resist compliance to top-down safety rules, regulations, and quick-fix safety programs (referred to by many as "flavor-of-the-month"). Some work cultures seem to restrict the perception of personal control and creativity when it comes to safety. And in these situations, contrary behavior or noncompliance becomes a personal statement of "freedom." Employee involvement is much more likely with top-down support of safety processes that are developed, owned, and continuously improved by work teams educated to understand relevant rationale and principles.

The metrics used to evaluate the safety performance of individuals, teams, and the organization as a whole establish a context that can increase or decrease employee commitment, involvement, and ownership. Injury statistics provide an overall estimate of distance from a vision of "injury-free" but such numbers are not diagnostic with regard to proactive action plans. If used as the only index of safety achievement (or failure), injury-related outcome numbers can do more harm than good from the perspective of empowering people to take control of safety. Numbers that measure the quantity and quality of process activities related to ongoing safety-related activities provide the context needed to motivate individual and team responsibility and to direct continuous improvement. Clearly this is not rocket science, nor is it common sense. It merely reflects some basic research-based principles of social dynamics applied to occupational safety and health.

References

Asch, S. E. 1951. Effects of group pressure upon the modification and distortion of judgments. In *Groups, leadership, and men.* ed. H. Guetzkow. Pittsburgh, PA: Carnegie Press.

———. 1955. Opinions and social pressure. *Scientific American,* 193:31–35.

Berkowitz, L., and L. R. Daniels. 1964. Responsibility and dependency. *Journal of Abnormal and Social Psychology*, 66:429–436.

Brehm, J. W. 1966. *A theory of psychological reactance*. New York: Academic Press.

Cialdini, R. B. 2001. *Influence: Science and practice. 4th* ed. New York: Harper Collins College Publishers.

Cialdini, R. B., J. E. Vincent, S. K. Lewis, J. Catalan, D. Wheeler, and B. L. Darby. 1975. Reciprocal concessions procedure for inducing compliance: The door-in-the-face technique. *Journal of Personality and Social Psychology*, 1:206–215.

Cialdini, R. B., J. T. Cacioppo, R. Bassett, and J. A. Miller. 1978. Low-ball procedure for producing compliance: Commitment then cost. *Journal of Applied Social Psychology*, 15:492–500.

Crutchfield, R. S. 1955. Conformity and character. *American Psychologist*, 10:191–198.

Deming, W. E. 1986. *Out of the crisis*. Cambridge, MA: Massachusetts Institute of Technology, Center for Advanced Engineering Study.

———. 1993. *The new economics for industry, government, education*. Cambridge, MA: Massachusetts Institute of Technology, Center for Advanced Engineering Study.

Freedman, J. L., and S. C. Fraser. 1966. Compliance without pressure: The foot-in-the-door technique. *Journal of Personality and Social Psychology*, 4:195–203.

Geller, E. S. 1996. How to give quality recognition. *Industrial Safety and Hygiene News*, 30(12):12.

———. 1997a. Hey, I'm trying to give you a compliment. *Industrial Safety and Hygiene News*, 31(1):12.

———. 1997b. How to celebrate safety success. *Industrial Safety and Hygiene News*, 31(2), 12–13.

———. 2001. *The psychology of safety handbook*. Boca Raton, FL. CRC Press.

Geller, E. S., and G. R. Lehman. 1991. The buckle-up promise card: A versatile intervention for large-scale behavior change. *Journal of Applied Behavior Analysis*, 24:91–94.

Isen, A. M., and P. F. Levin. 1972. Effect of feeling good on helping: Cookies and kindness. *Journal of Personality and Social Psychology*, 21:384–388.

Maslow, A. H. 1954. *Motivation and personality*. New York: Harper.

Our Need for Competence Fuels Safety Participation[3]

E. Scott Geller

Do you assume people will not want to participate in your safety process? Why? Because it is inconvenient or time consuming? Do you assume that safety participation represents altruistic or sacrificial behavior? Why? Because the obvious consequences are rewards for other people and inconvenience for the participant?

I would like you to consider an alternative perspective. I propose that people really want to participate—to play a role in keeping people safe. Consider that the opposite of active participation is incompetence, apathy, or helplessness. No one wants to be in those states.

Furthermore, I propose that we stop talking about participation in a safety process as if it requires personal sacrifice—as if such behavior requires a relinquishing of personal gain so others might benefit. How many people can get excited about enduring negative consequences so others in the workplace might receive positive consequences?

Instead, let's assume that helping others reaps personal gains for *everyone* involved. Such behavior is not sacrificial or even unselfish. It rewards a basic human need to make a difference—to serve others. In other words, serving others is self-rewarding not self-sacrificing.

The Need for Competence

Several researchers of human motivation have proposed that people naturally enjoy being able to solve problems and successfully complete worthwhile tasks. In other words, we are motivated to learn, to explore possibilities, to understand what is going on, and to participate in achieving worthwhile goals. The label for this fundamental human motive is "competence."

Motivational psychologists assume that the desire for competence is self-initiating and self-rewarding. In other words, behavior that increases feelings of effectiveness is self-directed and does not need extrinsic reinforcement to keep it going. Furthermore, behavioral skills that increase competence are readily learned and practiced. What does this mean for safety leadership and participation?

The Need for Self-Efficacy

The belief in self-efficacy is critical. In fact, self-efficacy implies competence. Having self-efficacy means the participant believes he or she can organize and perform the procedures needed to achieve a desired goal. When people learn specific procedural steps for carrying out a particular safety process, they gain self-efficacy. They feel competent.

The Need for Response-Efficacy

It is not enough to feel competent at a task; we need to believe our participation contributes to making a beneficial difference. Thus, effective training not only teaches step-by-step procedures, but it also includes a rationale for why the process is useful.

In safety, this translates to convincing the participants that applying the lessons learned will help people remain injury-free. Thus, the need to feel competent is satisfied by self-efficacy (believing one has the skills to implement a particular intervention) and response-efficacy (believing the intervention will have a desirable impact).

The Power of Feedback

So how do we know we are competent at something? How do we know our competence makes a valuable difference? You know the answer—feedback. Feedback about our ongoing behavior tells us how we are doing and enables us to do better. We hone our skills through practice and behavior-based feedback. Sometimes this feedback process comes naturally, like when we see our behavior produce a desired result. But, often behavioral feedback requires careful and systematic observation by another individual—a trainer or coach—who later communicates his or her findings to the performer. In each case, feedback enables the development of self-efficacy and fulfillment of the need for competence.

Outcome feedback supports the need for competence further by showing desirable effects of participation. For example, a display of the percentage of safe behaviors among a work group indicates whether an interpersonal coaching process is working. When these percentages are graphed daily or weekly, a work team can track their progress at improving their coaching competence.

Objective evidence of a reduction in injuries, property damage, or "near-miss" reports is, of course, the most rewarding feedback we can get in safety. This is the ultimate outcome feedback we work to achieve. Unfortunately, this feedback does not change rapidly enough to affect our competence. Also, this feedback can be invalid due to under-reporting. That is why we need ongoing behavior-based feedback to continuously build our self- and response-efficacy and to satisfy our need for competence.

I have heard some behavior-based safety consultants argue that feedback is not a reinforcer. Technically, a reinforcer is a behavioral consequence that increases the frequency of the behavior it follows. So, if behavior does not improve after feedback, then the feedback was not a reinforcer. Likewise, praise, reprimands, bonus pay, or frequent flyer points are not reinforcers when they do not increase the frequency of the behavior they target; they often do not.

Consider, however, that feedback delivered well, whether supportive, corrective or both, feeds the basic human desire to be competent. It is not a payoff for doing the right thing. Rather, it is objective information a person uses to feel competent or to learn how to become more competent. I believe there is no other consequence with greater potential to improve performance and thereby become a reinforcer.

Conclusion

This article calls for a paradigm shift—a change in perspective about participation in safety efforts. Instead of calling on guilt or sacrifice to get people involved in procedures to eliminate hazards or decrease at-risk behavior, I suggest we assume that people are naturally motivated to make beneficial differences. People hate feeling incompetent or helpless. They want to learn, to discover, to become more proficient at worthwhile tasks. People want opportunities to ask questions, to study pertinent material, to work with people who know more than they, and to receive feedback that can increase their competence.

Participation in a safety-related process is not a thankless job requiring self-sacrifice or a special degree of altruism. Safety participation puts people in control of environmental and human factors that can cause serious injury or death. Safety participation avoids one of the most aversive human states—the feeling of incompetence or helplessness. Participation in an effective safety process provides opportunities to satisfy a basic human need—the need for competence. And, the effective and frequent delivery of behavior-based feedback provides a mechanism for improving the quality of safety participation as well as cultivating feelings of competence throughout a work culture.

Total Safety Culture in Two Real World Companies

Case Study 1: E. Scott Geller, Bryan Krueger, Anne French, and Joshua H. Williams[4]

Case Study 2: Mary Foss, Joshua H. Williams, and Steve Roberts

The following article offers two companies as examples of the process of working toward a Total Safety Culture. The process of behavior-based observation and feedback differs in each company—as it should—to meet the requirements and needs of each industry, each company, and the employees of each organization.

Total Safety Culture

Achieving safety excellence requires going beyond the traditional safety focus of engineering and regulation. Because human behavior is a contributing cause to most incidents and injuries, safety excellence can only be achieved by addressing the human dimensions of safety. Reducing at-risk behavior and increasing safe behavior requires understanding how such behaviors are motivated by system and cultural influences. Developing and sustaining a supportive safety culture hinges on understanding and applying the principles of behavioral science and person-based psychology to build tools and methods that encourage personal responsibility and interpersonal interaction about safety. We call this "actively caring," and it is an integral part of a Total Safety Culture.

In a Total Safety Culture, employees not only feel responsible for their own safety, but they also feel responsible for their peers' safety, and the organizational culture supports them acting on that responsibility. Individuals have the necessary tools and methods, as well as the appropriate person states (e.g., self-esteem, group belonging, and personal control), to actively care for the safety of co-workers. Additionally, the organization's formal management systems and leaders' informal management practices facilitate actively caring behavior by encouraging, recognizing, and reinforcing appropriate actions. Achieving a Total Safety Culture involves the development and implementation of a behavior-based observation and feedback process, the assessment and refinement of existing safety management systems, and the improvement of management and employee practices influencing the safety culture.

Behavior-Based Safety (BBS) Observation and Feedback

To achieve a Total Safety Culture where employees feel as strong a sense of responsibility for the safety of their co-workers as they do for themselves, it is necessary to increase the openness and frequency of safety communication. A critical observation checklist encourages interpersonal behavioral observation and feedback, which improves this communication. The checklist has categories of relevant safety behaviors with corresponding columns to mark behaviors as either safe or at risk. The checklist also has a "Comments" section to provide specific information about an observation. Employees provide one-on-one feedback regarding the safe and at-risk behaviors observed. To ensure anonymity, the observee's name is never recorded.

The feedback which follows the observation provides positive social support for doing a task safely even when at-risk alternatives are easier, faster, and more convenient. Information about at-risk behaviors is given to avoid potential injury. The data from the observation cards are collected, compiled, and then shared with the employees as group feedback. This provides specific information and enables the DO IT process to improve safety performance, as defined in the first paper in this book.

The **DO IT** process involves four sequential steps: a) **D**efine relevant target behaviors to increase or decrease in frequency, b) **O**bserve target behaviors during a baseline phase and set specific goals for achievement, c) **I**ntervene to improve the target behaviors, and d) **T**est the impact of the intervention by continuing to observe the target behaviors. If the desired results are not achieved, other interventions are implemented. When improvement goals are met, other target behaviors are selected for improvement.

Case Study 1: Wisconsin Electric Fleet Services

Background

Fleet Services employees of Wisconsin Electric have experienced significant safety improvements since 1997 after implementing their behavior-based safety process. In fact, these employees, who repair vehicles and equipment used to build and maintain an extensive electric and natural gas infrastructure, have the best safety record at Wisconsin Electric. This is consistent with their safety mission statement: "Fleet Services, working together by actively caring for each other, will strive for a Total Safety Culture to create a safe working environment."

The central shop for Fleet Services is located in Milwaukee, with a series of surrounding remote satellite locations. Overall, Fleet Services includes 74 employees of which 70 are union members, the largest union being the International Brotherhood of Electrical Workers Local 2150.

Self-Observation

The Wisconsin Electric Fleet Services chose a unique, creative approach to provide observation and feedback. In addition to peer observations, Fleet Services employees also conduct self-observations for safety. Self-observations are performed because many of the Fleet Services employees often work alone at a remote repair center or in the field at a job site.

Not surprisingly, the average number of self-observations per month with Fleet Services personnel is 50 versus 30 per month with peer observations. Also, the self-observations are usually of very high quality (i.e., numerous comments and suggestions for improvement).

Self-observations are an important safety improvement process whereby individuals:

* identify target behaviors for improvement;

- establish a behavioral baseline for those target behaviors using self-observation;
- select a self-management strategy to promote desired behavior change and chart progress;
- select a goal which is specific, motivational, attainable, recordable, and trackable;
- record self-observation with target behaviors to measure progress toward the goal; and
- administer self-rewards that are accessible, immediate, individualized, and valued.

Identification of target behaviors is accomplished by reviewing injury data, analyzing "near-misses," and evaluating previous behavioral observations to identify individual problem areas.

Social cognitive theory suggests that self-observations are one of the primary components of the self-regulation of behavior change. Generally, when employees conduct self-observations, they are more likely to perform their jobs safely. This is explained by the principle of consistency. When our behaviors are different from our beliefs, feelings of discomfort are produced. Performing a job at risk when safety beliefs are triggered (by the use of a self-observation form) causes internal conflict. To eliminate this discomfort, individuals using a checklist are more likely to perform a task safely.

BBS Process

The Fleet Safety Solutions Team (FSST) is facilitated by Bryan Krueger, the lead technician for the night shift. In 1997, the FSST team developed the observation checklist (see Figure 1.4.1) which Fleet Services employees still use to observe one another and themselves. All Fleet Services employees received an eight-hour training course on how to use the observation checklist to provide co-workers with one-on-one feedback to improve safety. Typically, an observation and feedback session lasts between 10 and 15 minutes and may or may not be scheduled in advance.

Every month, the FSST team analyzes the data from the observation forms, documents trends in at-risk behaviors, and communicates the results

Observer ____________________ Date ________________

Observation Type: Self or Peer Time ________________

Location ____________ Day: M T W T F (circle one)

Housekeeping

	Safe	At Risk	N.A.
Floor – Clean and Dry			
Floor – Oil Free			
Doors Not Blocked			
Clear Walkways			
Clean Work Areas – Orderly			
Emergency Equipment Accessible			
Trash Containers – Marked/Used Correctly			
Tools – Organized/Clean			
Hoist – Clean/Accessible			
Parts Area – Clean			
Fire Cabinets – Used Properly			
PPE – Required as Necessary			

Body Mechanics

	Safe	At Risk	N.A.
Pushing/Pulling			
Proper Lifting			
• Bend at Knees			
• Back Straight			
• Load Close to Body			
• No Twisting			
• Using Assist/Help			
Pinch Points			
Line of Fire			
Working Above Ground			
• Climbing/Walking			
Cramped/Extended Conditions			

Figure 1.4.1 Fleet Services' Critical Observation Checklist

to the workforce. Also, the FSST team meets monthly to discuss the process and follow up on action items resulting from the "Comments" and "Suggestions" sections of the observation forms. Although a formal incentive process is not in place, periodic rewards (e.g., polo shirts, pens) are presented to employees for their participation in the BBS process and for actively caring behavior (going beyond the normal routine for the safety of someone else).

While not directly involved in the observation and feedback process, front-line supervisors (who also received BBS education and training and are members of the FSST team) are actively involved in the process. Although supervisors do not conduct observations, they regularly attend team meetings and provide the time and resources needed by hourly employees to effectively implement the new safety process. In order to foster employee ownership of the process, participation is encouraged (not mandated). Each employee is expected to perform two observations per month.

BBS Success

Not only have safety attitudes and beliefs improved following BBS implementation, more than 70 percent of all Fleet Services employees regularly perform both peer and self-observations. Since BBS implementation, there has been a steady increase in the number of interpersonal behavioral observations, at-risk behaviors identified, quick-fixes turned in, and comments and suggestions on the observation forms. In fact, the number of reported incidents has actually increased. Krueger, the FSST team facilitator, attributes this increase to employees feeling more comfortable discussing safety concerns openly and without fear of reprimand. He believes the *actual* number of incidents has not increased, but employees are more willing to report injuries because they trust their BBS process.

Overall, the number of *disabling* injuries has significantly dropped following BBS implementation. Despite two injuries near the end of 1999, employees were praised for their continued efforts in making safety a value.

Case Study 2: Koch Petroleum Group

Background

Koch Petroleum Group in Pine Bend, Minnesota has experienced significant safety improvements since the Fall of 1997 after implementing its behavior-based safety process. Pine Bend has 750 full-time employees, with 385 union employees represented by PACE International.

BBS Process

The behavior-based safety effort at Koch Pine Bend is lead by Marv Foss, the Health and Safety Representative. In 1997, the BBS implementation team designed the behavior observation checklist (Figure 1.4.1). Koch employees were then introduced to the voluntary, non-punitive nature of BBS through 8-hour training classes. Every month, the BBS team analyzes the data from the observation forms, documents trends in at-risk behaviors, and communicates the results to the remaining workforce.

The focus of the BBS at Pine Bend is on proactive process measures of safety. Small incentives ranging from free meals to gift certificates are provided for employees conducting behavior observations, safety audits, safe work permit audits, pre-work checklists, near-miss investigations, and safety meetings.

BBS Success

Foss points out, "The important part is lowering the number of at-risk behaviors and conditions we have in our refinery. By using the observation process along with all of the other proactive tools we have introduced to everyone, we are doing just that." In fact, the number of behavior observations increased from 6,000 in 1998 to 19,000 in 1999. There were also 5,000 safety audits, 700 incident analyses, and 22,000 "What-if" safety meetings in 1999.

Since BBS implementation, the number of injuries dropped from 27 to 15, and TRIR rates dipped from 3.4 to 1.7. Koch Pine Bend received three safety awards at the National Petrochemical Refinery Association conference last April for a million man hours without a lost time injury (currently at 2.4 million man hours), a calendar year without a lost time (currently at 514 days), and a 25% improvement in their recordable rate compared to the last three years.

While improvement efforts continue and some obstacles remain, BBS implementation has greatly helped Koch Pine Bend improve their safety efforts and reduce injuries.

Notes

[1] Based on a paper that originally appeared in *Proceedings for the 35th Annual ASSE Professional Development Conference and Exposition* (Des Plaines, IL: American Society of Safety Engineers, 1996):121–40.

[2] The original version of this paper appeared in *Proceedings for the 36th Annual ASSE Professional Development Conference and Exposition* (Des Plaines, IL: American Society of Safety Engineers, 1997):421–37.

[3] Originally appeared in *Industrial Safety and Hygiene News* (August 2001).

[4] Case Study 1 originally appeared as "The Human Factor" in *Utility Safety* (April 2000).

2

Actively Caring and Self-Management

"Actively caring" is more than a concept; it is a way of life. The famous psychologist Abraham Maslow taught that human beings have a "hierarchy of needs," including the need for food, shelter, social approval, and self-esteem. The highest "level" of attainment in this "needs hierarchy" was originally labeled "self-actualization," and Maslow cited individuals like Abraham Lincoln, and Allen Ginsberg, as being self-actualized.

In his later years, he revised his hierarchy to reflect one more "level" of attainment: "self-transcendence." At this level, individuals forsake their own needs for the greater good of other people. We call this actively caring, which involves systems thinking, an interdependent perspective, and an ongoing willingness to go beyond the call of duty for the safety of co-workers.

This chapter provides examples, empirical data, and strategies for spreading actively caring principles throughout a work culture, as well as in one's own home. We start with two research papers that initally defined the actively caring model and demonstrated significant validity for the basic concepts.

The third paper extends the notion of actively caring to a set of principles and procedures essential for safety improvement in numerous situations.

Specifically, the essence of self-management is described as it can be applied to occupational safety.

People who reach the actively caring level of Maslow's "hierarchy of needs" will not only use these self-management techniques on themselves, they will also teach these strategies to others—co-workers and family members—so the possibility of personal injury is reduced at work, at home, and throughout the community.

An Actively Caring Model for Occupational Safety[1]

Steve Roberts and E. Scott Geller

Abstract

"Actively caring" refers to individuals caring enough about the health and safety of others to act accordingly. Actively caring behavior in an industrial context can take the form of continually looking for environmental hazards and unsafe work practices and implementing appropriate corrective actions when unsafe conditions or behaviors are observed. Individuals presumed most likely to actively care are those high in self-esteem (those who feel valuable), optimism (those who feel they can make a difference), and group belonging or cohesiveness (those who feel close to members of their work group). In order to test the actively caring model, the following study assessed the relationship between self-esteem, group cohesion, and optimism with employees' self reports of willingness to actively care.

In addition to self-reported data, we assessed the occurrence of certain actively caring behaviors in the work setting. Specifically, actively caring behaviors were measured by counting the number of actively caring thank-you cards given or received for actively caring behaviors. Self-esteem, group cohesion, and optimism scores predicted significant and independent variance in self-reports to actively care. Furthermore, those workers who either gave or received thank-you cards scored significantly higher on measures of self-esteem and group cohesiveness than those workers who did not give or receive thank-you cards. Implications for future research and the application of the actively caring concept are discussed.

Introduction

Employees are often given mixed signals concerning safety issues. On the one hand, employees are told to work safely; on the other hand, they are also told to produce more, faster, or better, which often entails risky behavior. Furthermore, safety equipment is often uncomfortable to use, and safe operating procedures are often inconvenient and time consuming to follow. Statistics show that the chances of an employee being involved in an accident are relatively low. On average, only about four employees in 100 are involved in lost work time accidents per year (National Safety Council 1991). Therefore, when workers fail to use safety equipment or do not follow safety procedures, they are often rewarded by increased comfort, convenience, or speed of work without experiencing any aversive consequences. In other words, rewards for working at risk are often soon and probable, whereas the penalties for working at risk are usually delayed and improbable.

One way to increase the supportive consequences for safe behaviors and the aversive consequences for at-risk behaviors would be for management to make rewards contingent on following safe work practices and penalties contingent on at-risk work behaviors. However, managers are often absent when dangerous work is accomplished. Although managers and supervisors make rounds to check on employees, they are typically in an office during a large percentage of the workday. In most work situations, a person's co-workers are the ones most likely to be present when a work process warrants certain safety precautions.

Some employees work safely because of mandates or policy directives from management, but other individuals require more intrusive interventions to motivate their compliance with safety rules. After some individuals achieve the desired behaviors, it would be useful to enlist them as intervention agents to influence the behavior of others (Geller 1992; Geller et al. 1990). In other words, instead of "preaching to the choir," the choir should be sent out to enroll converts. In fact, Geller and Roberts (1993) found a direct relationship between the number of intervention agents and the impact of intervention programs designed to increase the use of vehicle safety belts.

From a brainstorming session with safety leaders at Exxon Chemical Company, Geller (1991) coined the term "actively caring" to refer to an ultimate goal in occupational safety—namely, that employees care enough about the safety of their co-workers to act accordingly. In other words, employees actively caring for safety would continually look for environmental hazards and at-risk work practices and then would implement appropriate corrective actions when unsafe conditions or behaviors are observed. Geller (1991) hypothesized that three individual difference factors increase the propensity for an employee to actively care (AC) for the safety or health of a co-worker. Individuals presumed most likely to AC are those high in self-esteem (those who feel valuable), optimism (those who expect the best), and group belonging or cohesiveness (those who feel close to members of their work group).

Empirical Support for the Actively Caring Model

Self Esteem. Coopersmith (1967) defined self-esteem as the evaluation an individual makes and usually maintains about oneself. This self-evaluation process indicates the extent to which the individual feels capable, significant, successful, and worthy. Michelini, Wilson, and Messe (1975) and Wilson (1976) measured subjects' self-esteem with a sentence completion test and then measured whether subjects helped another individual in a bystander intervention paradigm (Darley and Latané 1968). High self-esteem subjects were significantly more likely than low self-esteem subjects to help another person pick up dropped books (Michelini et al. 1975) and to leave an experimental room to assist a person in another room who screamed that he had broken his foot following a mock "explosion" (Wilson 1976). Similarly, subjects with higher self-esteem scores were more likely to help a stranger (i.e., a confederate) by taking his place in an experiment where they would presumably receive electric shocks (Batson, Bolen, Cross, and Newinger-Benefiel 1986).

Optimism. Optimism is the learned expectation that life events, including personal actions, will turn out well (Scheier and Carver 1985; Seligman 1991). Researchers have manipulated optimistic states (or moods) among individuals by giving them unexpected rewards or positive feedback and then observing the occurrence vs. nonoccurrence of AC behaviors. Isen and Levin (1972) showed that individuals finding a dime in the coin return slot of a public phone (placed there by researchers) were more likely to help a confederate who dropped a folder of papers than were individuals who did not find a dime. Similarly, students given a cookie while studying at the university library were more likely to agree to help another student by participating in a psychology experiment than those not given a cookie.

Isen, Clark, and Schwartz (1976) delivered free samples of stationery to people's homes and then called them later to request an AC behavior. The caller said he had dialed a wrong number but since he had used his last dime, he needed the subject to call a garage to tow his car. Subjects who had received the gifts of stationery were more likely to make the AC phone call than were subjects who had not received a gift.

Carlson, Charlin, and Miller (1988) reviewed these and other studies that showed direct relationships between mood (optimism) and AC behavior. They reported that the following pleasant experiences increased AC behavior, purportedly by inducing a positive mood or optimistic outlook: finding a dime, receiving a packet of stationery, listening to soothing music, being on a winning football team, imagining a vacation in Hawaii, and being labeled a charitable person.

Group Belonging /Cohesion. The social psychological construct most analogous to the AC concept of belonging is group cohesion—the sum of positive and negative forces attracting group members to each other (Wheeless, Wheeless, and Dickson-Markman 1982). Staub (1978) reviewed studies which showed that people were more likely to help victims who belonged to a relevant group determined by race, nationality, or an arbitrary distinction defined by preference of an artist's paintings. Similarly, Batson et al. (1986) found subjects more likely to help confederates if the subjects rated them as similar to themselves.

In a bystander intervention study, pairs of friends intervened faster to help a female experimenter who had fallen from a chair than did pairs of strangers. Thus, with friends as subjects, the bystander intervention effect (i.e., an inverse relationship between group size and victim-helping behavior) may not occur because group cohesiveness (or a sense of belonging) counteracts the diffusion of responsibility that presumably accounts for the bystander intervention effect (Latané and Nida 1981). In a similar vein, Rutkowski, Gruder, and Romer (1983) manipulated group cohesion in groups of two and four experimentally and found the most AC behavior among subjects in the high-cohesion conditions.

In a retrospective study, Blake (1978) studied real-world relationships between group cohesion and the ultimate in AC behavior—altruistic suicide. His data was gathered from official records of Medal of Honor awards given during World War II and Vietnam. The independent variable was the cohesiveness of combat units (estimated by group training and size) and the dependent variable was percentage of "grenade acts"—voluntarily using one's body to shield others from exploding devices. Results revealed that the smaller, more elite, specially trained combat units (e.g., the Marine Corps and Army airborne units) accounted for a substantially larger percentage of "grenade acts" than larger, less specialized units (e.g., Army non-airborne units), thus supporting the hypothesis that group cohesion increases AC behavior.

The helping behaviors previously discussed are somewhat different from the AC behaviors described by Geller (1991). The previously discussed helping behaviors usually occurred as a reaction to an accident which has already happened (e.g., falling off a ladder, explosion) or to an event that will inevitably lead to serious injury (e.g., the introduction of a live grenade). The AC behaviors most relevant to occupational safety and proposed by Geller (1991) help people *avoid* an injury that is only possible or even unlikely in any given situation.

In order to test the AC model, this study assessed the relationships between self-esteem, group cohesion, and optimism with employees' self reports of willingness to AC. In addition to self-reported data, we assessed the occurrence of certain AC behaviors in the work setting. Specifically, AC was measured by counting the number of actively caring thank-you cards given or received for AC behaviors, defined generally as: 1) recognizing and correcting an at-risk condition, 2) reminding a co-worker not to perform an at-risk act, 3) removing or cleaning at-risk objects or debris from a work area, 4) giving positive feedback to a co-worker for working safely, 5) reporting a near-miss, and 6) making a task safer.

Method

Subjects

The subjects were 65 hourly workers from one department of one division of a large fiber-manufacturing plant located in a rural section of southwestern Virginia. The subjects ranged in age from early twenties to sixties with an average job tenure of 18 years. The plant operates 24 hours per day, 365 days per year and employs approximately 2,000 workers, most of whom are production workers. Most of the hourly workers in the plant (approximately 1,800) belong to the local union, which has been represented at the plant since it first opened.

Personality Measures

The personality measures were all in the format of a five-point Likert Scale. The ten-item Self-Esteem Scale (Rosenberg 1965) was used to measure self-esteem. The Self-Esteem Scale is an established personality measure with reliability estimates ranging from .70 to .90 (Levy and Baumgardner 1991; Knight and Wadel 1986). The 12-item Life Orientations Test (LOT) was used to measure optimism. Scheier and Carver (1985) reported an internal consistency coefficient of .76 and a test-retest coefficient of .79 for the LOT. They also reported acceptable convergent and divergent validity with a number of other personality measures. The 18-item Group Cohesion Measure was used to measure belonging. Wheeless et al. (1982) reported this test to have a split-half reliability coefficient of .90. This measure was modified slightly in order to fit the industrial worker population. For example, the words "work group" were substituted for "group" in each item.

Actively Caring

Three questions used to measure willingness to AC were embedded within a test battery containing the personality measures and items regarding adequacy of safety training, attitudes toward other safety-related issues, and management response to various safety issues: 1) If I know a co-worker is going to do a hazardous job, am I willing to remind him/her of the hazards (even if the employee is familiar with the job); 2) Am I willing to warn my peers about working unsafely; and 3) Am I willing to do whatever I can to improve safety, even confronting my peers about their unsafe acts? The responses to these questions, measured on a five-point Likert scale, were added to attain an AC score.

Procedure

The subjects were given the test battery by their supervisors and told the questionnaire was a measure of the "safety climate" within their department and that their answers would be anonymous. However, a special code known only to respondents was used to match the surveys with information taken at a later date. The codes were formed by each employee writing the first letter of the city where he or she was born, the first letter of his or her mother's maiden name, and the number of the month when he or she was born. This process yielded a separate two-letter, one-number code for each employee.

Thank-you Cards

During a departmental meeting, area superintendents introduced the concept of AC with the hourly workers. Examples of AC behaviors were discussed as well as the need to increase AC behaviors. At the end of the meeting subjects were given five actively caring thank-you cards (see Figure 2.1.1). The employees were told to give the cards to their co-workers

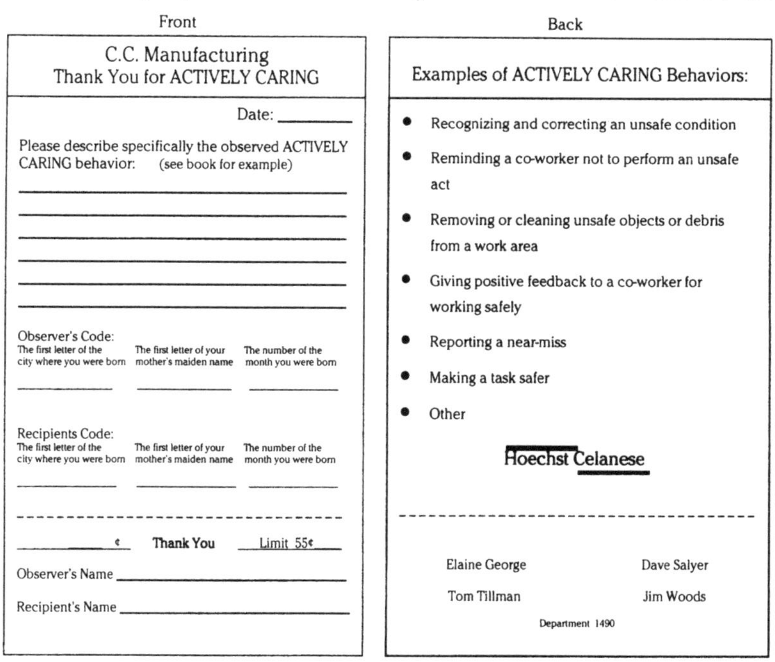

Front

C.C. Manufacturing
Thank You for ACTIVELY CARING

Date: ________

Please describe specifically the observed ACTIVELY CARING behavior: (see book for example)

Observer's Code:
The first letter of the city where you were born / The first letter of your mother's maiden name / The number of the month you were born

Recipients Code:
The first letter of the city where you were born / The first letter of your mother's maiden name / The number of the month you were born

¢ Thank You Limit 55¢

Observer's Name ________

Recipient's Name ________

Back

Examples of ACTIVELY CARING Behaviors:

- Recognizing and correcting an unsafe condition
- Reminding a co-worker not to perform an unsafe act
- Removing or cleaning unsafe objects or debris from a work area
- Giving positive feedback to a co-worker for working safely
- Reporting a near-miss
- Making a task safer
- Other

Hoechst Celanese

Elaine George Dave Salyer
Tom Tillman Jim Woods
Department 1490

Figure 2.1.1 Actively Caring Thank-You Card

whenever they saw an example of AC behavior for safety. If they ran out of cards, they could obtain more from their supervisors. The thank-you cards included examples of AC behaviors and an area to record an anonymous code for the observer and the card recipient as described above for the Safety Climate Questionnaire. This made it possible to match individual questionnaire results with the number of cards given and received. On the bottom of the cards was a perforated stub that could be torn off and redeemed by the recipient of the card for food in the company cafeteria (value 55¢). These stubs also included a space for the observer and recipient to write their names so that management could check for abuses in the system (e.g., to make sure two friends did not always give cards to each other).

Results

Survey Findings. Relationships between self-esteem, group cohesion, optimism and self-reports of willingness to actively care (AC) were tested using a stepwise multiple regression procedure. For this analysis all workers who completed the questionnaire (n = 31) were included even if they did not complete the identifying code (n = 6). To test whether each personality factor accounted for unique variance in AC scores, each personality factor was entered separately as the last step of the multiple regression. As shown in Table 2.1.1, the partial correlation for optimism with AC with

Variables	Partial R.	Model R	R^2	F to Enter
Group Cohesion and Self-Esteem Forced First				
Group Cohesion + Self-Esteem		.520	.271	
Optimism	.354	.601	.362	3.856 *
Group Cohesion and Optimism Forced First				
Group Cohesion + Optimism		.516	.266	
Self-Esteem	.361	.601	.362	4.034 *
Optimism and Self-Esteem Forced First				
Optimism + Self-Esteem		.456	.208	
Group Cohesion	.354	.601	.362	6.520 *

*Table 2.1.1 Multiple Stepwise Regression Analysis with Group Cohesion, Self-esteem, and Optimism Each Entered as the Last Step and the Celco Actively Caring Sub-scale Score as the Dependent Variable (N = 31). *p<.05.*

the effects of group cohesion and self-esteem partialed out was significant, $r_{12.34} = .354$, $p<.05$; the partial correlation for self-esteem with AC with the effects of group cohesion and optimism partialed out was significant, $r_{42.13} = .361$, $p<.05$; and the partial correlation for group cohesion with AC with the effects of optimism and self-esteem partialed out was significant, $r_{23.14} = .441$, $p<.05$. The terminal multiple regression model including group cohesion, self-esteem, and optimism accounted for a significant amount of variation in AC, $R^2=.362$.

Thank-you Card Findings

Eight different workers gave 23 thank-you cards to 15 different employees. However, only six of these cards from four different workers had codes that matched the codes from the Safety Climate Survey. Nevertheless, the hypothesis that workers who gave or received actively caring thank-you cards will score higher on the self-esteem, group cohesion, and optimism scales was partially supported. In order to ensure that the comparison group did not include any questionnaires from workers who gave or received thank-you cards, the analysis of the thank-you card data included only those questionnaires which included an identifying code (N = 25). Independent t-tests were used to compare the four participants in the thank-you card program with the 21 employees who did not participate.

Scale	Mean	SD	t-Score
Group Cohesion (G/R)	70.25	8.54	1.99 *
Group Cohesion (NG/NR)	54.19	15.54	
Self-Esteem (G/R)	34.75	3.77	1.71 *
Self-Esteem (NG/NR)	30.85	4.25	
Optimism (G/R)	21.25	1.26	.65
Optimism (NG/NR)	20.71	1.55	

*Table 2.1.2 Means, Standard Deviations, and One-Tailed t Scores for Group Cohesion, Self-Esteem, and Optimism Measures of Workers Who Gave or Received (G/R) Actively Caring Thank-You Cards (N = 4) and Workers Who Did Not Give or Receive (NG/NR) Actively Caring Thank-You Cards (N = 21). *p<.05*

As can be seen in Table 2.1.2, the one-tailed independent t-tests indicated that the self-esteem scores for the four workers who either gave or received thank-you cards (M = 34.75, SD = 3.77) was significantly higher than the self-esteem scores of those who did not give or receive thank-you cards (M = 30.85, SD = 4.25), $t(23) = 1.71$, $p = .05$; the group cohesiveness scores for those workers who either gave or received thank-you cards (M = 70.25, SD = 8.45) were significantly higher than those workers who did not give or receive thank-you cards (M = 54.19, SD = 15.54), $t(23) = 1.99$, $p = .029$. However, no significant difference in optimism scores was found between the participants (M = 21.25, SD = 1.26) and nonparticipants (M = 20.71, SD = 1.55), $t(23) = .65$, p = .262.

Discussion

Self-esteem, group cohesion, and optimism scores predicted significant and independent variance in AC, thereby supporting the AC model proposed by Geller (1991). Furthermore, those workers who either gave or received thank-you cards scored significantly higher on measures of self-esteem and group cohesiveness than those workers who did not give or receive thank-you cards.

These results are promising given the small sample size and the small number of individuals from whom both thank-you cards and questionnaire data were available. A problem with this study, however, was the low number of items to measure AC behaviors. Another problem was that 17 thank-you cards (from four different people) were returned without codes to match the codes on the survey. Therefore, these people could not be used in the thank-you card analysis. However, even if the additional four people who turned in thank-you cards could have been identified, the overall number of cards exchanged was well below what was expected.

Although relatively few people completed the AC thank-you cards, a great deal was learned from this first attempt at a novel intervention to increase employee involvement regarding industrial safety. Two of the major reasons given for the low rate of card exchange, as discussed at follow-up meetings with the experimental group, were embarrassment and inconvenience. Potential solutions to these problems were suggested by some of the workers. One suggestion was to be able to give the cards indirectly. For example, the cards could be placed in a sealed box on the operating floor where they could be collected later by a supervisor. A supervisor or other leader could then individually recognize each employee who exhibited the AC behavior or publicly post the names on a feedback board of those observed actively caring. Another suggestion was to have the cards available on the operating floor rather than from the supervisors. In this way, those employees who are uncomfortable about interacting with their supervisors would be more likely to acquire the cards. It is possible, however, that these conveniences could decrease the predictive utility of the AC scale.

It was also learned that a 55¢ food item in the cafeteria was not perceived as very valuable and may have actually diminished the intrinsic value of looking out for the safety of a co-worker. As an alternative method for allocating the same rewards, it was suggested that each time a card was given, 50¢ be credited to a departmental "goodwill fund" that could be spent by the employees as they saw fit. A popular suggestion was to donate the pool of money to a charity chosen by those employees who gave or received the most cards.

Although person factors (such as knowledge, intelligence, personality, motives, and attitudes) are neither objectively measurable nor directly controllable, they certainly influence acceptance of changes in the environment and safe operating procedures. Indeed, Skinner (1971) wrote of the indirect person effects (i.e., perceptions of "freedom") resulting from different behavior change contingencies (i.e., positive vs. negative reinforcement). Differential perceptions of self-esteem, ownership, teamwork, commitment, empowerment, and optimism can result from objective manipulations to environments, behaviors, and environment-behavior contingencies. For example, to help employees achieve a Total Safety Culture, we teach employees that the individual difference (or person) factors represented in the AC model are states or expectancies, not traits. These difference factors influence one's propensity to get involved in a safety process to benefit other employees and can be influenced by environmental and behavioral manipulations (Geller and Roberts 1993; Geller, Gilmore, and Roberts 1992). Indeed, a critical group exercise involves listing specific situations and incidents in the employees' particular work setting that increase (or facilitate) and decrease (or inhibit) the relevant person characteristics related to AC behavior.

Variables consistently listed as determinants of self-esteem include communication strategies, reinforcement and punishment contingencies, and leadership styles. Our discussions have led to a number of suggestions for building self-esteem: a) soliciting and following up employee suggestions, b) providing opportunities for personal learning and peer mentoring, c) increasing management and peer attention to the occurrence of safe behaviors as well as unsafe behaviors, and d) increasing recognition of personal competence and accomplishments.

Common suggestions for increasing a sense of group cohesion among employees in a corporate culture have included decreasing the frequency of top-down directives and "quick-fix" programs obtained from other facilities and increasing team-building discussions, group goal-setting and feedback, group celebrations for both process and outcome achievements, and the use of self-managed (or self-directed) work teams (Geller 1991).

Suggestions for increasing a sense of optimism include:

- breaking down overwhelming tasks into discrete smaller ones that are more easily managed;
- setting short-term goals and tracking accomplishments;
- offering frequent rewarding and correcting feedback for process activities, not just for outcomes;
- providing opportunities for employees to set their own goals;
- teaching co-workers to record and chart "small wins" (Weick 1984);
- teaching employees to define, observe, and record desired and undesired environments and behaviors, and giving them time and resources to conduct environmental and behavioral audits; and
- teaching employees basic behavior change and behavior support strategies and providing them time and resources to implement and evaluate the impact of their corrective and supportive actions.

Many of the above techniques are commonly used as part of organizational behavior management interventions. This research further validates the appropriateness of such techniques and suggests ways of assessing the needs of a particular work group, ways of increasing employee involvement in industrial safety programs, and ways of evaluating these programs. Future research in this area should focus on whether interventions that increase relevant personality characteristics lead to increases in AC behaviors and whether increases in AC behaviors lead to reduced occupational injuries.

References

Batson, C. D., M. H. Bolen, J. A. Cross, and H. E. Neuringer-Benefiel. 1986. Where is altruism in the altruistic personality? *Journal of Personality and Social Psychology*, 1:212–220.

Blake, J. A. 1978. Death by hand grenade: Altruistic suicide in combat. *Suicide and Life Threatening Behavior*, 8:46–59.

Carlson, M., V. Charlin, and N. Miller. 1988. Positive mood and helping behavior: A test of six hypotheses. *Journal of Personality and Social Psychology,* 55:211–229.

Coopersmith, S. 1967. *The antecedents of self esteem.* San Francisco: Freeman.

Darley, J. M., and B. Latané. 1968. Bystander intervention in emergencies: Diffusion of responsibility. *Journal of Personality and Social Psychology*, 8:377–383.

Geller, E. S. 1991. If only more would actively care. *Journal of Applied Behavior Analysis*, 24:607–612.

———. 1992. *Applications of behavior analysis to prevent injuries from vehicle crashes.* Monograph published by Cambridge Center for Behavioral Studies, Cambridge, MA.

Geller, E. S., T. D. Berry, T. D. Ludwig, R. E. Evans, M. R. Gilmore, and S. W. Clarke. 1990. A conceptual framework for developing and evaluating behavior change interventions for injury control. *Health Education Research: Theory and Practice*, 5:125–137.

Geller, E. S., M. R. Gilmore, and D. S. Roberts. 1992. *Achieving total safety management through employee involvement.* Newport, VA: Make-A-Difference, Inc.

Geller, E. S., and D. S. Roberts. 1993. *Achieving a total safety culture through employee involvement*. Newport, VA: Make-A-Difference, Inc.

Isen, A. M., M. Clark, and M. Schwartz. 1976. Duration of the effect of good mood on helping: "Footprints in the sands of time." *Journal of Personality and Social Psychology*, 34:385–393.

Isen, A. M., and P. F. Levin. 1972. Effect of feeling good on helping: Cookies and kindness. *Journal of Personality and Social Psychology*, 21:384–388.

Knight, P. A., and P. A. Wadel. 1986. Humility revisited: Self-esteem, information search, and policy consistency. *Organizational Behavior and Human Decision Processes*, 38:196–206.

Latané, B., and S. Nida. 1981. Ten years of research on group size and helping. *Psychological Bulletin*, 89 (2):308–324.

Levy, P. E., and A. H. Baumgardner. 1991. Effects of self-esteem and gender on goal choice. *Journal of Organizational Behavior*, 12:529–541.

Maslow, A. H. 1970. *Motivation and personality.* New York: Harper and Row.

Michilini, R., J. Wilson, and L. Meese. 1975. The influence of psychological needs on helping behavior. *The Journal of Psychology*, 91:253–258.

National Safety Council. 1991. *Accident facts.* Washington, D.C.

Rosenberg, M. 1965. *Society and the adolescent self-image.* Princeton, NJ: Princeton University Press.

Rutkowski, G., C. Gruder, and D. Romer. 1983. Group cohesiveness, social norms, and bystander intervention. *Journal of Personality and Social Psychology*, 44:545–552.

Scheier, M., and C. Carver. 1985. Optimism, coping and health: assessment and implications of generalized outcome expectancies. *Health Psychology*, 4:219–247.

Seligman, M. E. P. 1991. *Learned optimism.* New York: Alfred A. Knopf.

Skinner, B. F. 1971. *Beyond freedom and dignity.* New York: Alfred A. Knopf.

Staub, E. 1978. *Positive social behaviors and morality: Social and personal influence.* New York, NY: Academic Press.

Wilson, J. P. 1976. Motivation, modeling and altruism: A person X situation analysis. *Journal of Personality and Social Psychology*, 34:1078–1086.

Wheeless, L. R., V. E. Wheeless, and F. Dickson-Markman. 1982. The relations among social and task perceptions in small groups. *Small Group Behavior*, 13:373–384.

Weick, K. E. 1984. Small wins: Redefining the scale of social problems. *American Psychologist*, 39:40–49.

Predicting Propensity to Actively Care for Occupational Safety[2]

E. Scott Geller, Steve Roberts, and Mike Gilmore

Abstract

A 154-item safety culture survey (SCS) was administered to the employees of two industrial plants to test a model designed to predict individuals' willingness to "actively care" (AC) for the safety of co-workers. A total of 530 surveys were completed at a mean return rate of 89 percent. The regression analysis at each site ($R^2 = .267$ and .466) showed the same four subscales (measures of personal control, group cohesion, extroversion, and reactance) on the SCS to predict independent variance in employees' reported willingness to AC. Furthermore, a higher percentage of scales hypothesized to predict AC were significantly correlated with AC (90 percent) than the scales hypothesized not to predict AC (50 percent). Therefore, the results were largely consistent with theory (i.e., our AC model), except for the prominent impact of reactance on AC propensity. Thus, the results showed convergent and divergent validity of the AC model, while also suggesting specific theory refinements.

Introduction

Numerous researchers have reported significant improvement in individual and group work performance following a behavior-based feedback process (e.g., Geller, Eason, Phillips, and Pierson 1980; Komaki, Heinzmann, and Lawson 1980; Sulzer-Azaroff and De Santamaria 1980). This approach to organizational behavior management whereby workers receive specific feedback from systematic observation and recording of designated target behaviors has been applied frequently and successfully to reduce work injuries (Krause, Hidley, and Hodson 1996; Sulzer-Azaroff 1982, 1987). However, the research demonstrating the beneficial impact of behavioral observation and feedback on occupational safety has usually been short-term and small-scale, requiring outside agents (or consultants) to help implement the process. Large-scale and long-term application of behavior-change techniques requires the employees themselves to apply the interventions (for example, systematic behavioral observation and feedback) throughout the workplace.

Geller (1991) addressed the need to get employees actively involved in implementing behavior-change processes with the introduction of an "actively caring" model. In the context of occupational safety, actively caring behavior (AC) was operationally defined as employees acting to benefit the safety of other employees (e.g., observing and recording the safe and unsafe behaviors of co-workers and then giving them constructive behavioral feedback). From a review of the social/personality literature, Geller (1991) proposed that three personality states or expectancies, modified by workplace situations and interactions, influence employees' propensity to AC for another person's health or safety.

Specifically, individuals most likely to AC were presumed to have relatively high self-esteem ("I am valuable"), group cohesion ("I belong to a group"), and optimism ("I expect the best"). Support for this AC model was found in research showing that subjects were more likely to intervene in a bystander intervention paradigm when their self-esteem was relatively high (Michelini, Wilson, and Meese 1975; Wilson 1976), they felt close to group members (Rutkowski, Gruder, and Romer 1983), and they were in an optimistic mood after finding a dime, receiving a packet of stationery, listening to soothing music, being on a winning football team, imagining a vacation in Hawaii, or being labeled a charitable person (Carlson, Charlin, and Miller 1988).

Roberts and Geller (1995) tested the AC model by giving 65 hourly workers from one department of a large fiber-manufacturing plant a survey with subscales measuring self-esteem (Rosenberg 1965), group cohesion (Wheeless, Wheeless, and Dickson-Markman 1982), and optimism (Scheier and Carver 1985, 1993). Embedded in this questionnaire were three five-point agree/disagree questions assessing willingness to AC ("I am willing to warn my co-workers about working unsafely"). About two months later, these employees were introduced to an "AC thank-you program" whereby they could distribute AC thank-you cards to their co-workers for certain AC safety behaviors. Self-esteem, optimism, and group cohesion scores predicted significant and independent variance in self-reported willingness to actively care ($\underline{R}^2 = .362$), and those workers who either gave or received a thank-you card scored significantly higher on measures of self-esteem and group cohesion than employees who did not give or receive a thank-you card.

The present study followed up the research of Roberts and Geller by giving an expanded and refined questionnaire to a much larger sample of employees (n = 592). By replicating the survey administration across two companies, it was possible to conduct a cross validation analysis, and by including subscales predicted to be unrelated to AC propensity (i.e., measures of risky lifestyles, cognitive failures, and psychological reactance), we could obtain estimates of convergent and divergent validity. In addition, an expanded survey was administered in the present research because we modified our AC model to include four rather than three predictors of AC propensity. More specifically, we hypothesized propensity to AC to be a function of self-esteem, belonging (or group cohesion), and empowerment (Geller 2001). The empowerment construct (implied by the self-affirmation "I can make a difference") was presumed to vary directly with perceptions of optimism (Scheier and Carver 1985, 1993) and personal control (Rotter 1966; Nowicki and Duke 1974).

Whereas Roberts and Geller (1995) used only three survey questions to assess propensity to AC for safety, the survey for the present research included nine AC items. These AC items addressed employees' willingness to go beyond the call of duty to give co-workers feedback about their safe and at-risk behaviors and to look out for environmental safety hazards and take appropriate corrective action when warranted.

Actively caring (as we have operationalized the term in our survey) usually requires a person to interact with other people on behalf of their behaviors or an environmental risk. It is possible more outgoing individuals will score higher on our measures of AC behaviors than employees less sociable and more reserved; therefore, we included a measure of extroversion (Eysenck and Eysenck 1985) in our survey and hypothesized a direct relationship between extroversion and propensity to AC for occupational safety.

Method

Subjects and Setting

Employees at a plastics manufacturing plant (n = 374) in Texas (Site 1) and a textiles manufacturing plant (n = 218) in North Carolina (Site 2) were administered a Safety Culture Survey (SCS) by area supervisors, with assistance by one of the authors. The majority of the employees were either maintenance or operations personnel, although all supervisory staff also took the survey. The survey return rate was 86 percent for Site 1 (n = 328) and 92 percent for Site 2 (n = 202).

The Safety Culture Survey (SCS)

The SCS included measures of person factors hypothesized to predict one's propensity to actively care (AC) for the safety of others interspersed with questions regarding workers' perceptions of plant safety and measures of psychological reactance and cognitive failures. Each of the 154 items of the SCS was answered with a five-point Likert scale ranging from Strongly Disagree (1) to Strongly Agree (5). Five subscales were included and hypothesized to predict willingness to AC:

- self-esteem (16 questions) from Rosenberg's (1965) Self-Esteem Scale;
- group cohesion (20 questions) from the Wheeless, Wheeless, and Dickson-Markman (1982) Group Cohesion Measure;
- optimism (8 questions) from Scheier and Carver's (1985) Life Orientations Scale;
- personal control (25 questions) from the Nowicki-Strickland Internal-External Scale (Nowicki and Duke 1974; Strickland 1989); and
- extroversion (9 questions) from the Extroversion/Introversion scale of the Eysenck Personality Questionnaire (Eysenck and Eysenck 1985).

Each subscale was refined slightly from the original to fit a corporate culture.

The SCS included three personality measures hypothesized not to predict AC propensity: 1) psychological reactance (12 questions) based on the Merz Psychological Reactance Scale (Merz 1983; Tucker and Byers 1987); 2) cognitive failures (26 questions) from the Cognitive Failures Scale developed by Broadbent et al. (1982); and 3) a Risky Lifestyles Scale (29 questions) developed by the authors to predict personal injury rate on the job (Geller, Roberts, and Gilmore 1992).

The SCS also included a subscale to measure willingness to AC (i.e., the AC subscale). The three items used in the original AC subscale (Roberts and Geller 1995) were expanded to include the following nine statements:

- If I know a co-worker is going to do a hazardous job, I am willing to remind him or her of the hazards, even if the employee is familiar with the job;
- I feel comfortable praising my co-workers for working safely;
- I am willing to warn other co-workers about working unsafely;
- I am willing to do whatever I can to improve safety, even confronting other co-workers about their unsafe acts;
- I am willing to observe the work practices of a co-worker and record his or her safe and unsafe behaviors;
- I am willing to pick up after another employee to maintain good housekeeping;
- When I see a potential safety hazard (e.g., oil spill), I am willing to correct it myself if possible;
- I am willing to pick up work-place litter that I did not cause myself;
- If I notice an unsafe feature in the equipment outside my work area, I am willing to take corrective action (e.g., notify my supervisor or complete appropriate paperwork).

All 154 items of the SCS were intermixed randomly with no two items of the same subscale occurring serially.

Procedure

During area safety meetings, the subjects from both sites received the SCS from their supervisors who introduced it as an information-gathering tool to find both good and bad aspects of the safety climate at their plant. Although codes were used to categorize the surveys according to department or work group, names were not included on the SCS to identify subjects, and the employees were told their answers would be completely anonymous. The survey took approximately one hour to complete. Site 1 received the SCS in November, 1992; Site 2 received the SCS in December, 1992.

Results

Inter-item Reliability

An internal consistency analysis was used to estimate the reliability of each SCS subscale (Murphy and Davidshofer 1988). Table 2.2.1 lists each subscale, the number of scale items, and Chronbach's alpha for both Site 1 and Site 2.

Subscales	Number of Items	Site 1 Alpha (n = 328)	Site 2 Alpha (n = 202)
Self-Esteem	16	.80	.79
Group Cohesion	20	.90	.89
Optimism	8	.75	.69
Personal Control	25	.73	.73
Extroversion	9	.69	.72
Reactance	12	.74	.75
Cognitive Failures	26	.90	.92
Risky Lifestyle	29	.80	.77
Actively Caring	9	.79	.83

Table 2.2.1 Scales from the Safety Culture Survey, the Number of Scale Items, and the Scale Alpha for Sites 1 and 2

Relationships between Subscales

Table 2.2.2 depicts the correlations between each SCS subscale for Site 1 and Site 2. At both sites, the highest inter-subscale correlations occurred between the three subscales measuring self-esteem, personal control, and optimism. Also for both sites, the AC subscale correlated significantly ($p < .01$) with each of the subscales predicted to influence AC propensity (i.e., self-esteem, group cohesion, optimism, personal control, and extroversion). The measure of cognitive failure correlated negatively with self-esteem, optimism and personal control at both sites, and at one site cognitive failures correlated significantly (and inversely) with the AC subscale. Table 2.2.2 also depicts significant negative relationships between reactance and the AC subscales at both sites. All of the reliable correlations were notably higher at Site 2 than Site 1.

Site 1

	SE	GC	Op	PC	Ex	Rx	CF	RL	AC
Self-Esteem (SE)	-								
Group Cohesion (GC)	0.21	-							
Optimism (Op)	0.60	0.27	-						
Personal Control (PC)	0.48	0.20	0.44	-					
Extroversion (Ex)	0.23	0.08	0.18	0.02	-				
Reactance (Rx)	-0.18	-0.24	-0.29	-0.30	0.00	-			
Cognitive Failures (CF)	-0.30	-0.13	-0.27	-0.21	0.13	0.28	-		
Risky Lifestyle (RL)	-0.02	0.01	-0.09	0.02	0.08	-0.03	0.17	-	
Actively Caring (AC)	0.25	0.29	0.28	0.39	0.13	-0.38	-0.09	0.01	-

Site 2

	SE	GC	Op	PC	Ex	Rx	CF	RL	AC
Self-Esteem (SE)	-								
Group Cohesion	0.15	-							
Optimism (Op)	0.64	0.23	-						
Personal Control (PC)	0.58	0.23	0.58	-					
Extroversion (Ex)	0.30	0.26	0.21	0.30	-				
Reactance (Rx)	-0.25	-0.31	-0.36	-0.26	0.04	-			
Cognitive Failures (CF)	-0.50	-0.16	-0.43	-0.39	-0.03	0.31	-		
Risky Lifestyles (RL)	-0.01	-0.16	-0.09	-0.01	0.14	0.12	0.04	-	
Actively Caring (AC)	0.36	0.41	0.41	0.45	0.41	-0.47	-0.29	-0.07	-

Table 2.2.2 Correlations Between Each Sub-scale of the Safety Culture Survey at Site 1 (n = 328) and Site 2 (n = 202). Correlations of .18 or Higher are Significant at the .01 Level.

Regression Analyses

Site 1 regression. A stepwise multiple regression (Norûsis 1990) was performed with the Site 1 AC subscale scores as the dependent variable and self-esteem, optimism, personal control, group cohesion, psychological reactance, extroversion, risky lifestyles, and cognitive failures as the independent variables. Table 2.2.3 presents the independent variables for each step, the partial *r*, model *R*, R^2, and *t* values for the partial *r*. The model R^2 (.27) was significantly different than zero after four steps $F(4, 323) = 27.7$, $p < .001$. The four variables included in the regression equation were personal control, reactance, group cohesion, and extroversion.

Cross validation. Because a relatively large number of variables (8) was used to predict propensity to AC, the potential for capitalizing on chance could lead to an inflated R. Therefore, the Site 1 regression equation was

Step	Variable	Partial R.	Model R	R^2	t	
1	Personal Control		.39	.15	7.57	**
2	Reactance	-.30	.48	.23	-5.67	**
3	Group Cohesion	.19	.51	.26	3.54	**
4	Extroversion	.13	.52	.27	2.35	*

*Table 2.2.3 Multiple Stepwise Regression with Site 1 AC Subscale Scores as the Dependent Variable and Self-Esteem, Optimism, Personal Control, Group Cohesion, Psychological Reactance, Extroversion, Risky Lifestyles, and Cognitive Failures as Independent Variables (n = 28). *p<.01*

cross validated (e.g., Stephens 1992) using the SCS data from Site 2. As shown in Table 2.2.4, the variance explained in AC scores at Site 2 ($R^2 = .31$) from applying the Site 1 regression equation to the Site 2 data was actually higher than the variance explained in Site 1 AC scores predicted with Site 1 data ($R^2 = .27$).

Site 2 stepwise regression. To determine whether the same variables included in the Site 1 regression equation would also be included for Site 2, a stepwise multiple regression was performed with the Site 2 AC subscale as the dependent variable and the other SCS subscales as independent variables. As depicted in Table 2.2.5, the same four factors (i.e., personal control, reactance, group cohesion, and extroversion) that entered into the Site 1 regression equation also entered into the Site 2 regression equation, although not in the same order. The model R^2 (.47) was significantly different from zero after four steps, $F(4, 179) = 43.1$, $p < .001$.

	R	R^2	F	
Site 2 AC Predicted by Site 1 Regression Equation †	.55	.30	84.3	**

*Table 2.2.4 Site 1 Regression Cross Validated Using Site 2 Data. ** p< .01*

† $\hat{y}$ =21.6 + .272 (Personal Control) + -.259 (Reactance) + .168 (Group Cohesion) + .11 (Extroversion)

Step	Variable	Partial R.	Model R	R^2	t	
1	Reactance		.47	.22	-7.52	**
2	Extroversion	.48	.63	.40	7.75	**
3	Personal Control	.27	.67	.45	3.98	**
4	Group Cohesion	.20	.68	.47	2.80	**

*Table 2.2.5 Multiple Stepwise Regression Analysis with the Site 2 AC Subscale as the Dependent Variable and Self-Esteem, Optimism, Personal Control, Group Cohesion, Psychological Reactance, Extroversion, Risky Lifestyles, and Cognitive Failures as Independent Variables (n = 202).** p< .01*

Convergent and Divergent Validity

Convergent validity was estimated by calculating the correlations between each of the scales hypothesized to predict AC (i.e., self-esteem, optimism, personal control, group cohesion, and extroversion) and the Actively Caring Subscale (ACS) for Sites 1 and 2. All scales hypothesized to correlate with the ACS except one (i.e., extroversion at Site 1) were significantly correlated with the ACS. In other words, nine of the 10 correlations (five measures at two sites) hypothesized to be significantly correlated with the ACS were significantly correlated with the ACS.

Divergent validity was estimated by calculating the correlations between each of the scales hypothesized not to predict AC (i.e., psychological reactance, risky lifestyles, and cognitive failures) and the ACS for Sites 1 and 2. Three of the six correlations (risky lifestyles at both sites and cognitive failures at Site 1) were not significant.

Discussion

Geller et al. (1990) introduced a system for categorizing behavior change techniques and developing a more systematic approach to designing intervention programs to fit a particular behavioral problem, target audience, and organizational culture (see also Geller 1998). Interventions were categorized into multiple levels or tiers, with each level defined by its intrusiveness and cost-effectiveness. At the top of the "multiple intervention hierarchy" (Level 1), the interventions are least intrusive and target the maximum number of persons for the least cost per person. At this level, intervention techniques (e.g., attempts to activate behaviors through signs, billboards, and public service announcements) are designed to have maximum large-scale appeal with minimal personal contact between target individuals and intervention agents. Geller et al. hypothesized that those individuals uninfluenced by initial exposure to Level 1 types of interventions will be uninfluenced by repeated exposures to interventions at the same level of cost-effectiveness. These individuals require a more intrusive and costly (i.e., higher level) intervention.

Higher level (and more influential) intervention processes require increased costs in terms of materials and personnel. Compared to signs, lectures, and policy statements, for example, an observation and feedback process changes the behavior of more individuals, but such programs are much more costly to implement with regard to personnel, materials, and effort. These programs are wasted on individuals who already emit the target behavior, but they are necessary for "hard-core" resistant (perhaps reactant) individuals who are not influenced by behavior change techniques less intensive, less intrusive, and less costly. In the job setting, it is important to develop and implement higher-order (more intrusive and costly) intervention processes for the more resistant employees. These higher-level intervention processes require the active assistance of other employees.

A key proposition in the multiple intervention level model proposed by Geller et al. (1990) and refined by Geller (1998) is that individuals influenced by an intervention program should not be targeted for further intervention; they should instead be enrolled as intervention agents for the next level of behavior change intervention. In other words, "preaching to the choir" is not as beneficial as enlisting the "choir" to preach to others (Katz and Lazarfeld 1955).

Indeed, employee involvement is a key issue in almost every recent publication addressing the human element of occupational safety (Carder 1994; Montante 1994; Shields 1994). The actively caring (AC) model introduced by Geller (1991, 1994) and refined by Geller, Roberts, and Gilmore (1992) was designed as a heuristic to identify those employees most likely to become intervention agents for organizational behavior change and to guide the development of intervention strategies for increasing the probability that employees will become intervention agents. This latter purpose has substantial potential for beneficial application in real-world settings. For example, if certain personality states or experiences reliably predict propensity to actively care (AC) or to get involved in an extra safety effort, then a case could be made for implementing intervention techniques to increase these states or expectancies.

The present research demonstrated that certain person factors predicted by the AC model did predict employees' willingness to AC at both industrial sites. Three of the five personality factors hypothesized to influence AC behavior predicted independent and significant variance in employees' reported willingness to AC for the safety of other co-workers. The two factors which did not enter into the regression equations (self-esteem and optimism) correlated highly with personal control and the AC subscale. Thus, the measures of self-esteem, optimism, and personal control predicted overlapping variance in propensity to AC, with personal control having the most predictability.

The use of the same self-report measure to provide both the independent and dependent variables leads to two potential problems. First, there was no direct measure of safety-related behavior; therefore, the extent to which these self-reports correspond to actual safety-related behaviors could be questioned. However, Roberts and Geller (1995) found self-reports of AC behaviors (i.e., the number of AC thank-you cards given and received) in a similar group of workers to match exactly with the number of AC thank-you cards turned in to the experimenters.

A second potential problem with the use of the same self-report measure to provide both the independent and dependent variables is that common method variance could lead to significant correlations among predictor and criterion variables. For example, subjects could have simply answered all the questions in the most socially desirable way. It is likely there were low demand characteristics for socially desirable responses, however, because names were not included on the SCS, and it was stressed that the SCS was an information gathering tool to find both good and bad aspects of the safety climate. Estimates of both convergent and divergent validity were obtained by examining the correlations between the scales hypothesized to predict and hypothesized not to predict AC to further rule out the possibility of common method variance as the cause of the current results.

The correlations in Tables 2.2.6 and 2.2.7 suggest both convergent and divergent validity. In other words, a higher percentage of scales hypothesized to predict AC were significantly correlated with the ACS (90 percent) than the scales hypothesized not to predict AC (50 percent). The correlations between reactance and the ACS, however, were substantial (giving the highest average across the two sites), thus weakening the evidence for divergent validity and suggesting certain modifications in our AC model as discussed below.

	Self-Esteem	Group Cohesion	Optimism	Personal Control	Extroversion
Site 1	.25	.29	.28	.39	.13
Site 2	.36	.41	.41	.45	.41

Table 2.2.6 Convergent Validity Estimate: Variables Hypothesized to Predict Actively Caring Correlated with the AC Subscale

	Reactance	Cognitive Failures	Risky Lifestyles
Site 1	-.38	-.09	.01
Site 2	-.47	-.27	-.07

Table 2.2.7 Discriminate Validity Estimate: Variables Hypothesized Not to Predict Actively Caring Correlated with the AC Subscale

The cross validation analysis resulted in reasonably high predictability of Site 2 AC propensity (R^2 = .31) with the regression equation derived from Site 1 data. The four person states that consistently predicted significant and independent variance in the AC subscale were personal control, group cohesion, extroversion, and reactance. Reactance was not hypothesized to be a significant predictor of AC behavior, but the negative relationship between reactance and willingness to go beyond the call of duty for occupational safety is reasonable from post hoc considerations.

Occupational safety is typically perceived as following top-down policies and mandates, and compliance with safety rules is usually managed with fault-finding investigations, interpersonal confrontations, discipline sessions, and letters of condemnation. Given a top-down, rule-enforcement perception of corporate safety, it seems reasonable that persons scoring high on reactance would be relatively unwilling to go beyond the call of duty and AC for the safety of other employees. Thus, this post hoc interpretation of our findings warrants an extension of our AC model. Our next test of the AC model will include reactance as a predictor of propensity to AC.

The high correlations between self-esteem, optimism, and personal control suggests substantial redundancy in these constructs as they were assessed in this research, at least with regard to influencing propensity to AC. Actually, when we have asked employees to operationalize these person factors by listing procedures and situations that increase and decrease these states, we have found substantial overlap between lists (Geller et al. 1992). Likewise, when we ask employees to suggest action strategies for independently increasing self-esteem, optimism, and personal control, the same basic strategies are suggested for each concept. Employees suggested:

- increasing positive recognition programs,
- listening actively and with empathy,
- setting realistic and trackable goals,
- treating others with respect,
- considering safety as achievement rather than as loss-control,
- holding people accountable for processes to reduce injuries rather than for the company's injury rate, and

⁎ celebrating "small win" accomplishments of safety achievement goals.

The results of the present research supported this observation from employee workshop discussions. Follow-up research is needed to study the unexpected negative relationship between reactance and reported willingness to AC and to determine whether the characteristics or skills of extroverts that make them more willing to AC could be taught at employee training sessions.

References

Broadbent, D., P. F. Cooper, P. Fitzgerald, and K. Parkes. 1982. The cognitive failures questionnaire (CFQ) and its correlates. *British Journal of Clinical Psychology*, 21:1–16.

Carder, B. 1994. Quality theory and the measurement of safety systems. *Professional Safety*, 39(2):23–28.

Carlson, M., V. Charlin, and N. Miller. 1988. Positive mood and helping behavior: A test of six hypothesis. *Journal of Personality and Social Psychology*, 55:211–229.

Eysenck, H. J., and M. W. Eysenck. 1985. *Personality and individual differences: A natural science approach*. New York, NY: Plenum.

Geller, E. S. 1991. If only more would actively care. *Journal of Applied Behavior Analysis*, 24:607–612.

———. 1994. Ten principles for achieving a total safety culture. *Professional Safety*, 39(9):18–24.

———. 1998. *Applications of behavior analysis to prevent injuries from vehicle crashes. 2nd ed.* Monograph published by the Cambridge Center for Behavioral Studies, Cambridge, MA.

———. 2001. Actively caring for occupational safety: Extending the performance management paradigm. In *Organizational performance: Behavior analysis and management*. C. M. Johnson, W. K. Redmon, and T. C. Mawhinney, eds. New York: Springer.

Geller, E. S., S. Eason, J. Phillips, and M. Pierson. 1980. Interventions to improve sanitation during food preparation. *Journal of Organizational Behavior Management*, 2:229–240.

Geller, E. S., T. D. Berry, T. D. Ludwig, R. E. Evans, M. R. Gilmore, and S. W. Clarke. 1990. A conceptual framework for developing and evaluating behavior change interventions for injury control. *Health Education Research: Theory and Practice*, 5:125–137.

Geller, E. S., D. S. Roberts, and M. R. Gilmore. 1992. *Achieving a total safety culture through employee involvement*. Newport, VA: Make-A-Difference, Inc.

Katz, E., and P. E. Lazarfeld. 1955. *Personal influence: The part played by people in the flow of mass communication*. Glencoe, IL: Free Press.

Komaki, J., A. T. Heinzmann, and L. Lawson. 1980. Effect of training and feedback: Component analysis of a behavioral safety program. *Journal of Applied Psychology*, 65:261–270.

Krause, T. R., J. H. Hidley, and S. J. Hodson. 1996. *The behavior-based safety process. 2nd ed.* New York: Van Nostrand Reinhold.

Merz, J. 1983. Fragebogenzur messung der psychologischen reaktanz (Questionnaire for measuring psychological reactance). *Diagnostica*, 29:75–82.

Michelini, R. L., J. P. Wilson, and L. A. Meese. 1975. The influence of psychological needs on helping behavior. *The Journal of Psychology*, 91:235–258.

Montante, W. M. 1994. An ergonomic approach to task analysis. *Professional Safety*, 39(2):18–22.

Murphy, K. R., and C. O. Davidshoher. 1988. *Psychological testing*. New Jersey: Prentice Hall.

Norûsis, M. J. 1990. *SPSS base system users guide*. Chicago: SPSS, Inc.

Nowicki, S., and M. P. Duke. 1974. A locus of control scale for college as well as non-college adults. *Journal of Personality Assessment*, 38:136–137.

Roberts, D. S., and E. S. Geller. 1995. An "actively caring" model for occupational safety: A field test. *Applied and Preventive Psychology*, 4:53–59.

Rotter, J. B. 1966. Generalized expectancies for internal versus external control of reinforcement. *Psychological Monographs*, 80:1.

Rosenberg, M. 1965. *Society and the adolescent self-image*. Princeton, NJ: Princeton University Press.

Rutkowski, G. K., C. L. Gruder, and D. Romer. 1983. Group cohesiveness, social norms, and bystander intervention. *Journal of Personality and Social Psychology*, 44:545–552.

Scheier, M. F., and C. S. Carver. 1993. On the power of positive thinking: The benefits of being optimistic. *Current Directions in Psychological Sciences*, 2:26–30.

———. 1985. Optimism, coping and health: Assessment and implications of generalized outcome expectancies. *Health Psychology*, 4:219–247.

Shields, M. A. 1994. Human relations in safety. *Professional Safety*, 39(2):40–42.

Stephens J. 1992. *Applied multivariate statistics for the social sciences.* New Jersey: Lawrence Erlbaum Associates.

Strickland, B. R. 1989. Internal-external control of expectancies: From contingency to creativity. *American Psychologist*, 44:1–12.

Sulzer-Azaroff, B. 1982. Behavioral approaches to occupational health and safety. In *Handbook of organizational behavior management.* L. W. Frederiksen, ed. New York: John Wiley and Sons. 505–538.

———. 1987. The modification of occupational safety behavior. *The Journal of Occupational Accidents*, 9:177–197.

Sulzer-Azaroff, B., and M. C. De Santamaria. 1980. Industrial safety reduction through performance feedback. *Journal of Applied Behavior Analysis,* 13:287–295.

Tucker, R. K., and P. Y. Byers. 1987. Factoral validity of Merz's psychological reactance scale. *Psychological Reports*, 61:811–815.

Wheeless, L. R., V. E. Wheeless, and F. Dickson-Markman. 1982. The relations among social and task perceptions in small groups. *Small Group Behavior*, 13:373–384.

Wilson, J. P. 1976. Motivation, modeling, and altruism: A person X situation analysis. *Journal of Personality and Social Psychology*, 34:1078–1086.

Does Everything Happen for a Reason?[3]

E. Scott Geller

It is not unusual to blame the victim of unfortunate circumstances—including workplace safety incidents. Psychological research has shown that victims of crimes and vehicle crashes are often perceived as causing their fate.

For example, many people believe rape victims are somewhat to blame for the rape. "If she had not dressed and acted that way, she would not have been sexually assaulted." In other research, battered wives were viewed by some as being responsible for their abusive husbands' behavior.

Social psychologists believe this thinking is a form of self-defense that helps us maintain our belief that life is safe, orderly, and predictable. If life were unfair and random, then innocent people like ourselves could also be victimized and injured.

In psychological terms, this attributional bias is referred to as the "just world" hypothesis. This is the assumption that people get what they deserve and deserve what they get. How do we deal with the constant reminder that tragedies not only happen, but they are also relatively common? One way is to believe in a "just world": Bad things happen to bad people and good things happen to good people.

Since most of us see ourselves as decent and basically good, our belief in a just, safe world means bad things won't happen to us. In the remainder of this article, I would like to point out four ways that this defense mechanism can hurt our efforts to protect workers on the job.

1. Increased Risk-taking

The phrase "It won't happen to me" is familiar to all of us. This belief enables us to face situations where a serious injury is possible—from the drive to and from work to numerous situations while we work. As it turns out, most days this conviction is supported. Nothing happens to us.

This belief, obviously consistent with the "just world" hypothesis, increases risk-taking. We take a chance—a calculated risk—because we think we are safe and secure in a "just world."

Maybe we would not take as many risks if we realized that our co-workers who suffered injuries also believed in a "just world." They did not expect anything bad to happen to them—only to the other guy. Now they know what it is like to be that other guy.

2. Poor Incident Analysis

Let's be honest—haven't you at one time or another tried to feel better about someone getting hurt or killed by looking for things the victim could have done to prevent the incident? Do you feel more sympathy for the passengers than the driver killed in a vehicle crash?

We feel safer when we can somehow blame the victim of misfortune. Then the tragedy does not seem so tragic. We feel more secure knowing we could prevent such a misfortune from happening to us by doing what the victim did not do.

But the actual or potential behaviors of a victim do not tell the whole story. Any analysis of what happened and why can't begin and end here. But sometimes it does, and for this we can blame our belief in a "just world."

3. Lack of Involvement

It is not easy to get people to actively care for each other's safety. Not only must we be educated and trained to know what to do, we must feel comfortable intervening and believe it is the right thing to do.

We might be less likely to step in or speak up if we believe that people simply get what they deserve. Why should we bother? Well, you can also believe it is your responsibility to help make the world safer just by intervening. This includes removing environmental hazards and barriers to safe behavior, as well as helping others become more mindful of their behaviors that put themselves and others at risk for injury.

4. Lack of Public Support

When hazards and injuries seem unfair, such as when a child is molested or inflicted with a deadly disease, the public pays special attention. It is comparably easy to obtain contributions or voluntary assistance for programs that target vulnerable populations, like physically challenged individuals or learning-disabled children.

Victims of workplace injuries are not viewed as weak and defenseless. Occupational injuries are indiscriminately distributed among employees who take risks, and the "just world" perception is that they get what they deserve. This helps explain the low public and political support for agencies like OSHA and NIOSH that are charged with protecting employees.

More productive lives of Americans are lost through occupational mishaps and vehicle crashes than through cancer and heart attacks combined. Yet the amount of government support for cancer and cardiovascular disease is many times greater than the amount allocated for research and interventions to prevent injuries and fatalities in the workplace and on the road. I believe the "just world" hypothesis is partly responsible for this unjust situation.

What to do

So what can we do about the "just world" hypothesis and its impact on workplace safety? We can turn it in our favor. Through education and regular conversations, we can convince people that a "just world" actually depends on our actions. We can help make our world safer and more just by doing our part to remove environmental hazards, suggesting refinements that will make equipment more "user friendly," and helping people improve their safety-related behavior.

A "just world" can be an ultimate vision, made possible by our attempts to actively care for the safety and health of our co-workers and ourselves.

Flawed Thinking

Believing in a "just world" in which people get what they deserve can~~

~~Lead to more risk-taking—because in a "just world" we do not believe we deserve to get hurt.

~~Stifle incident investigations and getting at root causes—because we blame the victim and leave it at that.

~~Prevent us from getting involved in safety efforts—because in a "just world" everything happens for a reason and people get what they deserve, so why should we bother?

~~Reduce public funding for government agencies that target occupational safety (OSHA, MSHA, and NIOSH)—because workers injured on the job get what they deserve. In comparison, EPA gets a bigger budget because it protects children and families unfairly exposed to pollution threats.

Safety Self-Management: A Key Behavior-Based Process for Injury Prevention[4]

E. Scott Geller and Steve Clarke

Many safety professionals believe behavior-based safety is a one-to-one coaching process whereby one employee (the coach) completes a behavioral checklist while watching another employee work. Then the coach gives the employee feedback by reviewing the "safe" and "at-risk" checks on the list. This observation and feedback process is certainly an important behavior-based safety method, and it can reap great benefits (Geller 1996; Krause, Hidley, and Hodson 1996; McSween 1995). Safe behaviors will increase, at-risk behaviors will decrease, and workplace injuries will be prevented. Plus, interpersonal trust and group cohesion will increase throughout the culture as workers experience interdependency with regard to personal health and safety (Geller 1998c).

It is useful, however, to consider the application limitations of a one-to-one coaching process. While it is a powerful approach to improving behavior and attitude, one-to-one safety coaching requires the presence of another person (the coach) as well as a situation in which critical behaviors are likely to occur. What about those numerous times when people are working alone? And, what about those important safety-related behaviors that occur infrequently and are not likely to be occurring during a scheduled observation and feedback session? Figure 2.4.1 illustrates such a situation.

The worker in Figure 2.4.1 is performing a critically important task. The various components of this task, from body positioning to using certain personal protective equipment, are inconvenient or somewhat uncomfortable. And, the safe procedures are not the most efficient. Why is this worker following safe operating procedures? He is working alone, so there is no one around to hold him accountable. He could take a shortcut and get to the breakroom faster. Someone could hold him accountable for completing the lock-out procedure by checking out the equipment later, but no one can know *how* this lone worker locked out the equipment. And the probability of incurring an injury while locking out equipment with the wrong body language must be minuscule.

Figure 2.4.1 Critically Important Safe Behaviors Are Relatively Infrequent and Often Performed Alone.

So what motivates the worker in the figure to follow a certain list of lock-out procedures? How did this person get to that level of commitment to safety? And how can an organization help others hold themselves accountable for going out of their way for safety when working alone? That is the theme of this article. Interpersonal observation and feedback are not relevant, but principles from behavior-based safety are applicable. In fact, the very essence of behavior-based safety is used for the process described here. We call it safety self-management.

What Is Safety Self-Management?

The methods and tools of effective self-management have been derived from behavioral science research and are perfectly consistent with the principles of behavior-based safety (Geller 1998b, c). Self-management techniques are clearly relevant for situations beyond safety. In fact, the procedures and tools of this behavior-based intervention process were developed and evaluated by behavioral scientists interested in helping people with the variety of personal adjustment issues seen in the clinical psychology clinic; few if any of these problems relate to safety. For example, a popular and comprehensive review of self-management technology first published in 1970 and now in its seventh edition (Watson and Tharp 1997) never mentions safety-related behavior. Personal improvement domains addressed thoroughly by Watson and Tharp with self-management techniques include: anxiety and stress; weight loss and overeating; studying and time management; depression and low self-esteem; smoking, drinking, and drugs; exercise and athletics; and relations with others, including interpersonal anxieties, social skills, and dating.

Although specific therapy procedures vary considerably across the various personal adjustment areas targeted with self-management techniques, the basic principles are the same and parallel those used in behavior-based safety (as reviewed by Geller, 1998a, b). Therefore, it is not difficult to apply self-management principles to safety-related problems. And the best news is that over 30 years of rigorous evaluative research has shown that when these basic principles of behavior-based psychology are applied correctly to the variety of personal improvement areas listed above, self-management works (as reviewed by Watson and Tharp 1997).

From Outside to Inside Direction

Watson and Tharp (1997) emphasize that all behavior passes through three sequential stages: control by others, control by self, and automatization. It is useful to consider control by others as "accountability" and control by self as "responsibility" (Geller 2001a). Automatization occurs when behavior becomes a habit. If the habit is good, appropriate, or safe, we say the individual is "unconsciously competent." If the habit is bad, inappropriate, or unsafe, the person is "unconsciously incompetent."

Other-directed behavior is at-risk when the instructions or the example followed are not safe. Self-directed behavior is at-risk when self-regulation does not match the safe directive, as in a calculated risk. Automatic behaviors or habits are at risk when the performer is unaware of a mismatch between the safe directive and the current behavior, as in most human error. Thus, safety self-management is a process of either matching self-directed behavior to a safety standard or bringing at-risk habits into conscious awareness so a mismatch between safe and at-risk behavior can be realized and adjusted.

Getting Started in Safety Self-Management

The first step of safety self-management is to recognize a problem or a need for improvement. And that is easier said than done, especially in the realm of safety. Most of us believe we are safe enough. If we were not, we would be getting hurt. That sounds like good common sense, but we know it is not true. From an individual perspective, the self-statement "It's not going to happen to me" makes sense and is verified everyday. Teachers and consultants who ask us to stop believing or saying this phrase are making a mistake. You cannot discredit a self-statement that is verified consistently by personal experience. If people really thought it *was* going to happen to them, they would have a good excuse to stay home.

From a group or workplace perspective, the statement "It's not going to happen" is not true and can be verified by industrial injury reports. So it is fair to say "It's going to happen to someone." Then ask the relevant follow-up question. "Do you care?" We have never heard "No" to this question, meaning most people want to help reduce the probability of co-workers getting hurt. But people might not know how to "actively care." One sure way is to always demonstrate the safe way, because if you do not, you are inadvertently teaching others to take the same calculated risks and put themselves in some degree of danger. If everyone takes the same risk, someday someone will get hurt. From an individualistic viewpoint, the probability of an injury might be small, even minuscule; from a big picture, collective outlook, the probability approaches certainty.

Now we have a reasonable self-statement to use as justification for engaging in safety self-management. We need to practice the safety standards consistently, because people learn by watching us. And if we demonstrate an at-risk shortcut, we could be teaching others to make an at-risk decision. And they will teach others, and the others will teach others, and so on until at-risk behavior becomes the norm. With an entire work force exposing themselves daily to a particular risk, someone will eventually get hurt, regardless how small the probability of injury is at the individual level.

The Techniques of Safety Self-Management

Safety self-management starts with defining one or more target behaviors that need improvement. These might be safe behaviors you need to perform more often to meet a safety standard, or they could be at-risk behaviors you need to perform less frequently. You probably know where your weaknesses are in regard to complying with ideal safety procedures. When starting out, however, you should not develop a long list of behaviors to change. Start with a few target behaviors that need improvement and are critical with regard to injury reduction.

Do not rely completely on common sense to select a critical behavior to work on. If many of your safety-related behaviors need improvement, then start with behaviors that are most critical with regard to preventing serious injury to yourself or someone else. Start with behaviors that are relatively easy to change. Then when you experience success, you will be ready to add more critical behaviors to your safety improvement list.

Self-Observing and Recording

Self-knowledge is key to successful self-management. Most of us believe we understand ourselves, and we place great faith in our ability to remember our own behaviors. You may feel it is not necessary to systematically observe your own safety-related behaviors. Unfortunately, our memories of behavior are much less accurate than we think. Research indicates that people have much more confidence in their memories than they should.

In order to improve your safety-related work practices, you need to obtain a baseline record of your safe and at-risk behaviors. In fact, before selecting a few critical safe behaviors to work on, you might need to sample several safety-related behaviors in various situations. You will, of course, pay more attention to these behaviors when you begin self-observing and recording certain behaviors, so your record will likely be an overestimation of your personal safety. That's good, of course, and shows the benefit of bringing behavior to the self-directed stage for evaluation and self-improvement. This observing and recording process is facilitated with the right kind of behavioral checklist.

Developing a Self-Observation Checklist

A good self-observation checklist (SOC) has a number of qualities. The most effective SOCs:

- ※ are portable and accessible when the target behaviors occur;
- ※ are easy and convenient to use;
- ※ are noticeable, so they can serve as an activator for making observations; and
- ※ record both safe and at-risk behaviors, allowing for a computation of percent safe behaviors.

An SOC with these characteristics will benefit self-management by making the recording process easy and convenient. The SOC depicted in Figure 2.4.2 has all of these qualities, and is simple to use on a daily basis. Space is provided for five target behaviors. Every time an opportunity for a particular behavior occurs, you merely judge whether the behavior was safe or at-risk and mark the checklist accordingly. The observation procedures are exactly as detailed elsewhere for interpersonal coaching

Self-Observation Checklist

Target Behaviors:	Safe	At-Risk	% Safe
1) ________			
2) ________			
3) ________			
4) ________			
5) ________			
Comments: Totals	________	________	

____

____

________ **Overall % Safe Behaviors** ____

Figure 2.4.2 A Self-observation Checklist Can Be Simple

(Geller 1996, 1998b), except here the observers hold themselves rather than another person accountable for certain behaviors.

Multiple opportunities to perform each target behavior can be tabulated on the SOC. At the end of a day or week, the number of safe and at-risk occurrences of each behavior can be totaled to calculate a percent safe score (number of safe occurrences divided by number of safe and at-risk occurrences) for each behavior. And an overall percent safe can be calculated from the totals. These percent safe scores can be charted on a graph and used to assess daily or weekly progress. The mere recording and charting of percent safe scores will lead to significant improvement because the process is a means of holding yourself accountable. Later we discuss additional intervention strategies you can use to make success easier and greater.

Using a self-observation checklist. Delays in recording occurrences of the target behavior(s) can weaken the improvement process. Our memories for specific behaviors become less accurate over time. So it is not a good idea to wait until the end of the day and then try to remember how many times you engaged in a particular target behavior. Your count will not be accurate. Obviously, the most accurate accountability occurs when you record occurrences of behavior soon after they occur.

Taking note of specific activators and consequences of your safety-related behaviors can provide valuable insight into the events that control your other-directed and habitual behaviors. Record these observations in the comment section of your SOC (see Figure 2.4.2). This information can be useful when developing a self-management plan. For example, comments about activators that set the occasion for at-risk behaviors tell you what situations require special attention and might suggest ways to manage certain activators. Information related to the various consequences motivating specific safe and at-risk behaviors can help you devise a plan for making it more rewarding to choose the safe behavior and less rewarding to work at risk.

The fewer critical behaviors targeted for self-observing and recording, the faster and greater the impact. Thus, targeting only a few behaviors at a time usually improves success. Also, the greater the frequency of self-observing and recording per behavior, the greater the beneficial impact.

The length of time needed to make self-observations of a specific target behavior will vary across situations and behaviors. In most cases, however, self-observations should continue for at least three to four weeks.

In summary, the process of self-observing and recording of critical safe versus at-risk behaviors has a number of benefits:

- increasing awareness of personal safety performance,
- identifying activators that influence other-directed and automatic behaviors,
- identifying consequences that motivate certain safe and at-risk behaviors, and
- providing information that can be used to develop an action plan for improving behavior and tracking progress toward reaching safety goals.

Intervening for Self-Improvement

In this phase of safety self-management you derive a plan to improve certain critical behaviors that do not meet your safety standards. This essentially means that a safe behavior needs to occur more often and/or an at-risk behavior should occur less often. You might need to conduct the observing and recording stage several times on various behaviors before pinpointing critical behaviors that require your intervention attention. As indicated earlier, the observing and recording process alone will bring other-directed and habitual behaviors under self-control. This might be sufficient for the desirable self-improvement.

Your self-observing and recording processes will likely point out ways, in particular activators and consequences, to manage the situation in order to improve safety performance. With this process you pinpoint behaviors that need self-improvement. Select a "manageable number" of these behaviors to focus on first. While there is no specific prescription for the optimal number of behaviors to address at once, it is better to start with a few behaviors and build. Once the target behavior(s) are selected, a self-improvement plan can be developed. Let's turn to a discussion of the behavior change strategies from which you can choose when designing a self-management intervention.

Activator management

Activator management involves identifying environment, behavior, and person factors that precede the occurrence of safe and at-risk behaviors. Strategies are then employed to eliminate activators that precede at-risk behaviors and add activators that can increase the probability of safe behaviors. Referring to the baseline information collected during your self-observations can identify safety-related activators. The comment section on your self-observation checklist should include useful insight about activators.

The self-observation process draws our attention to relevant activators in the workplace. This is good. We might become aware of an existing safety activator and use this reminder to bring at-risk habits under safe, self-directed control. Or, our attention to a particular activator might discourage us from taking a self-directed calculated risk—if only to set the right example. Sometimes employees have opportunities to design and post activators in their work areas. In fact, you can place a safety reminder almost anywhere—on your locker door, lunch box, workbench, computer screen, or on the dashboard of your personal vehicle.

If you have occasion to design your own safety activator for yourself or others, consider the following guidelines:

※ specify the behavior,
※ maintain salience with novelty,
※ vary the message over time,
※ activate close to the response opportunity,
※ imply consequences, and
※ involve other employees if the activator is for more than self-direction.

If you want an activator to motivate action, you need to specify the desired behavior and imply relevant consequences. The most powerful activators make the observer aware of positive consequences to gain or negative consequences to avoid by performing the target behavior. You can make a weak safety activator stronger by writing a mental script that specifies a behavior to perform for a particular consequence. Such internal scripts or self-statements make a safety message more personal and influential.

Self-statements

Self-statements are activators. They can cue certain safe or at-risk behaviors. When getting in your vehicle you might tell yourself, "I'm only going a few blocks, so I really don't need to put on my safety belt." Then you probably will not buckle up. But you will likely buckle up if you say to yourself, "I need to buckle up to set a safe example; my behavior could influence anyone who sees me, especially a family member." Notice that this safety script includes specific self-instructions, while also expressing a belief about example-setting which provides strong rationale for performing the behavior.

Self-directed messages come in three basic forms:

- self-instruction,
- belief, and
- interpretation.

Self-instruction

Self-instruction can be obvious and manifested as a clear inner voice saying, "I need to use my legs when lifting this box." Or the instructions might be suppressed and occur only as a "faded voice," so weak that only careful attention can reveal it. In this case, your challenge is to turn up the amplifier on this small voice and hear yourself reiterating the safe way to do something. It is a matter of bringing your inner voice into conscious awareness—the self-directed phase of behavioral development.

Beliefs

Beliefs are the underlying assumptions that can provide useful rationale or motivating consequences for a certain behavior. They are often not clearly available in our inner speech, but they certainly influence what we say to ourselves. They also determine whether particular self-instructions are loud and clear or weak and ambiguous.

It is useful to reflect on your beliefs and evaluate how they are influencing your self-statements. Do you really believe safety should be a core value, uncompromised by fluctuating priorities? Do you sincerely believe your behavior can influence the behavior of onlookers, thus making you responsible to always set the safe example? Do you honestly believe the behavior-based approach to safety is more effective than the more traditional attitudinal perspective?

You can see how your answers to these belief questions will influence the nature and strength of your self-statements. This critical role of belief also explains why it is important to include powerful rationale or intuitive principles with any safety training. If people hear instructions as only directions without personal meaning and significance, their behavior will be only other-directed compliance and will not reach the self-directed stage needed for personal responsibility.

Interpretations

Interpretations influence our beliefs and our self-instructions. And, of course, our beliefs influence our interpretations. People often misperceive and misinterpret information in order to protect their self-esteem. This "self-serving bias" (Geller 2001b) is likely to occur in situations when admitting an error or a calculated risk leads to punishment or feelings of embarrassment and inadequacy. Under these circumstances we are likely to lie or make excuses about our at-risk behavior and harbor misinterpretations, beliefs, and self-statements that inhibit safety self-management. On the other hand, situations that enable us to talk about our "near-misses" and calculated risks provide impetus for safety self-management and self-improvement.

Mental imagery

Mental imagery is using your "mind's eye" to picture situations without actually being there. It is a way to anticipate and prepare for events. It can be used to direct behavior (as an activator) and to motivate behavior (as a consequence). It is often most productive to create a mental picture of positive consequences resulting from your safe actions. By focusing on positive outcomes from safe behaviors, you anticipate achievement and the good feelings it brings. This can increase your confidence in being successful as well as increase your desire to reach your goals. Imagery can also help clarify what you need to do in order to accomplish your goals.

When using mental imagery for safety self-management you should:

- see yourself performing the appropriate safe behavior with ease and convenience,
- visualize avoiding specific negative consequences with the safe behavior,
- imagine feelings of accomplishment following the safe behavior,
- take an active rather than passive perspective, and
- share your imagery with others.

It is important to be active in your image. Do not see yourself as a passive observer watching a movie with you in it. Rather, imagine yourself acting within the complete activator-behavior-consequence (ABC) framework. See the activators in the situation which cue the desired behavior, visualize yourself actually performing the safe act, and then imagine positive feeling states from setting the safe example and acknowledging safety as a value.

It is often useful to share your motivating imagery with others. One reason personal testimonies of injuries or near-misses are powerful motivators is because the listeners can get a mental image. They can readily visualize the speaker in the precarious situation described, especially if the presenter gives a passionate and realistic delivery. Even more motivating is when listeners can put themselves or family members in the situation that caused negative consequences for the speaker. Of course, it is essential to specify the behaviors that can be performed to avoid the injury or near-miss discussed.

Self-rewards

In addition to imagining pleasant consequences after performing a desired target behavior, you can arrange to actually experience them. These are referred to as self-rewards in the self-management literature. Research has shown that individuals who use self-rewards are more likely to sustain their self-management process and optimize self-improvement. Consider three factors when using self-rewards:

- selection of the reward,
- delivery of the reward, and
- timing of the reward.

Individualize your reward. People like different things. The variety of possibilities is endless. However, it is important for the rewards to be accessible and convenient to use following the target behavior. Some potential rewards may not be practical. It is useful to think of self-rewards as behaviors or activities. You give yourself an opportunity to do something you really like to do after doing something less fun but important for self-improvement. Self-rewarding activities can be internal (as in self-dialogue) or external (as in spending money for food, clothes, or entertainment). Often, the strongest self-reward is simply the personal pride associated with objective and charted accomplishments resulting from self-management.

Rewards should be based on behavior, not affirmation or intention. First, self-observe your behavior systematically and self-deliver rewards according to the criteria you define. The reward should come after the target behavior, not before. And, they should also be administered as soon as possible following the target behavior.

During the early stages of self-management, it is often useful to reward yourself for participating in the process. For example, in one self-management program, the individual rewarded her involvement in self-management by adding $1 to her vacation fund each day she completed her self-observing and recording. Remember: the most successful self-rewards are readily accessible, individualized, valued, varied, and follow the desired behavior as soon as possible.

Goal-setting

Setting personal goals is one of the most powerful ways to self-manage safety. Behavior-based improvement goals should be set high yet be achievable. They specify expectations for improvement and provide for the tracking of progress. The acronym **SMART** provides a way to remember the qualities of the most influential goals: **S**pecific, **M**otivational, **A**chievable, **R**ecordable, and **T**rackable.

The goal should specify the target behavior(s) and the amount of change desired. Sometimes a time deadline is declared. Often goals are either unrealistically difficult or too vague to be meaningful and measurable. When goals are realistic and under the individual's personal control to achieve, they will be reached. Then the individual feels empowered to reach more challenging goals in the future.

Goal-setting helps you focus on improving specific critical behaviors. The target behavior(s) should be observed and recorded and charted on a graph to track progress. Sometimes it is useful to post one's self-management results publicly as a way to gain social support from co-workers. This is especially beneficial when everyone on a work team is working on their own safety self-management project. Then social support becomes a powerful mechanism for rewarding progress, activating additional improvement efforts, and setting expectances for continuous improvement.

Social support

People in control of their self-management programs create a supportive social context for themselves. They look around for people who might appreciate their personal commitment to improve personal safety. They enlist support from supervisors, co-workers, friends, or family members whom they know care.

Social support can take many forms and can be either formal or informal. For example, social support can include simple words of encouragement from co-workers, positive feedback from supervisors for personal successes, and/or formal recognition of accomplishments from your workgroup. One way to cultivate opportunities for social support is to display your self-observation data so others can see your accomplishments. You can also create a supportive social environment by signing a behavior-based promise card or devising a behavior-change contract and then sharing your commitment with co-workers, friends, and family members.

Commitment

As detailed elsewhere (Geller 1998b, 2001b), once people make a commitment, they encounter internal and external pressures to think and act consistently with their position. That is why people can act themselves into thinking differently as well as think themselves into acting differently. This is the Consistency Principle in action (Cialdini 2001). Our point here is that you will provide impetus to your self-management project if you make a formal commitment to improve your behavior. You can do this by:

※ signing and publicly displaying a simple promise card indicating your commitment to participate in a self-management process or reach a certain safety-improvement goal (see Figure 2.4.3),

I PROMISE
To Do ____________ Self-Observations
By __________.

Yes No
Met Goal ☐ ☐

Signature (optional) – I would like this posted
☐ Yes
☐ No

Figure 2.4.3 A Simple Promise Card for Getting People Started on a Behavior-based Self-management Project

※ signing a contract with a co-worker or friend that specifies the safety-related behavior(s) you wish to improve and includes a brief outline of your plan for accomplishing this goal (see Figure 2.4.4).

You might, for example, draw up a safety contract that specifies the behavior(s) you wish to change, sets a specific behavior-based goal, and indicates a time frame for reaching your goal. The contract should also include a place for both your signature and that of a person you can call on to help you reach your goal. Then share your contract with someone you would like to support your behavior improvement efforts. Once you have someone who is willing to support you, you should both sign the contract. A sample contract for promoting safe driving is shown in Figure 2.4.4.

Safe Driving Contract

During the next four weeks I, Joe Peterson, will increase my safety-belt use from 50% to 95%. During that time I will self-observe and record my safety-belt use each day and graph the results. This graph will be placed on my locker. Each day I reach my goal I will reward myself by placing $1 in my vacation fund.

____________________ __________

Signature Date

During the next four weeks I, Sally Stevens, will help Joe increase his safety-belt use by providing him encouragement on days when he does not reach his goal, and recognize his accomplishments on those days when he does reach his goal. In addition, to show my support of his commitment to increase his safety-belt use, I will set the right example and use my safety belt whenever I travel in a vehicle during this four-week period.

____________________ __________

Signature Date

Figure 2.4.4 A Sample Safety Contract between Co-workers for Increasing Safe Driving Practices

Conclusion

Safety self-management means taking personal responsibility for doing the right things to prevent injury both to self and others. When you are out there by yourself with no one to hold you accountable, do you choose the safe behavior or a more efficient or convenient at-risk behavior? In other words, do you take a calculated risk when you think you can get away with it? Safety self-management techniques can help you choose the safe way consistently, but you need to choose to use these techniques. And this takes commitment and personal responsibility.

Of course, not everyone feels personal responsibility for safety. Safety self-management will only work for behaviors that are self-directed; for some people, safety-related behaviors are only other-directed. In other publications the senior author addresses organizational and personality factors that influence whether safety-related behavior is other-directed or self-directed (Geller 2001a). A significant challenge for safety professionals is to help people transition from other-directed to self-directed for industrial safety. Then safety self-management will be relevant for more people and have greater impact on injury prevention.

This article reviewed seven self-management intervention strategies applicable individually or in various combinations to improve safety-related behaviors:

※ managing environmental activators or conditions antecedent to a target behavior;

※ developing and using self-statements to encourage desired behavior and discourage undesirable behavior;

※ picturing situations in your mind's eye to direct and motivate target behavior;

※ administering self-rewards strategically to support desired behaviors, including the completion of certain self-management processes;

※ setting specific, motivational, achievable, recordable, and trackable goals regarding both the completion of certain self-management techniques and the achievement of a certain level of behavioral improvement;

※ making an explicit voluntary commitment to complete particular aspects of a self-improvement project and/or to reach a designated level of behavioral improvement; and

※ enlisting the support of one or more friends or co-workers for social encouragement and support.

Researchers have shown each of these techniques is effective at improving target behaviors beyond levels obtained during initial self-observing and recording.

References

Cialdini, R. B. 2001. *Influence: Science and practice.* 4th ed. New York: Harper Collins College Publishers.

Geller, E. S. 1998a. Principles of behavior-based safety. In Proceedings of *Light Up Safety in the New Millennium: A Behavioral Safety Symposium*. Des Plaines, IL: American Society of Safety Engineers.

———. 1998b. *Understanding behavior-based safety: Step-by-step methods to improve your workplace.* 2nd ed. Neenah, WI: J. J. Keller & Associates.

———. 2001a. *Beyond safety accountability.* Rockville, MD: Government Institutes.

———. 2001b. *The psychology of safety handbook.* Boca Raton, FL: CLC Press.

Krause, T. R., J. H. Hidley, and J. J. Hodson. 1996. *The behavior-based safety process* 2nd ed. New York: Van Nostrand Reinhold.

McSween, T. E. 1995. *The values-based safety process: Improving your safety culture with a behavioral approach.* New York: Van Nostrand Reinhold.

Watson, D. L., and R. B. Tharp. 1997. *Self-directed behavior: Self-modification for personal adjustment.* 7th ed. Pacific Grove, CA: Brooks/Cole Publishing Company.

Notes

[1] Revised from a research article that originally appeared in *Applied and Preventive Psychology* 4 (1995):53–59.

[2] Originally appeared in *Journal of Safety Research* 27 (1996):1–8.

[3] Originally appeared as "Everything Happens for a Reason?: It's Not Safe to Believe in a Just World" in *Industrial Safety & Hygiene News* (June 1997).

[4] Updated from the original which appeared in *Professional Safety*, 47 (1999):29–33.

3

Behavior-Based Observation and Feedback

Achieving and maintaining a Total Safety Culture involves more than just good intentions and motivation. A *system* must be in place to encourage safety discussion, objectively assess safety behaviors and improvement, and sustain and reinforce actively caring behavior over time. A behavior-based observation and feedback process provides the tools and structure to accomplish this. Specific guidelines for establishing and maintaining a successful behavior-based observation and feedback process are provided in this chapter.

The first article compares athletic coaching with coaching for safety, and provides guidelines for effective delivery of both rewarding and correcting feedback. The second paper discusses the importance of integrating behavioral observation and feedback processes into other management systems.

Next, social comparison information is introduced as a powerful but under-used type of behavioral feedback to improve safety behaviors. This article describes important behavior-based field research and serves as an exemplar for the type of cost-effective research safety leaders can readily accomplish. We need more research like this to explore and objectively evaluate various ways to improve the impact of behavior-based safety (BBS).

Motivating Personal Protective Equipment (PPE) use is the topic of the next article followed by a discussion of the importance of addressing ergonomic hazards through behavior-based observation and feedback.

Safety Coaching through Observation and Feedback[1]

E. Scott Geller and Anne French

Safety coaching is essentially an interpersonal process of one-on-one observation and feedback. One person (the coach) systematically observes the behaviors of another person and then provides constructive feedback related to these observations. Safety coaches support the safe behaviors they see, and they offer useful and caring feedback regarding any at-risk behaviors they observe. This paper outlines the steps of safety coaching, points out basic skills needed to accomplish the process, and reviews methods for increasing the benefits of a safety coaching process. Most of this information is based on earlier writing by Geller (1995, 1996, 1998).

Safety coaching versus athletic coaching

The term "coach" is familiar to us from competitive or recreational sports. The coaches of winning athletic teams practice the basic observation and feedback processes needed for effective safety coaching. They follow most of the guidelines reviewed here. The most successful athletic coaches record the ongoing behaviors of individual players in a systematic fashion on a behavioral checklist, team roster, or videotape. Later, they deliver specific and constructive feedback to team members with hopes of increasing desirable behavior and decreasing undesirable behavior. This behavioral feedback might be given in a group meeting, or it might be given individually in a one-on-one conversation.

The most successful athletic coaches communicate feedback so team members learn from the exchange and increase their motivation to continuously improve. The same goals are true for safety coaches; the basic principles and methods for effective coaching are the same. It is more challenging, however, to coach for safety than for sports.

People are usually more receptive of advice to improve their performance in sports than in safety. Athletes actually show genuine appreciation for feedback that could improve their recreation-related behavior, such as a golf swing, a tennis stroke, a batter's stance, or a basketball maneuver. But how often do individuals offer sincere gratitude for being corrected on an at-risk work behavior? Unfortunately, coaching for safety is often seen as a personal confrontation. In the workplace, feedback regarding work production and quality is usually more accepted and appreciated than feedback about work safety.

One reason safety coaching is especially difficult is because people view safety as a personal matter, whereas work production or quality is more readily considered a group or team issue. If you get hurt, you suffer the consequences. But, if you slow production or reduce quality, the whole team suffers the negative consequences. Safety leaders understand the fallacy in this reasoning, but many others do not. They often do not consider the negative team consequences resulting from an individual injury.

When we adjust our behavior following instructive athletic coaching, we might notice an immediate improvement in our performance. We often see a direct connection between the coach's feedback and our improvement. We might even note an improvement in a team score as a result of individual feedback. Athletic coaching is frequently reinforced by natural consequences occurring soon after the behavior targeted by the coach; thus, the value of athletic coaching is evident.

The value of safety coaching is usually not as obvious. On a day-to-day basis there is rarely a clear connection between safety coaching and the ultimate purpose of coaching—a reduction in injuries. When you follow the advice of a safety coach you usually do not observe an immediate difference in your own safety or in the company's injury rate. People do not expect an injury to happen to them, and since their daily experiences support this belief, they do not perceive a personal need for advice from a safety coach.

Changing the way we keep score for safety can increase acceptance and appreciation of safety coaching. Although a reduction in injuries is the ultimate purpose of coaching, behavior change is the immediate goal of coaching. At-risk behavior is involved in almost every workplace injury; therefore, an increase in safe work behavior or a decrease in at-risk behavior as a result of safety coaching should be appreciated as a proactive, upstream approach to injury reduction. This requires a behavior-based approach to safety accountability and evaluation. Some tools and methods for accomplishing this are reviewed in this paper.

The safety coaching process

The five letters of the word **COACH** reflect the essential aspects of effective coaching—whether coaching to improve athletic or safety performance. This acronym is especially instructive because it not only contains the components of successful coaching, but it also gives them in the order in which they should occur.

"C" for Care

Know I care and you will care what I know. Safety coaching starts with caring. Safety coaches genuinely care about the health and safety of their co-workers, and they act on such caring. In other words, they "actively care" (Geller 1991, 1994, 1996). People are more likely to accept a safety coach's advice when the coach's words and body language reflect caring.

Many people are unwilling to coach or to be coached for safety because they consider safety a matter of individual or personal responsibility. This is reflected in the statement "If Susan and Chuck want to put themselves at risk, that's their problem, not mine." For some individuals a change in personal perspective or attitude is needed to facilitate the occurrence and acceptance of safety coaching. Everyone needs to consider safety coaching a shared responsibility or an interdependent opportunity to prevent injuries throughout their work culture. This calls for a shift from an individualistic to a collectivistic frame of reference (Triandis 1977, 1985).

Most employees are willing to adopt a collectivistic or team approach when it comes to work quantity and quality. In other words, people are more willing to give and receive feedback about work performance than safety performance. Coaching for production or quality is part of the job; coaching for personal safety is meddling. Workers need to realize that safety-related behaviors need as much (probably more) interpersonal observation and feedback as any other job activity.

Pointing out a company's injury record for a certain period of time can convince people to accept safety coaching as a shared responsibility. Although they might not have been injured, others were hurt on the job, and everyone obviously cares about that. With an over-arching caring attitude, people are ready to learn what they can do to make a difference. The answer is to become an actively caring safety coach. The coaching must be behavior-based, and the process begins with behavior-based observation.

"O" for Observe

Safety coaching involves the objective and systematic observation of other people's behavior. The focus is on supporting safe behavior and correcting at-risk behavior. Safety coaches are on the lookout for those who go beyond "the call of duty" for the safety of another person. This actively caring behavior contributes significantly to safety improvement and can be increased through proper recognition or rewarding feedback, and safety coaches know how to do this.

The safety coaching process starts with the coach asking the person for permission to be observed. If the person says "no" to the request, the coach honors the refusal and politely retreats. Some people are puzzled by this permission step. Giving permission for an observation means people are aware they are being observed. If workers know they are being observed, won't they be on their "best behavior"? Won't the results look "safer" than they really are? Sure they will, but that can be a good thing. People will feel a sense of achievement or accomplishment when the results look favorable, and they will strive to achieve more. "Working to achieve success" is a far more desirable state than "working to avoid failure."

Even when people know they are being observed and are practicing their "best" safety-related behaviors, they can still be at risk. This is unconscious incompetence and is exactly the kind of behavior that requires observation and feedback. People are usually willing to receive this kind of feedback, especially when they give permission for the observation process and are first recognized for certain safe work practices.

Consider the potential problems of not asking permission before observing. You might catch the person avoiding some basic safe work practice, and feedback would likely be primarily negative. If people choose to perform at-risk behavior when they think no one is watching them, catching them at it will not improve anything. A temporary change in behavior will be more than offset by a negative attitude.

So the purpose of a behavior-based coaching process is not to catch people performing at-risk behavior. Rather, this safety coaching process is designed to help people know what behaviors are safest in a particular situation and to help them think about those behaviors and practice them on a regular basis. When people make a conscious effort to perform a task safely while a coach is observing them, they are practicing ways to prevent personal injury. The coach is there to point out their safety-related accomplishments and to show them how they might achieve even more.

Observing behavior in order to give supportive and constructive feedback is easy if the coach knows what particular behaviors are appropriate and inappropriate and takes the time to look for occurrences of these behaviors in relevant work situations. It is advantageous to develop a checklist of those safe and at-risk behaviors which are most related to injury or injury prevention in the various work settings of an industrial complex. Ownership and empowerment increase when workers develop their own behavioral checklists.

Developing a critical behavioral checklist. Observation checklists can be either generic or job-specific. A generic checklist is used to observe behaviors that might occur during several jobs, whereas a job-specific checklist is designed for a particular job. Deciding what to include on a critical behavior checklist (CBC) is an important ingredient of the coaching process.

Critical behaviors can be safe or at-risk. Critical safe behaviors are those needed to avoid injury, and critical at-risk behaviors are those which put people at risk for personal injury. A critical at-risk behavior is one that

- has led to a large number of injuries or near-misses in the past,
- could potentially contribute to a large number of injuries or near-misses because many people perform the behavior,
- has previously led to a serious injury or a fatality, or
- could lead to a serious injury or fatality.

Critical safe behaviors are those desirable behaviors that contribute to injury prevention. Some are as basic as using the appropriate personal protective equipment and many represent the alternative to the at-risk behaviors referred to above. Other safe behaviors can be classified as special actively caring behaviors or behaviors that reflect going beyond the call of duty of another person's health and safety. Conducting an observation and feedback process is such a behavior, as is cleaning up a spill in another person's work area. The standard CBC does not include a category for these actively caring behaviors, but these behaviors need to be considered as targets for special one-on-one recognition in a safety coaching process (see Geller 1997 for guidelines for giving quality recognition).

Several sources can be consulted to obtain behaviors for a CBC:

- injury records,
- near-miss reports,
- job hazard analyses,
- standard operating procedures,
- rules and procedural manuals, and
- the workers themselves.

People know a lot about the critical safe and at-risk behaviors occurring in their work areas. They know which safety rules are frequently ignored, and they often know when a near-miss has occurred because of at-risk behavior. Also, it is useful to seek advice from the plant doctor, nurse, safety director, or anyone else who maintains injury statistics for the work setting.

It is not often a good idea to start a safety coaching process with a lengthy and exhaustive CBC. A long list of behaviors to check at the initiation of one-on-one coaching can seem overwhelming at first and can inhibit the process. As with anything new and in need of voluntary support, it is beneficial to start small and build. With experience, people find a CBC easy to use and later accept additions to the list. Expect them to contribute to continual improvement of the CBC, from refining behavioral definitions to suggesting behavioral additions and substitutions. The development and application of a CBC is an ongoing improvement process. Continuous development and refinement of a behavioral checklist enhances coaching observations and vice versa.

Operating Procedures	Safe Observation	At-Risk Observation
BODY POSITIONING/PROTECTING Positioning/protecting body parts (e.g., avoiding line of fire, using PPE, equipment guards, barricades, etc.).		
VISUAL FOCUSING Eyes and attention devoted to ongoing task(s).		
COMMUNICATING Verbal or nonverbal interaction that affects safety.		
PACING OF WORK Rate of ongoing work (e.g., spacing breaks appropriately, rushing).		
MOVING OBJECTS Body mechanics while lifting, pushing/pulling.		
COMPLYING WITH LOCKOUT/TAGOUT Following procedures for lockout/tagout.		
COMPLYING W/ PERMITS Obtaining, then complying w/ permit(s). (e.g., Confined space entry, hot work, excavation, open line, hot tap, etc.).		

Figure 3.1.1 A Worksheet for Developing a Generic Critical Behavior Checklist (CBC)

The worksheet depicted in Figure 3.1.1 can be used to begin the development of a CBC. Through interactive discussions, a work team defines safe and at-risk behaviors in their own work areas relevant to each operating procedure in Figure 3.1.1. For example, the first category on body positioning and protecting should include job-specific methods for protection from environment or equipment hazards, from using certain personal protective equipment to positioning body parts in particular ways.

Some categories in Figure 3.1.1 may be irrelevant for certain work groups, such as locking or tagging out equipment or complying with certain permit policies. Moreover, a work group might add another general operating procedure to cover a distinct set of work behaviors. It is noteworthy that the process of defining safe and at-risk behaviors is safety training in the best sense of the word. The participants learn exactly what safe behaviors need to occur and what at-risk behaviors need to be avoided in a designated work process.

A list of behaviors covering all of the operation categories in Figure 3.1.1 can be quite lengthy. While such a list provides numerous opportunities for coaching feedback, it can also appear overwhelming and can reduce voluntary participation in the coaching process if used at the start. It takes time and practice for people to observe behaviors reliably, and for people to get used to being observed while they work. We have found it useful to begin with a brief CBC of four or five behaviors and then to build on this list after practice, group discussion, checklist refinement, and more practice.

Each operating procedure in Figure 3.1.1 could be used as a separate checklist. The first category, for example, could be developed into a CBC for the observation of personal protective equipment (PPE). In this case, you would list the PPE behaviors for the work area in the left-hand column with columns on the right to check safe versus at-risk observation. This type of CBC could be used to record the PPE use of several workers by adding checks in the safe or at-risk columns for each observation of an individual's use or non-use of a particular PPE item.

A CBC for personal protective equipment. Figure 3.1.2 depicts a CBC for recording a number of PPE-related behaviors, including the use of gloves, safety glasses, ear plugs, safety shoes, and hard hat. With this

Critical Behavior Checklist for Personal Protective Equipment

Observation period (dates): __________

Observer: ______________________________

TOTAL NUMBER OF EMPLOYEES OBSERVED	NUMBER OF EMPLOYEES OBSERVED USING ALL REQUIRED PPE

PPE (For Observed Area)	SAFE OBSERVATION (Proper Use of PPE)	AT-RISK OBSERVATION (Improper or No Use of PPE)
Gloves		
Safety Glasses/Shield		
Hearing Protection		
Safety Shoes		
Hard Hat		
TOTAL		

$$\% \text{ Safe Observations} = \frac{\text{Total Safe Observations}}{\text{Total Safe Observations} + \text{Total At-Risk Observation}} \times 100 = \quad \%$$

Figure 3.1.2 A Checklist for Monitoring the Use of Personal Protective Equipment (PPE)

sort of CBC you can carry out an audit of PPE use much the way you or your colleagues currently conduct an environmental audit. After observers know what PPE is required in a certain situation, they watch people work throughout the work area and record what they see regarding PPE use. For every PPE item required in the work area and for every employee on a work team, an observer checks the safe (use) or at-risk (improper or non-use) column. This information can be easily transformed into two meaningful summary statistics, which can then be graphed and displayed openly for individual and group accountability.

The check marks in the individual PPE categories of the CBC in Figure 3.1.2 are totaled. Then by dividing the total number of safe checks by the total safe and at-risk checks, the percentage of safe PPE observations can be assessed. See the formula at the bottom of Figure 3.1.2.

The box at the top of Figure 3.1.2 can be used to calculate another summary statistic: the percentage of employees protected appropriately. Specifically, if you put a tally mark in the left-most cell each time you observe an employee's PPE use and check the right cell every time the employee is using all the PPE called for in the situation, you have all you need to calculate the "percentage of safe employees." This percentage will usually be lower than the "percent safe observations" (bottom of Figure 3.1.2) because with the "percent safe employee" statistic, it takes only one improper or non-use of PPE to classify an employee as "at-risk."

Besides calculating summary percentages for group display and team accountability, the CBC in Figure 3.1.2 can also be used to assess problems with individual PPE devices. The percentage of safe behaviors can be calculated separately for each PPE category. In other words, if several employees were observed regarding all PPE items in the CBC of Figure 3.1.2, the percentage of safe observations could be calculated for each of the six PPE items. And, of course, other PPE items could be added to this checklist and evaluated.

A generic critical behavior checklist. Figure 3.1.3 depicts a CBC for recording the results of a coaching observation. This kind of CBC recording sheet should be used after the participants (optimally, everyone in a particular work area) have derived precise behavioral definitions for each category and have practiced rather extensively with shorter CBCs. Practice with a CBC enables careful refinement of the behavioral definitions and builds confidence and trust in the process.

There is a place on the CBC for the observer's name, but obviously no name is recorded for the person(s) observed. The CBC in Figure 3.1.3 can be used to record the observations of two individuals. The number "1" could be used to categorize the safe vs. at-risk judgments of one person, and "2" could be used for the second person.

The "percent safe observations" formula at the bottom of this CBC could be used for each person individually or summed across all observations. Individual feedback can be given by reviewing individual classifications for each operating procedure, along with specific comments which might be recorded on the back of the CBC. The "percent safe observations" can

A Generic Critical Behavior Checklist

Observer: ________________

Person 1

Date: ________________
Time: ________________
Department: ________________
Building: ________________
Floor: ______ Area: ________

Person 2

Date: ________________
Time: ________________
Department: ________________
Building: ________________
Floor: ______ Area: ________

Operating Procedures	Safe Observation	At-Risk Observation
BODY POSITIONING/PROTECTING Positioning/protecting body parts (e.g., avoiding line of fire, using PPE, equipment guards, barricades, etc.).		
VISUAL FOCUSING Eyes and attention devoted to ongoing task(s).		
COMMUNICATING Verbal or nonverbal interaction that affects safety.		
PACING OF WORK Rate of ongoing work (e.g., spacing breaks appropriately, rushing, etc.).		
MOVING OBJECTS Body mechanics while lifting, pushing/pulling.		
COMPLY WITH LOCKOUT/TAGOUT Following procedures for lockout/tagout.		
COMPLYING WITH PERMITS Obtaining, then complying w/ permit(s). (e.g., confined space entry, hot work, excavation, open line, hot tap, etc.).		

1 = observations for first person; 2 = observations for second person

% Safe Behavior: $\frac{\text{Total Safe Observations}}{\text{Total Safe Observations + At-risk Observations}} \times 100 =$ ________ %

Feedback with Person 1

Time Begin ______
Date ________ Time End ______

Feedback with Person 2

Time Begin ______
Date ________ Time End ______

Figure 3.1.3 Generic Critical Behavior Checklist

be combined with the results of other "one-on-one" observations to obtain group results for public posting.

Scheduling coaching sessions. When you are setting up a one-on-one coaching process, you need to develop a protocol. Where and when will the coaching take place? How long will each observation last? How often will each person be observed? Now, you do have a lot of flexibility. For example, I worked with one group who decided to do infrequent but detailed observations that lasted 30 minutes. On the other end of the spectrum, another group used frequent, one-minute observations. They said, "Hey, if you can have a one-minute manager, why not a one-minute coach?"

You can make just about any observation protocol work, as long as people feel it is reasonable. Nevertheless, there are a few points you need to keep in mind. First, as we have said earlier, participation in a safety coaching process should always be voluntary. Remember that a lot of people are just not comfortable being observed one-on-one. Some people start out feeling that a person's safety is his or her own business. They do not want to be observed and coached by their co-workers. It is best to give these people time to watch the process work, and encourage them to join in when they are ready. If people do not believe in the one-on-one coaching process, they are not going to improve their behavior. Even worse, they might make the process unpleasant for others, and that is the last thing you want to happen.

By the way, some people might agree to be coached but NOT to observe others, and that is okay. Not everyone has to coach. We have found that once the process is underway, if there are not enough coaches to go around, more people will volunteer. They learn the value of coaching, and they want the process to continue.

The second point to remember is that the name of the person being observed should never be recorded. Unfortunately, many people have come to expect a top-down, enforcement approach to safety, and they are concerned that observation data might be used against them. You need to make it perfectly clear that coaching sessions are private and personal. Only the combined data from individual CBCs will be shown to others.

Feedback is essential. Each CBC includes a provision for tallying results and graphing them as "percent safe behavior" on a group feedback chart. The CBCs shown in Figures 3.1.2 and 3.1.3 include a formula at the bottom for calculating "percent safe behavior" per coaching session. In this case, all checks for safe observations are added and divided by the total number of checks (safe plus at-risk behaviors). The result is multiplied by 100 to yield "percent safe behavior."

The feedback percentages can be readily displayed on a group feedback chart or graph. While feedback percentages are valuable, it is vital to realize that the process is more important than the numbers. The real value of the coaching process is not in the behavioral data (which are no doubt biased by uncontrollable factors) but in the behavior-based communication between employees. At times we have seen observers get so caught up in recording the numbers that they let co-workers continue to perform an at-risk behavior while they observe and check columns on a CBC.

An individual's safety must come before the numbers. In other words, when coaches observe an at-risk behavior that is immediately threatening to a person's health or safety, they should intervene at once. They can usually continue the observation process after the intervention. On the other hand, if the CBC was partially completed before the intervention, it might be most convenient to communicate other observations, especially if there are some safe behaviors to report. This "deposit" will help compensate for the "withdrawal" that was probably implicated by the intervention to stop an at-risk behavior.

"A" for Analyze

When evaluating their observations, safety coaches understand and appreciate the three-term contingency or ABC model (for Activator, Behavior, and Consequence). They appreciate there are usually environmental or system reasons for behaviors occurring, whether safe or at risk. Safety coaches realize, for example, that certain at-risk behaviors occur because they are directed by activators such as work demands, at-risk example-setting by peers, and inconsistent or mixed messages from management. They also understand that at-risk behaviors are often motivated by one or more natural consequences, such as comfort, convenience, work breaks, and approval from peers or work supervisors.

An ABC analysis helps people discover the reasons behind at-risk behaviors and facilitates the design of interventions to decrease at-risk behavior and increase safe behavior throughout the workplace. Such an analysis can be accomplished before giving feedback to the person observed, or it can be done during the one-on-one feedback process. Talking about the activators and consequences that influence certain work practices can lead to environment or system improvements for decreasing at-risk behavior. These are explored in the next component of safety coaching—the heart of the process.

"C" for Communicate

Effective coaching requires basic communication skills, including strategies for active listening and persuasive speaking. Communication training sessions that include role-play exercises can be invaluable in developing individuals' confidence and competence for sending and receiving behavioral feedback. This training should illustrate the need to separate behaviors (actions) from attitudes and feelings. This enables corrective feedback to occur at the behavioral level without "stepping on" feelings.

People need to believe that anyone can be at risk without even realizing it, and performance can only improve with behavior-specific feedback. When this is understood, behavior-based feedback is accepted and even appreciated. Of course, for feedback to benefit behavior and attitude, it needs to be delivered correctly. Let's review some basic guidelines for giving behavior-based feedback.

Individual feedback. Whether supportive or corrective, feedback needs to be specific (with regard to a particular behavior) and timely (occurring soon after the target behavior). It should be given in a private, one-on-one situation to avoid potential interruption or embarrassment from others. In addition, corrective feedback is most successful when the appropriate safe behavior is specified and possible solutions for eliminating the at-risk behavior are discussed.

Whether giving supportive or corrective feedback, the safety coach must actively listen with sincerity when the feedback recipient reacts to support or correction. By giving empathetic attention, the safety coach shows genuine concern for the feelings and self-esteem of the feedback recipient. The best listeners paraphrase what they hear to check understanding, ask for more details, accept statements without judgmental interpretation, and avoid autobiographical reactions (such as, "When I worked in your department, I always worked safely") that can stifle self-disclosure.

The critical characteristics of effective rewarding and correcting feedback are listed in Figure 3.1.4. We suggest you use this figure as a guideline for group practice sessions. It is not easy to give safety feedback properly, and many people feel awkward or uncomfortable initiating a one-on-one feedback conversation about personal health or safety. It is important for people to practice giving rewarding and correcting feedback. In this regard, we have found it educational and engaging when workers demonstrate the characteristics of effective rewarding and correcting feedback in group training sessions.

Use Rewarding Feedback to Support Safe Behavior

- Give the feedback one-on-one and privately.
- Give the feedback as soon as possible after the observation process.
- Identify the safe behavior(s) observed.
- Be sincere and genuine.
- Express personal appreciation for setting the right example for others.

Use Correcting Feedback to Decrease At-Risk Behavior

- Give the feedback one-on-one and privately.
- Give the feedback as soon as possible after the observation process.
- Begin with acknowledgement of safe behavior(s) observed.
- Identify the at-risk behavior(s) observed.
- Specify the safe alternative to the at-risk behavior(s).
- Indicate concern for the person's welfare.
- Request commitment to avoid the at-risk behavior(s).
- Thank the person for their commitment to continuous improvement.

Figure 3.1.4 Key Characteristics of Rewarding and Correcting Feedback

To facilitate the practice of communicating one-on-one safety feedback, we ask groups of three to seven individuals to develop a brief skit to illustrate rewarding or correcting feedback. Each skit includes at least three participants: a safety coach, a worker who receives feedback, and a narrator who sets the scene. We have found this exercise particularly enjoyable when groups show the wrong way to give feedback first and then demonstrate the correct feedback procedure (following the guidelines given in Figure 3.1.4).

After the right way to provide safety feedback is illustrated in a role-play exercise, the audience provides feedback to the presenters regarding particular strengths of their demonstration and places where the demonstration could be improved. An effective facilitator gleans important lessons from this communication exercise while maintaining a cordial and gratifying atmosphere. It takes practice and peer support for participants to feel comfortable and effective at giving rewarding and corrective feedback on the job, so role-play exercises need to be offered on several occasions. Since people learn much by observing others (Bandura 1969), even those who do not perform a skit learn vicariously by observing their peers.

Group feedback. As discussed above, the CBC records can be summarized as a calculation of "percent safe behavior." Such percentages can be calculated per day, per week, or per month and then displayed on a feedback graph. Posting this graph in a prominent place, perhaps next to the plant's statistical process control charts, provides employees with an opportunity to monitor their progress and to be rewarded intrinsically for their efforts. This puts safety on the same status level as production and quality. It promotes a group achievement approach.

The percentage of safe observations can be calculated separately for various categories of workplace activities (as shown in Figure 3.1.3). Then feedback graphs can be posted for individual behavioral categories. For example, behavior observation graphs could show the percentage of employees

- wearing safe apparel,
- using appropriate personal protective equipment,
- avoiding the line of fire, lifting or moving objects safely,
- keeping work areas neat and free of trip hazards,
- using tools and equipment safely, and
- complying with lockout/tagout procedures or work-permit requirements.

By monitoring behavioral categories separately you can identify certain activities needing special attention. A particular intervention approach could be indicated (e.g., a special training session, an activator or consequence technique, or an ergonomic rearrangement of a particular work area). Observation and feedback can provide invaluable diagnostic information. Behavioral feedback graphs hold people accountable for process numbers they can control on a daily basis (in contrast to outcome numbers like total recordable injury rate or workers' compensation costs). When you improve upstream numbers (percent safe behavior), you will eventually reduce the outcome numbers (work injuries).

"H" for Help

The last letter of COACH reflects the essential mission of a safety coach. The purpose of safety coaching is to help people prevent injury to themselves and others by correcting at-risk behaviors and supporting safe behaviors. It is essential that a coach's helping communication is accepted. The four letters of HELP offer a mnemonic for remembering four words which suggest strategies for increasing the likelihood that a safety coach's advice, directions, or feedback will be appreciated.

***"H"** for humor*. Obviously, safety is a serious matter, but sometimes a little humor can be added to our safety communications as a way of increasing interest and acceptance. Researchers have found that laughter reduces distress and even improves a person's immune system (Goodman 1995; *National Safety News* 1985). Also, we know from personal experience that humor can enhance the constructive gains from a communication by increasing our attention and decreasing our resistance to an appeal for change.

***"E"** for esteem*. Esteem, or more precisely self-esteem, is the personal perception of self-worth ("I am valuable") and is critical to increasing safe work practices. People who feel inadequate, unappreciated, or unimportant in a particular work setting are not as likely to go beyond "the call of duty" to benefit the safety of themselves or others as are people who feel capable and valuable (Geller 1991, 1996; Schroeder, Pennar, Dovidio, and Piliavin 1995). The most effective safety coaches emphasize the positive over the negative in an attempt to build or avoid lowering another person's self-esteem.

"L" *for listen*. Listening attentively to another person is one of the most powerful ways to support or increase self-esteem. Actively listening shows that the listener cares about the individual and his or her circumstances. This can be interpreted as reflecting self-worth with an internal script like "I must be valuable because my opinion is appreciated." After a safety coach listens actively, his or her feedback message is more likely to be heard and accepted. As Covey (1989) put it, "Seek first to understand, then to be understood."

"P" *for praise*. Another powerful way to build self-esteem is to praise others for their specific accomplishments. If the praise targets a particular behavior, the probability that the behavior will reoccur increases. In other words, behavior-focused praise is a powerful rewarding consequence that not only increases the behavior it follows but also increases a person's self-esteem. This in turn increases the individual's willingness to actively care for the safety of others (Geller 1996, 1997, 1998).

It is human nature to attend more to people's mistakes than successes. Errors stick out and disrupt the flow and therefore attract reaction and effort to correct them. However, when things are going well—when people are performing safely—there is usually no activator to signal success. Good performance is typically taken for granted. Effective safety coaches resist the tendency to go with the flow, and they find appropriate times to express genuine appreciation for the safe behavior they observe.

Conclusion

The safety coaching process reviewed here is founded on the basic premise of behavior-based safety. Injuries result from at-risk behaviors, and if these behaviors are decreased and safe behaviors increased, injuries will be prevented. Over the past two decades, a variety of behavior-based research studies have verified this premise by systematically evaluating the impact of interventions designed to decrease employees' at-risk behaviors and increase their safe behaviors. Behavior-based feedback from behavioral observations was a common ingredient in most of the successful intervention processes, whether the feedback was delivered through tables, charts, interpersonal communication, congratulatory notes, or the receipt of a reward following a particular behavior. (See, for example, the comprehensive review by Petersen 1989 or individual research articles by Chhokar and Wallin 1984; Geller, Eason, Phillips, and Pierson 1980; Komaki, Heinzmann, and Lawson 1980; Sulzer-Azaroff and De Santamaria 1980).

The behavior-based observation and feedback process reviewed here is analogous to the behavior-based safety process detailed earlier by Geller (1994), Krause, Hidley, and Hodson (1996), and McSween (1995) and taught in training videotapes and workbooks developed by Tel-A-Train, Inc. (1995). Besides the research referred to above, there are hundreds of real-world case studies which demonstrate the remarkable injury-prevention impact of behavior-based coaching. But, there is no effective step-by-step cookbook available for this process. And, the process is not a *quick* fix. The injury prevention achievements from safety coaching are a direct function of the effort put into the process. The guidelines presented here need to be customized into specific procedures for a particular work team in a particular plant. Who knows best which step-by-step coaching procedures will work in a given work area? The people who work there know best; therefore, they need to be empowered to develop their own safety coaching process from the principles and guidelines presented here.

Empowerment to implement a large-scale safety coaching process requires time and resources to develop materials, train necessary personnel, establish support mechanisms, monitor progress, and continually improve the coaching tools, methods, and support mechanisms. A behavior-based safety steering committee needs to address the following questions/issues before initiating a large-scale safety coaching process:

- Who will develop the critical behavior checklist (CBC)?
- How many behaviors will be included on the first CBC?
- What information will be used to decide which behaviors are critical?
- Who will train the safety coaches, facilitate their necessary practice, and give them behavior-based feedback?
- How many coaches will be trained initially, and how can additional people volunteer to participate as a safety coach?
- How will formal coaching sessions be scheduled, how often will people be coached, and how long will the coaching sessions last?
- Where will group feedback graphs be posted, and who will be

responsible for preparing the displays of safe behavior percentages?

⋇ Who will be on the steering committee to oversee the safety coaching process, answer questions about process implementation, maintain records, monitor progress, and refine procedural components whenever necessary?

The above list of implementation issues is certainly not exhaustive, yet it might appear overwhelming at first. Organizational cultures vary widely as a function of personnel, history, policy, work processes, environmental factors, and current contingencies. Therefore, implementation procedures need to be customized with significant input from the people protected by the coaching process and from whom long-term participation is needed.

It will take significant time, effort, and resources to achieve a plant-wide safety coaching process. With this end in mind, we recommend starting small to build confidence and commitment on small-win successes. Celebrate those achievements that reflect progress toward your ultimate vision—a Total Safety Culture in which everyone consistently coaches each other, formally and informally, to improve safety-related behavior.

References

Bandura, A. 1969. *Principles of behavior modification*. New York: Holt, Rinehart, and Winston.

Chhokar, J. S., and J. A. Wallin. 1984. A field study of the effects of feedback frequency on performance. *Journal of Applied Psychology*, 69:524–530.

Covey, S. R. 1989. *The seven habits of highly effective people*. New York: Simon and Schuster.

Geller, E. S. 1991. If only more would actively care. *Journal of Applied Behavior Analysis*, 24:607–612.

———. 1994. Ten principles for achieving a total safety culture. *Professional Safety*, 39(9):18–24.

———. 1995. Safety coaching: Key to achieving a Total Safety Culture. *Professional Safety*, 40(7):16–22.

———. 1996. *The psychology of safety: How to improve behaviors and attitudes on the job*. Radnor, PA: Chilton Book Company.

———. 1997. Key processes for continuous safety improvement: Behavior-based recognition and celebration. *Professional Safety*, 42(10):40–44.

———. 1998. *Understanding behavior-based safety: Step-by-step methods to improve your workplace*. 2nd ed. Neenah, WI: J. J. Keller & Associates, Inc.

Geller, E. S., S. L. Eason, J. A. Phillips, and M. D. Pierson. 1980. Intervention to improve sanitation during food preparation. *Journal of Organizational Behavior Management,* 2(3):229–240.

Goodman, J. B. 1995. Laughing matters: Taking your job seriously and yourself lightly. *Journal of the American Medical Association*, 267(13):1858.

Komaki, J., A. T. Heinzmann, and L. Lawson. 1980. Effect training and feedback: Component analysis of a behavioral safety program. *Journal of Applied Psychology,* 65(3):261–270.

Krause, T. R., J. H. Hidley, and S. J. Hodson. 1996. *The behavior-based safety process: Managing involvement for an injury-free culture*. 2nd ed. New York: Van Nostrand/Reinhold.

McSween, T. E. 1995. *The values-based safety process*. New York: Van Nostrand/Reinhold.

National Safety News. 1985. Laughter could really be the best medicine. 15.

Schroeder, D. A., L. A. Penner, J. R. Dovidio, and J. A. Pillavin. 1995. *The psychology of helping and altruism: Problems and puzzles.* New York: McGraw-Hill, Inc.

Sulzer-Azaroff, B., and M. C. De Santamaria. 1980. Industrial safety hazard reduction through performance feedback. *Journal of Applied Behavior Analysis,* 13:287–295.

Tel-A-Train, Inc. 1995. *Actively caring for safety*. Dallas, TX: Wescott Communications.

Triandis, H. C. 1977. *Interpersonal behavior*. Monterey, CA: Brooks/ Cole.

———. 1985. The self and social behavior in differing cultural contexts. *Journal of Personality and Social Psychology*, 96:506–520.

Beyond Observation and Feedback: Integrating Behavioral Safety Principles into Other Safety Management Systems[2]

Sherry R. Perdue

Abstract

A behavioral observation and feedback process is a very effective means of reducing injuries and incidents in the workplace. By observing and providing feedback, peers encourage one another toward safe rather than at-risk work practices. In addition, the data collected provide a leading metric predictive of downstream safety performance. A behavioral observation and feedback process, however, is just one tool that applies principles of psychology to encourage an improved safety culture. In fact, in absence of a positive (or improving) safety culture, an observation and feedback process is likely to meet limited success.

Traditional safety management systems and activities (e.g., incident and near-miss reporting, incident investigations, safety meetings, safety committees, safety accountability processes, safety reward and recognition programs) may undermine safety culture change efforts. Systems that are ineffective or counterproductive should be carefully examined and modified. Because as many as 95 percent of all injuries and incidents can be attributed, at least in part, to human behavior, any new approach should naturally include better ways to manage human behavior and better means to measure and reinforce safety-related behaviors.

This paper provides an overview of a behavioral observation and feedback process and discusses the tool's broader purpose of influencing an organization's safety culture. Next, the paper reviews some of the principles from the field of psychology which underlie the observation process and illustrates through case study how these principles should also influence the design of other safety management systems. Finally, the paper presents a strategy for assessing and modifying traditional safety systems so they support an organization's safety culture change efforts.

Behavioral Observation and Feedback Reviewed

A behavioral observation and feedback process is a tool designed to help encourage safe work practices and discourage at-risk work practices. Using simple but effective observation techniques, employees periodically observe each other. Following the brief observation period, the observer gives tactful one-on-one feedback regarding safe and at-risk behaviors observed.

The one-on-one dialogue that occurs between peers immediately following the observation can be instrumental in decreasing at-risk work practices. In some cases, the feedback provides information to the individuals, pointing out a risky behavior that they were unintentionally performing. Whether because they did not understand the proper way to perform the behavior or a lapse in attention caused them to operate at-risk, they were not purposefully violating safe work practices. Feedback can make the individuals aware of the risks they were taking so they can work to change their behavior in the future.

Alternatively, feedback sometimes provides social support to encourage peers to take the time and make the extra effort to perform behaviors in the safest manner. When employees purposefully choose to perform at-risk behaviors, it is usually because the at-risk behavior is easier, faster, and/or more convenient. Heavy emphasis on meeting schedules and production goals often further reinforces at-risk behaviors. Although the chance for injury is increased, it is still highly unlikely. Furthermore, many cultures support this "minor" risk-taking. At the time, the decision seems reasonable. In this case, peer feedback is providing the social support to encourage the safe work practice instead of the at-risk work practice.

In many cases, however, feedback alone is not sufficient to address the at-risk behavior. Often at-risk behaviors are encouraged or even required by the work environment. For example, improper lifting practices may be necessary because of the layout of a particular workstation. Using a metal ladder for electrical work may be encouraged if fiberglass ladders are needlessly inconvenient to get. In these cases, changes in the workplace should be made to increase the likelihood that safe work practices will occur. Therefore, the next step of the observation process is for the observer and observee to analyze the work situation to determine the contributing causes of any at-risk behaviors and determine improvement opportunities.

In addition to the one-on-one discussion, the data from an entire group's observations are periodically collected and compiled. The compiled data are shared with the employees as a second form of feedback, illustrating the areas in which the team is excelling and those that provide the greatest potential for injury. Again, feedback alone may be insufficient to encourage behavior change. Where the compiled data reveal frequent, repeated occurrences of a particular at-risk behavior, there are likely system-level influences affecting the behavior. These areas are analyzed in detail by the work team, which then develops relevant intervention strategies to reduce the likelihood of the at-risk behaviors recurring. For example, if the at-risk behavior of failing to use a mechanical hoist were observed repeatedly, the work team would investigate why the behavior occurs and then introduce appropriate changes. Perhaps the only hoist available in the area is in constant use, in which case the purchase of a second hoist may be justified.

A Total Safety Culture Defined

While the immediate goal of a behavioral observation and feedback process is to identify and increase critical safety-related behaviors, the ultimate goal of the process is much greater. A behavioral observation and feedback process is a key tool to influence an organization's overall safety culture. The likelihood that such feedback and problem solving will begin to occur more frequently on an informal basis is increased by providing a structured approach to encourage systematic peer observation and feedback. This frequent informal communication between co-workers about safety is a key characteristic of a Total Safety Culture.

A Total Safety Culture is defined as a culture in which individuals hold safety as a value, feel a sense of responsibility for the safety of their co-workers as well as themselves, and are *willing* and *able* to ACT on the sense of responsibility they feel. That is, individuals have the skills and tools necessary and are supported by the culture to go "beyond the call of duty" on behalf of the safety of others and themselves. In workshops conducted within dozens of organizations across the country, participants were asked to list specific components of a Total Safety Culture. Following are some common responses:

- ※ All employees comply with safety rules and regulations at all times;
- ※ Employees continuously search for safety hazards and take the initiative to correct hazards when found;
- ※ Employees are eager to participate in safety-related activities, which are promoted and encouraged through respect and positive recognition;
- ※ All safety-related issues are openly communicated. Fear of reprimand or discipline does not inhibit discussions;
- ※ Safety incidents are viewed as an opportunity to identify system failures and therefore improve the system. Individuals are not assumed to be, and are rarely found to be, at fault;
- ※ Training systems result in all employees having the needed knowledge, skills, and abilities to perform their jobs safely;
- ※ All employees fully understand and appreciate the potential hazards of the operations performed;
- ※ Employees do not consider taking unnecessary risks;
- ※ Management never (knowingly or otherwise) encourages employees to take unnecessary risks;
- ※ Regular safety-related feedback is a way of life. Corrective feedback is constructive and appreciated;
- ※ Peer pressure acts toward rather than against safety; and
- ※ All business activities are managed with a constant focus on injury prevention and occupational health.

Principles of Psychology That Underlie an Observation and Feedback Process

If designed and implemented carefully, an observation and feedback process is a key tool to help drive change in an organization's safety culture. When done right, it is built upon principles from psychological research. These behavioral science principles, fifty of which are compiled and reviewed in the first chapter of this book, include the following concepts:

- Behavior is directed by activators and motivated by consequences;
- People are motivated to maximize positive consequences and minimize negative consequences;
- Feedback can be a powerful consequence;
- Although negative consequences can lead to behavior change, they may have undesirable side effects;
- People compensate for increases in perceived safety by taking more risks;
- People view behavior as correct and appropriate to the degree they see others doing it;
- Long-term behavior change requires people to change "inside" as well as "outside";
- When people feel empowered, their safe behavior spreads to other situations and behaviors;
- People feel more personal control when working to achieve success than when working to avoid failure; and
- Choice increases commitment, ownership, and involvement.

The observation process cannot, however, succeed on its own in changing an organization's culture, no matter how well the observation process is designed, implemented, and executed. Other safety management systems, as well as individual management practices, must also support the desired culture. In fact, in the absence of a positive (or changing) safety culture, an observation and feedback process is likely to meet limited success.

Application of Principles of Psychology to Other Safety Management Systems

Organizations rely on a number of processes and procedures to manage risk and decrease the chance of incidents and injuries. These generally include systems such as discipline, safety training, safety suggestions, safety communications, safety rules and procedures, hazard identification and correction, incident reporting and investigation, and rewards and recognition. Each safety management system has an important contribution to make in terms of not only improving workplace safety, but also influencing an organization's safety culture.

When the system is poorly designed or operating ineffectively, its ability to accomplish its primary purpose will at best be compromised. At worst, a poorly designed, badly implemented, or ill-functioning system can also have a destructive influence on an organization's overall safety culture. The culture cannot advance when, for example, incident investigations create an air of mistrust and fault-finding, safety incentive programs discourage injury reporting, accountability processes fail to recognize individuals for their accomplishments, and performance evaluations only consider safety performance to be whether the individual was involved in an incident.

To compound the situation further, the systems are interactive and, in many cases, overlap. For example, hazard identification and correction require a climate that fosters employee participation, sufficient training so employees can recognize and correct hazards, and ample communication of the hazard and/or its solution. The poor features of one system can have negative influences on other systems.

When employee incentive programs and/or supervisor performance evaluations are based primarily on injury rates, it is unreasonable to expect those employees to embrace an open injury reporting and investigation system. Or when the incident analysis process is viewed as extremely blame-oriented, it is not reasonable to expect employees to feel comfortable having their safe and at-risk behaviors observed and recorded. Consider the following case study.

An Illustrative Case Study

Despite a long-standing, fairly structured protocol for investigating employee safety and process safety-related incidents, the management team of a mid-sized chemical manufacturing plant was concerned that the investigations were not very effective. In particular, when the incidents involved some form of human behavior (which, of course, most did) the analysis nearly always stopped at identifying the contributing behavior but failed to explain *why* the behavior had occurred.

An overall effort to improve the company's investigation process included process redesign and training. First, however, an attempt was made to better understand the employees' perceptions of the existing investigation process. Employees were selected randomly and asked to complete a questionnaire describing their experience in the reporting and analysis of incidents. The questionnaire focused on three areas. First, a series of questions addressed the individual's history of reporting near-misses, minor incidents, and major incidents. Those who indicated they had not reported (or would consider not reporting) an incident were asked to indicate their reasons. Next, the questionnaire asked for the person's history of participating in incident investigations. Finally, the questionnaire included several items regarding the person's opinion of the incident analysis process. The questionnaire was anonymous and was administered and collected in a manner that ensured the participant's confidentiality.

The results of the analysis revealed some interesting information. Of all employees surveyed (both hourly and salaried), 60 percent indicated they would *not* report an incident if they could avoid doing so; of those respondents, 60 percent suggested they would likely not report an incident because "they or someone else would be blamed," while 40 percent feared that discipline would result. Interestingly, company records showed very few instances of disciplinary actions.

In examining employees' opinions of the existing incident analysis process, the survey responses were separated by position (hourly and salaried) and by whether individuals had or had not been involved in an incident investigation within the previous 12 months. The results for two of the questionnaire items are shown in Figure 3.2.1. As illustrated, the salaried employees responded fairly positively. Most felt the investigations were of a "fact-finding" nature and resulted in accurate conclusions.

The hourly respondents' perceptions were not as favorable. Those who *had been involved* in an investigation had somewhat mixed opinions; some held fairly strong negative opinions, but most remained neutral. On the other hand, those who *had not* recently been involved in an investigation rated the investigations quite negatively. Unfortunately, this group accounted for most of the hourly respondents and was representative of a large majority of the plant population.

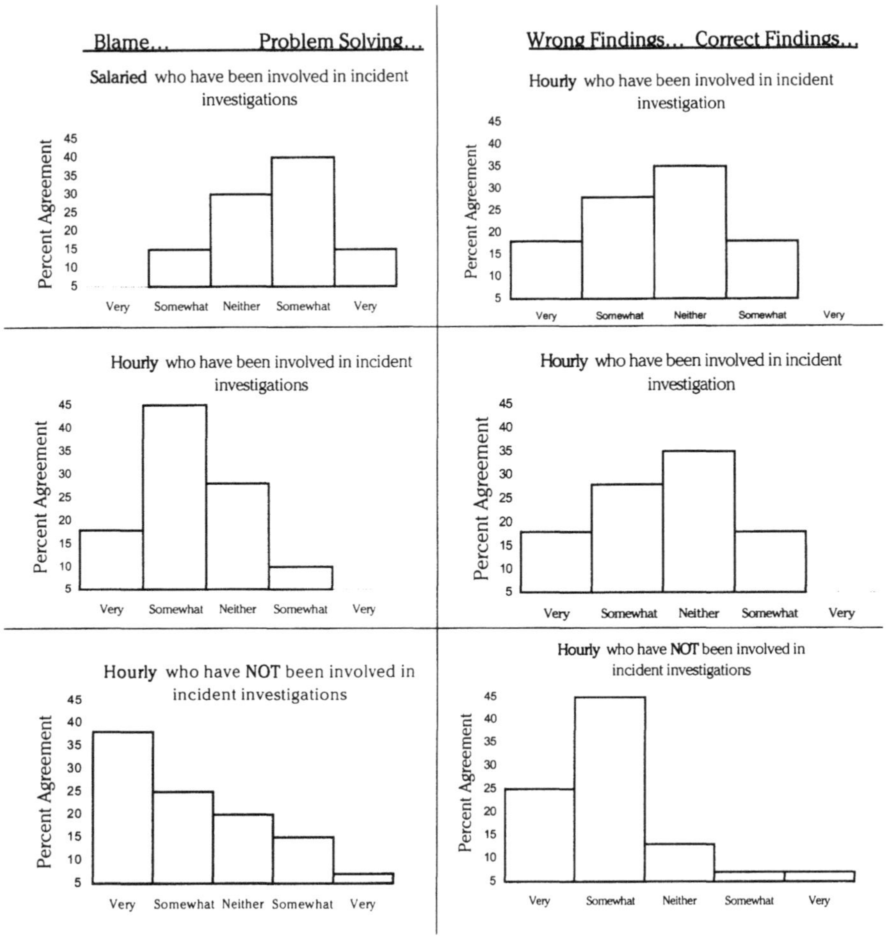

Figure 3.2.1 Sample Survey Results

So, although participation in the process enhances most people's opinion of it, the pervading opinion of the investigation process is overwhelmingly negative. Follow-up focus groups confirm these opinions. Overall perceptions of the investigation process is probably best described by the nickname given it: Kangaroo Court.

The existing investigation process was not only failing to provide the organization with information to allow them to make true improvements and therefore minimize the possibility for similar events to be repeated, but it also had an extremely destructive influence on the organization's culture. People were hiding incidents whenever they could manage to do so, and they were guarded in providing information to the investigation team whenever an event was reported and investigated. As a result, the investigations were carried out with less than adequate information, likely resulting in poor conclusions, further enhancing people's negative perceptions of the process and resulting in a spiraling situation likely to continue to get worse.

Interestingly, this same organization had attempted to introduce a peer observation and feedback process several years before. After having designed the details of the process and trained a pilot group of employees, the safety coordinators were surprised and disappointed to see extremely low participation rates within the pilot group. The employees did not seem interested in participating in this safety improvement process. Although the employees were assured the process was anonymous and the data would be confidential, they were still reluctant to observe and record the safety-related behaviors of their co-workers. Wonder why?

An Assessment Tool to Evaluate Safety Management Systems

The same principles of psychology which underlie the behavioral observation and feedback process are equally applicable for creating other safety management systems to motivate and reinforce safe work practices and create a culture which promotes true interdependency for safety. The central themes of a Total Safety Culture should serve as a ruler against which to measure and improve all organizational safety management systems so that they are effective not only in their primary missions but also have a positive influence on the organization's safety culture.

An assessment tool comprised of a set of "Maturity Paths" (one for each of several common safety management systems) was developed to help organizations assess and redesign existing safety management systems based on the principles of a Total Safety Culture. Each Maturity Path lists between ten and 20 key components of the system to be evaluated. Each component is evaluated on a four-point scale (i.e., *Beginning*, *Improving*, *Achieving*, *and Leading*), with each level describing a stage in the evolution of that particular safety management system. *Beginning* indicates that the component is nonexistent or poorly designed, and *Leading* indicates that the component is "cutting edge." The assessment tool is not intended to be a thorough evaluation of the system's effectiveness at accomplishing its primary mission but rather to assess the system's influence on the organization's safety culture as it operates. A portion of one of the Maturity Paths is shown in Table 3.2.1.

Beginning (1)		Leading (4)
All but the most serious incidents go unreported.	1 2 3 4	All incidents (e.g., near-misses, property damage, injuries) are reported in a timely fashion.
Analyses are conducted by safety professionals and/or supervisors only.	1 2 3 4	An incident analysis team is composed of members representing a cross-section of the organization and includes individuals involved in the event.
No training is provided for the incident reporting and analysis.	1 2 3 4	All employees receive training in the *philosophy* and *process* of investigating incidents. Investigation team leaders receive detailed training in investigation skills such as interviewing and root cause analysis.
Analyses often result in identifying "who's to blame." Corrective measures such as discipline or "counseling" are common.	1 2 3 4	Incident analyses focus on determining system-level roots causes and minimizing or eliminating them. Individuals are not assumed to be at fault. *Appropriate* behavior-based corrective actions are introduced where warranted.
Corrective actions and follow-up activities are handled by the safety department only.	1 2 3 4	Corrective actions and follow-up activities are handled by appropriate personnel (e.g., safety department, safety committee, area personnel).
Feedback concerning investigation results and corrective action implementation is haphazard.	1 2 3 4	Feedback concerning investigation results and corrective action implementation occurs without fail with all stakeholders.
Employees characterize the incident reporting and analysis systems as unsystematic, fault-finding, and/or ineffective.	1 2 3 4	Employees have confidence in the incident reporting and analysis systems to reduce the chance of future incidents.

Table 3.2.1 A Portion of the Maturity Path for Incident Analysis Process

Naturally, because the purpose is to assess the impact of the system on organizational culture, many of the components evaluated have to do with employee perceptions. Additional data are generally needed during the assessment to rate system components accurately. Survey data from focus group results and a comprehensive safety culture survey or safety-specific survey are ideal inputs.

The assessment can be completed in a variety of ways but is most effective when representatives from the entire organization participate. The use of experts from outside the organization to facilitate the evaluation may be useful. The assessment can be completed by a single team or through the compilation of several teams' results.

The results of the assessment readily reveal areas in which modifications are needed and provide direction for doing so. When planning system modifications, it is important to remember this assessment focuses on employee *perceptions*, not necessarily reality. That is, the results indicate how employees *feel* incident investigations are being handled, not necessarily how the actual process works. Similarly, items on the Maturity Path concerning discipline primarily measure what employees *perceive* as common practice. Perhaps only one incident in one hundred results in an employee being formally disciplined, but if that event is what employees remember, discipline may be viewed as the norm. Further complicating this scenario is the fact that perceptions often lag behind reality; therefore, recent changes to any management system may not be reflected in employees' current perceptions.

When considering modifications to any system, two primary options for action should be considered. First, look at the *actual process*. For example, safety training concerns may be caused by a variety of issues, each with different solutions. Safety training may be too short, too complicated, too boring, or too general for application on the job. Training may be given by employees who lack credibility, or it may be conducted on overtime either for the trainee or his counterpart back on the job. Alternatively, the training itself may be top notch, but the employee's supervisor may consider training a nuisance or secondary to "getting the work done." For these reasons, it is also important to consider how *perceptions* of the system are being managed (or not managed).

Consider another problem with the evaluation of safety suggestions. Only those whose suggestions have been implemented—and perhaps by those who have at least received feedback to their input—will be likely see a safety suggestion process as beneficial. Suggestions might be actively solicited, amply funded, quickly acted upon, and fairly evaluated by a cross-functional team of employees, but if they are poorly communicated to the rest of the workforce, employees may perceive the system as useless. Worse still, they may view the system as an indication of management's low concern for safety and lack of appreciation for their input. Here the situation can be addressed easily without revamping the entire safety suggestion program, just the communications aspects.

Employees at all levels of the organization must be an integral part of the design and implementation of a behavioral observation and feedback effort. Hourly employees not only know the hazards associated with their jobs, but they also know when and where at-risk behaviors tend to occur. By involving hourly employees in conducting observations, the process systematically introduces and encourages *peer* feedback, an extremely effective means for influencing human behavior as well as an effective mechanism for encouraging open, honest communication about safety-related issues in general. In addition, complete understanding and participation in any change effort is necessary to foster the buy-in and commitment needed for success. When employees are involved in creating a safety process, they are motivated to make it succeed; people care about the things they own.

Management, too, plays a crucial role in the behavioral observation and feedback process. In some cultures, managers participate just as the hourly employees do, serving both as observers and observees. In others, they serve as the administrators of the process, ensuring that time is allocated for observations, facilitating the analysis of the data generated, and initiating follow-up activities. Regardless of any additional role, management must support and encourage employee participation and ensure that any barriers to success are identified and addressed.

Remember, while the immediate goal of a behavior observation and feedback process is to identify and increase critical safety-related behaviors, the ultimate goal of the process is much larger. A behavior observation and feedback process is a key process to influence an organization's overall safety culture. By providing a structured tool to encourage peer feedback about safety-related behaviors, we increase the likelihood that such feedback will begin to occur more frequently and informally as well.

Conclusions

A behavioral observation and feedback process is not only an effective means of reducing workplace injuries; it can also be instrumental in driving change in an organization's safety culture. Other safety management systems also influence an organization's safety culture, and that influence may be either positive or negative. This paper has presented a tool to help apply the same psychological principles that underlie a behavioral observation and feedback process to these other systems so that every system is working together to support an organization's drive toward a Total Safety Culture.

References

Geller, E. S. 2001. *The psychology of safety handbook*. Boca Raton, FL: CRC Press.

Case Study: Behavior-Based Intervention for Occupational Safety: Critical Impact of Social Comparison Feedback[3]

Joshua H. Williams and E. Scott Geller

Abstract

The relative impact of global, specific, and social comparison feedback on safety behaviors was assessed at a large soft-drink bottling company. A Two Feedback Type (Specific vs. Global) x Two Social Comparison Feedback (Present vs. Not Present) analysis of covariance was used to test the hypothesis that specific, social comparison feedback would lead to the greatest improvement in percentage of safe behavior. Employees ($n = 97$) received behavioral safety training and then developed a critical behavior checklist (CBC) for their work areas. Researchers used the CBC to systematically categorize work behaviors throughout the plant as safe or at-risk and give employees written behavioral feedback each week. Social comparison feedback (SCF) led to significantly higher percent safe scores than did the No SCF conditions (mean percent safe was 78 percent vs. 68 percent, respectively). Follow-up Chi-Square analyses and practical considerations suggest global/SCF is optimal for improving safety performance. Limitations of the study and future implications for safety feedback research are discussed.

Introduction

More than 11.3 million U.S. employees are seriously injured and nearly 11,000 workers are killed on the job each year (Baker et al. 1992). It has been estimated that U.S. employers pay $155 billion in direct costs (e.g., workers' compensation, insurance premiums) associated with workplace injuries, amounting to over $1,400 per work-related injury (Baker et al. 1992; Miller 1997; The National Committee for Injury Prevention and Control 1989). Moving from at-risk to safe work practices is key to preventing injury (Califano 1979; Geller 1999).

Employees face numerous instances throughout the course of a workday when they choose to put themselves at some risk for injury. Unfortunately, workers are often rewarded for performing tasks in an at-risk manner because doing so is typically faster, easier, more comfortable, and more efficient or convenient than following the safe procedures (Geller 1996). Examples of at-risk choices include not wearing protective gear, failing to follow standard energy-controlled lock-out procedures, lifting a heavy object without a hoist, and standing on a machine instead of a ladder. In addition to such natural consequences like ease and comfort, some workers feel management pressures them to take safety short cuts for more efficient production.

The antecedent-behavior-consequence model of applied behavior analysis has been used frequently and successfully to prevent occupational injuries (e.g., Alavosius and Sulzer-Azaroff 1986; Komaki, Collins, and Penn 1978; Streff, Kalsher, and Geller 1993). Antecedents or activators (such as a safety sign) direct one's focus and attention on relevant safety behaviors needed for a given task. Consequences (e.g., an interpersonal "thank you") follow behavior and motivate the future occurrence of the recognized behavior.

Behavioral approaches to injury control have a number of advantages:

- they can be administered by people with minimal professional training (McSween 1995),
- they can reach people in the natural setting such as a particular work area where a problem occurs (Geller 1996),
- the leaders in these settings can be taught the behavior-change techniques most likely to work under specific circumstances (Baer, Wolf, and Risley 1968, 1987; Daniels 1989; Geller 1998b), and
- they can be extremely cost-effective (Daniels 1989; Sulzer-Azaroff and De Santamaria 1980).

Improvements in safety-related behaviors following specific feedback have been demonstrated in a number of organizational settings:

- a plastics manufacturing plant (Sulzer-Azaroff and De Santamaria 1980),
- a metal fabrications company (Zohar, Cohen, and Azar 1980),

- a bakery (Komaki, Barwick, and Scott 1978),
- a public works department (Komaki, Heinzmann, and Lawson 1980),
- a university chemical laboratory (Sulzer-Azaroff 1978), and
- a university cafeteria (Geller, Eason, Phillips, and Pierson 1980).

Fellner and Sulzer-Azaroff (1984) conclude that, "Informational feedback on performance has been shown to be a simple, effective, and durable method for promoting safety" (p. 7).

In the safety research literature, behavioral feedback has been delivered in two different forms—specific and global. Briefly, specific feedback is defined as the percentage of certain safe behavior occurrences over a given time period for certain target behaviors (e.g., bending knees when lifting is 45 percent safe for Week 1). Theoretical support for specific feedback is grounded in behavior analysis. Interventions to improve safety performance typically involve modifying or changing the salience of the antecedents and/or consequences of the specific safety behaviors targeted (Krause, Hidley, and Hodson 1996; McSween 1995). In fact, specific feedback has been demonstrated to be highly effective in various industry-based studies (e.g., Cohen and Jensen 1984; Hopkins et al. 1986; Komaki et al. 1980; Ludwig and Geller 1997).

In contrast, global behavioral feedback can be defined as an overall safety score based on the percentage of safe work practices occurring over a given time period across a certain number of behaviors (e.g., 76 percent safe for Week 1). Support for using global feedback is driven by theories of generalized responding. For a given set of behaviors, overall feedback without information regarding the specific behaviors contributing to the results should lead to more vigilant attention to *all* behaviors, resulting in generalized and long-term improvement (Boyce and Geller 2001; Stokes and Baer 1977). Global behavioral feedback has been effective at improving safety-related behaviors in numerous occupational safety studies (e.g., Austin et al. 1996; Cooper et al. 1994; Komaki et al. 1978; Reber, Wallin, and Chhokar 1990).

While there is ample evidence that both specific and global behavioral feedback lead to beneficial behavior change, no published research has systematically compared these two feedback approaches. The present study compared the impact of giving employees specific versus global feedback regarding certain safety-related behaviors. In addition, the impact of social comparison feedback was evaluated.

Social comparison feedback (SCF) displays specific information about the performance of an individual or group in comparison to a relevant comparison group. Wood (1989) maintains that, "...people not only wish to evaluate their abilities, they also feel pressure to continually improve them. When combined with the desire to compare with similar others, this drive upward leads the individual to strive toward a point slightly better than that of comparison others" (p. 232). Thus, SCF could serve a motivational function. In other words, the addition of SCF, also referred to as normative feedback, should increase the impact of both specific and global behavior-based feedback.

We could find no published evaluation of SCF in the safety literature. However, research has shown SCF to improve performance of a word completion assignment (Mathieu and Button 1992), a mental rotations task (Tindale, Kulik, and Scott 1991), and a work-related simulation (Mitchell, Rothman, and Liden 1985). Some researchers, however, have suggested that SCF will have either no affect or negative impact on performance because of group pressure to perform at the mean. Thus, exceptional performers may be motivated to perform less effectively. In fact, SCF led to diminished mean performance in a word search task (Johnson et al. 1996) and had no effect on actual job performance (Smither, Wohlers, and London 1995). Thus, there is contradictory evidence regarding the influence of SCF on task performance, and no published data on the influence of SCF on industrial safety practices.

In summary, there is a plethora of research suggesting behavior-based feedback leads to increases in safe work practices and reductions in occupational injuries. However, less is known about the specific types of feedback which lead to the greatest improvement in safety performance. More specifically, the existing body of literature does not address the relative impact of specific versus global feedback and SCF versus no SCF on safety-related behavior. The current research made these comparisons.

Method

Participants and Setting

The participants were 97 employees from both day shifts at a soft-drink bottling facility located in southwestern Virginia. The majority of employees were male (69 percent) and Caucasian (82 percent). The facility consists of four production lines, a warehouse, and a laboratory area. Both cans and plastic bottles are filled with the product and distributed via conveyers to large "labeling" machines, and then sent to large packaging machines (or palatizers). Because the daily activities of the laboratory are very dissimilar to the other areas, laboratory observations were not used in the study. Although no fatalities have occurred at this site within the last five years, 75 restricted workdays and 12 lost work days due to injury occurred in 1998. The majority of injuries were back strains/sprains and hand/head abrasions.

Site-Specific Observation Checklist

The behaviors included on a critical behavior checklist (CBC) were chosen by the safety manager and safety steering committee with guidance from the senior author. Given the high percentage of back injuries, specific lifting behaviors were chosen. Because of past near-hits and property damage involving forklift trucks, specific forklift truck-driving behaviors were also included on the CBC. In addition to protective gear, generic safety concerns were chosen for the checklist, including stacking (danger of pallets tipping over), using the handrail (slips occur on stairs with wet surface areas from the sanitation procedures), and avoiding the overhead conveyer (causes head abrasions).

Site-Specific Areas

The employees were categorized into four similarly sized groups of approximately 25 employees each: Group A = employees on Lines 1 and 2 of the plant; Group B = those on Line 3; Group C = employees in the warehouse; and Group D = employees on Line 4. The tasks and requirements of the employees in these four groups were relatively similar, although loading and lifting occur more frequently in the warehouse. According to the safety director at the facility, injury reports were similar across these groups.

Procedure

Behavioral safety education/training

The senior author delivered one hour of behavioral safety education and training to all employees with approximately 25 participants per session. All training sessions occurred on one day. Principles and practical applications of behavior-based safety were presented with an emphasis on behavioral observation and feedback. Participants were informed that behavioral observations would follow training for a period of ten weeks. Feedback was given individually to each employee, and participants were repeatedly asked not to discuss feedback information with co-workers. In other words, they were told to keep their feedback experiences personal and private. Also, participants were informed that observation data would not be shared with management and that only group data would be given.

Behavioral observations

After all employees received the behavioral safety education and training, behavioral observations occurred systematically for ten consecutive weeks. The same CBC was used for three consecutive phases: 1) baseline, 2) six weeks of intervention, and 3) four weeks of withdrawal. The CBC observations occurred every workday and for both shifts, Monday through Friday. The observers followed a specific route for all sessions. They were not informed of the four experimental conditions and three phases.

When an employee was observed to be engaged in a task, the observers marked individual behaviors as either safe or at-risk, based on the operational definitions given on the back of the CBC. When a judgment about a specific behavior was not possible (for instance, when long hair obscured a clear view of hearing protection), a line was drawn through the safe and at-risk box for that specific observation to indicate a "non-observation." The observations on both shifts lasted two to five minutes per target individual with the length of an entire observation session ranging from one to two hours.

Feedback intervention

At the end of each week of the intervention period, behavioral feedback was displayed on bar graphs and delivered in sealed envelopes addressed to each individual employee. The specific target and nontarget safety-related behaviors were listed on the feedback page above the graphs. Participants in the global feedback condition received a single percent safe score for the week, whereas participants in the specific feedback condition received a bar graph for each target behavior.

Participants in Groups A (n = 23) and B (n = 22) received their area's percent safe score(s) for the week compared with the other shift's performance in the same area of the plant. These percentages were global for Shift 1 and specific for Shift 2. Participants in Groups C (n = 28) and D (n = 24) only received their own group's percent safe scores.

Results

Reliability of Observations

For 27 percent of the 15,183 total behavioral observations balanced across phases, two trained and experienced research assistants independently observed the participants. These observers communicated with each other during the observation procedures only to identify the person being observed. After selecting a target individual, each observer completed the CBC independently.

A percent agreement score was calculated for both safe and at-risk behaviors for each behavioral category using the formula: (Total Number of Agreements/Total Number of Agreements + Total Disagreements) x 100. The mean agreement percentages for safe observations were 93 percent (n = 1725) for PPE, 80 percent (n = 3564) for lifting, 77 percent (n = 3880) for forktruck driving, and 91 percent (n = 975) for general safety behaviors. The mean agreement percentages for at-risk observations were 88 percent (n = 681) for PPE, 83 percent (n = 1502) for lifting, 78 percent (n = 2607) for forktruck driving, and 93 percent (n = 313) for general safety behaviors (cutting, conveyor avoidance, stacking).

Percent Safe Scores

Daily data from the CBC cards were converted to daily percent safe scores with the formula: Percent Safe = (Total Safe Observations/Total Safe Observations + At-Risk Observations) x 100. The percentages for each day were collapsed over consecutive three-day periods for the analysis of variance because there were large variations in the number of observations made per individual each day. This daily number ranged from 1 to 223, with a mean of 78 observations per individual per day and a standard deviation of 49.7. This three-day measurement percentage was used for the Two Feedback Type (Global or Specific) x Two SCF (Present or Not Present) analysis of covariance (ANCOVA) using baseline scores as the covariate.

The ANCOVA indicated that behavioral feedback was effective at increasing percent safe scores over baseline. Overall, mean percent safe scores across all feedback conditions rose from 64 percent at baseline to 73 percent at intervention (and fell only slightly to 72 percent when feedback was withdrawn). A plot of mean percent safe scores per feedback condition at baseline, intervention, and withdrawal is provided in Figure 3.3.1.

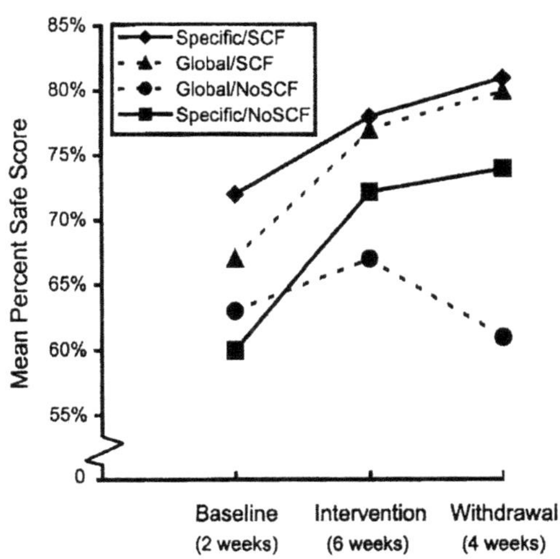

Figure 3.3.1 Consecutive Experimental Phases

Global and Specific Feedback

Mean percent safe scores at intervention were only slightly higher in the specific (M = .74, SD = 8.7) versus global (M = .72, SD = 9.7) feedback conditions. Specific/SCF led to slightly higher percent safe scores than Global/SCF (M = .79 and .77, respectively), and Specific/NoSCF led to higher percent safe scores than Global/NoSCF (M = .70 and .66, respectively). Overall, the main effect for feedback type was not statistically significant (F = .17, $p>.05$).

Social Comparison versus No Social Comparison Feedback

Percent safe scores were substantially higher in the SCF (M = .78, SD = 6.94) versus no SCF conditions (M = .68, SD = 8.43). Global/SCF led to a higher mean percent safe score than Global/NoSCF (M = .77 and .66, respectively). Similarly, Specific/SCF led to a higher mean percent safe score than Specific/NoSCF (M = .79 and .70, respectively). Not surprisingly, the main effect for SCF was statistically significant (F = 10.8, $p<.01$).

Feedback Type by SCF

Specific/SCF resulted in the highest percent safe score across feedback conditions (M = .79, SD = 6.7), followed by Global/SCF (M = .77, SD = 7.41), Specific/NoSCF (M = .70, SD = 8.48) and Global/NoSCF (M = .66, SD = 8.24). Overall, the Feedback Type by SCF interaction was not significant (F = .02, $p>.05$).

A series of Chi-Square analyses were used to test for significant differences between the four feedback conditions at intervention. Specific/SCF was not significantly higher than Global/SCF ($X^2_{(1)}$=.46, $p>.05$). Next, Global/SCF was significantly higher than Specific/NoSCF ($X^2_{(1)}$=14.3, $p<.05$). Specific/NoSCF was significantly higher than Global/NoSCF ($X^2_{(1)}$=12.81, $p<.05$). During the withdrawal phase, only the percent safe score for the Global/NoSCF dropped significantly after the intervention was removed ($X^2_{(1)}$=8.54, $p<.05$).

Discussion

Performance feedback research has seldom included experimental comparisons of different ways to deliver behavioral feedback (Balcazar, Hopkins, and Suarez 1985/86). This lack of research extends to the safety literature, which consistently demonstrates that behavior-based feedback improves safety performance without exploring the characteristics of feedback driving these results. To this end, the objective of the current study was to add to the existing literature by increasing our understanding of the relative impact of global, specific, and social comparison feedback on subsequent safety performance.

Similar to prior industrial safety research using behavior-based (BB) feedback as an intervention (Fellner and Sulzer-Azaroff 1984; Sulzer-Azaroff and De Santamaria 1980; Zohar, Cohen, and Azar 1980), BB feedback led to clear increases in safe work practices. Based on more than 15,000 observations over an eight-week period, overall percent safe scores increased from 64 percent to 73 percent following the introduction of BB feedback and was maintained during the withdrawal period (72 percent).

The results supported the use of social comparison feedback (SCF) to improve safety performance. Prominent differences in percent safe scores between the SCF and NoSCF conditions at intervention are reflected in Figure 3.3.1. Overall, SCF led to an overall ten-percentage point increase in percent safe scores over the NoSCF conditions. More specifically, the intervention impact was significantly greater for the Specific/SCF condition than Specific/NoSCF, and the feedback benefits were significantly higher for Global/SCF than Global/NoSCF.

Although SCF was beneficial at improving safety performance, the expected advantage of specific over global BB feedback was not found. Specifically, the impact of Specific/SCF and Global/SCF was similar. On the other hand, Specific/NoSCF led to a significantly higher mean percent safe score than Global/NoSCF. Plus, only the Global/NoSCF group showed a decrease in percentage of safe behaviors after the feedback intervention was withdrawn. This suggests specific feedback is superior to global feedback when SCF is *not* present.

While specific feedback may outperform global feedback in the absence of SCF, the similar impact of Global/SCF and Specific/SCF feedback has important practical implications. It took an extremely long time to code, aggregate (by week), calculate, and post percent safe scores for the nine specific behaviors targeted in this research, especially when SCF was added. It took the senior author five to six hours a week to prepare the charts for the specific feedback conditions. Conversely, computing one single overall score (i.e., global feedback) even with SCF took only about 30 minutes a week. The difference in response cost between the two feedback conditions for eight consecutive weeks cannot be overstated.

Most safety professionals have limited time and resources for implementing and maintaining the variety of interventions available to improve safety performance. In an effort to reduce injuries, many professionals choose BB feedback to supplement existing safety programs. The results of the current study suggest that Global/SCF is basically as effective as Specific/SCF in improving safety performance at a *fraction* of the time and effort required to provide it.

However, global feedback (without SCF) led to no meaningful change over baseline and a drop in safe work practices after removal of BB feedback. Simply put, global feedback by itself had almost no effect on subsequent safety performance and, given the follow-up results, may have actually done more harm than good. In contrast, substantial improvements from baseline to intervention were shown with the other three feedback conditions.

While providing employees one global score did not improve performance, adding a second overall score (from another shift) led to a feedback effect comparable to the two specific feedback conditions. Apparently, providing comparison information affected the workers' collective desire to outperform their peers and thereby served a motivational function.

The powerful influence of SCF is not surprising given past theoretical and empirical support, including Festinger's (1955) seminal article, "A Theory of Social Comparison Processes." Festinger argues that in Western cultures, individuals are driven to continually improve performance and seek out similar others with whom to gauge their performance. Thus, the critical factor in the evaluation of an individual's performance is his or her *relative* standing in comparison to others.

Furthermore, Festinger's (1955) concept of the unidirectional drive upward suggests that individuals are motivated not only to improve their own performance but also to outperform the comparison group (Wood 1989). The result of this comparison information is improved performance, as demonstrated in a number of empirical studies (e.g., Goltz et al. 1990; Mathieu and Button 1992; Mitchell, Rothman, and Liden 1985; Tindale, Kulik, and Scott 1991).

In the current study, SCF probably served as a motivational referent for comparison and triggered participants' collective desire to outperform a similar work group. The result is that participants given SCF demonstrated significant and substantial improvements over baseline performance.

Geller (1999) proposed that behavior-change interventions are instructional, motivational, or supportive and that the impact of a particular intervention depends upon the awareness versus motivational state of the target individuals. Since Global/SCF was just as effective as Specific/SCF, it can be assumed the employees did not need specific direction or instruction (as provided by specific feedback). Rather, the employees apparently knew how to perform their jobs safely but needed some extrinsic motivation to follow the nine safety policies implied by the nine different target behaviors. Equivalent motivation was provided by a global comparison score and by specific feedback. In this case, specific feedback was probably more motivational than directional, reflecting a precise accountability system (Geller 1998a).

Conclusion

Behavior-based feedback increased percent safe scores from baseline to intervention, and these improvements were maintained for three of four groups after the BB feedback intervention was withdrawn. Global/SCF provided the most promising results in terms of cost effectiveness, and Global/NoSCF was least effective. Unfortunately, Global/NoSCF is the most common type of behavioral feedback used in applied settings (Krause, Hidley, and Hodson 1996; McSween 1995), so the majority of organizations using BB feedback are probably not experiencing the full potential and benefits of this intervention.

References

Alavosius, M. P., and B. Sulzer-Azaroff. 1986. The effects of performance feedback on *the safety of client lifting and transfer. Journal of Applied Behavior Analysis,* 19:261–267.

Austin, J., M. L. Kessler, J. E. Riccobono, and J. S. Bailey. 1996. Using feedback and reinforcement to improve the performance and safety of a roofing crew. *Journal of Organizational Behavior Management*, 16(2):49–75.

Baer, D. M., M. M. Wolf, and T. R. Risley. 1968. Some current dimensions of applied behavior analysis. *Journal of Applied Behavior Analysis*, 1:91–97.

———. 1987. Some still-current dimensions of applied behavior analysis. *Journal of Applied Behavior Analysis*, 20:313–327.

Baker, S. P., C. Conroy, and J. J. Johnston. 1992. Occupational injury prevention. *Journal of Safety Research*, 23(2):129–133.

Balcazar, F., B. L. Hopkins, and Y. Suarez. 1985/86. A critical, objective review of performance feedback. *Journal of Organizational Behavior Management*, 7(3/4):65–89.

Boyce, T. E., and E. S. Geller. 2001. Applied behavior analysis and occupational safety: The challenge of response maintenance. *Journal of Organizational Behavior Management*, 21(1): 31–60.

Califano, J. A., Jr. 1979. *Healthy people: The Surgeon General's report on health promotion and disease prevention*. Washington, D.C.: U.S. Government Printing Office.

Cohen, H. H., and R. C. Jensen. 1984. Measuring the effectiveness of an industrial lift truck safety training program. *Journal of Safety Research*, 15:125–135.

Cooper, M. D., R. A. Phillips, V. J. Sutherland, and P. J. Makin. 1994. Reducing accidents using goal setting and feedback: A field study. *Journal of Occupational Psychology*, 67:219–240.

Daniels, A. C. 1989. *Performance management*. Tucker, GA: Performance Management Publications.

Fellner, D. J., and B. Sulzer-Azaroff. 1984. Increasing industrial safety practices and conditions through posted feedback. *Journal of Safety Research*, 15:7–21.

Festinger, L. 1955. A theory of social comparison processes. *Human Relations*, 7:117–140.

Geller, E. S. 1996. *The psychology of safety: How to improve behaviors and attitudes on the job*. Radnor, PA: Chilton Book Company.

———. 1998a. *Beyond safety accountability: How to increase personal responsibility*. Neenah, WI: J. J. Keller & Associates, Inc.

———. 1998b. *Understanding behavior-based safety: Step-by-step methods to improve your workplace*. 2nd ed. Neenah, WI: J. J. Keller & Associates, Inc.

———. 1999. Behavior-based safety: Confusion, controversy, and clarification. *Occupational Health and Safety*, 68(1):40–49.

Geller, E. S., S. Eason, J. Phillips, and M. Pierson. 1980. Interventions to improve sanitation during food preparation. *Journal of Organizational Behavior Management*, 2:229–240.

Goltz, S. M., M. Citera, M. Jensen, J. Favero, and J. L. Komaki. 1990. Individual feedback: Does it enhance effects of global feedback? *Journal of Organizational Behavior Management*, 10(2):77–93.

Hopkins, B. L., R. J. Conard, , R. F. Dangel, H. G. Fitch, M. J. Smith, and W. K. Anger, 1986. Behavioral technology for reducing occupational exposures to styrene. *Journal of Applied Behavior Analysis*, 19:3–11.

Johnson, D. S., D. B. Turban, K. F. Pieper, and Y. M. Ng. 1996. Exploring the role of normative- and performance-based feedback in motivational processes. *Journal of Applied Social Psychology*, 26(11):973–992.

Komaki, J., K. D. Barwick, and L. R. Scott. 1978. Behavioral approach to occupational safety: Pinpointing and reinforcing safe performance in a food manufacturing plant. *Journal of Applied Psychology*, 63:424–445.

Komaki, J. L., R. L. Collins, and P. Penn. 1978. The role of performance antecedents and consequences in work motivation. *Journal of Applied Psychology*, 67(3):334–340.

Komaki, J., A. T. Heinzmann, and L. Lawson. 1980. Effects of training and feedback: A component analysis of a behavioral safety program. *Journal of Applied Psychology*, 65:261–270.

Krause, T. R., J. H. Hidley, and S. J. Hodson. 1996. *The behavior-based safety process: Managing involvement for an injury-free culture.* 2nd ed. New York: Van Nostrand Reinhold.

Ludwig, T. D., and E. S. Geller. 1997. Assigned versus participative goal-setting and response generalization: Managing injury control among professional pizza deliverers. *Journal of Applied Psychology*, 82(2):253–261.

Mathieu, J. E., and S. B. Button 1992. An examination of the relative impact of normative information and self-efficacy on personal goals and performance over time. *Journal of Applied Social Psychology*, 22:1758–1775.

McSween, T. E. 1995. *The values-based safety process: Improving your safety culture with a behavioral approach.* New York: Van Nostrand Reinhold.

Miller, T. R. 1997. Estimating the costs of injury to U.S. employers. *Journal of Safety Research,* 28(1):1–13.

Mitchell, T. R., T. Rothman, and R. Liden, 1985. Effects of normative information on task performance. *Journal of Applied Psychology*, 70:48–55.

National Committee for Injury Prevention and Control. 1989. *Injury prevention: Meeting the challenge.* New York: Oxford University Press.

Reber, R. A., J. A. Wallin, and J. S. Chhokar. 1990. Improving safety performance with goal setting and feedback. *Human Performance*, 3(1):51–61.

Smither, J. W., A. J. Wohlers, and M. London. 1995. A field study of reactions to normative versus individualized upward feedback. *Group and Organizational Management*, 20(1):61–89.

Stokes, T. F., and D. M. Baer. 1977. An implicit technology of generalization. *Journal of Applied Behavior Analysis*, 10:349–367.

Streff, F. M., M. J. Kalsher, and E. S. Geller. 1993. Developing efficient workplace safety programs: Observations of response covariation. *Journal of Organizational Behavior Management*, 13(2):3–15.

Sulzer-Azaroff, B. 1978. Behavioral ecology and accident prevention. *Journal of Organizational Behavior Management*, 2:11–44.

Sulzer-Azaroff, B., and M. C. De Santamaria. 1980. Industrial safety hazard reduction through performance feedback. *Journal of Applied Behavior Analysis,* 13:287–297.

Tindale, R. S., C. T. Kulik, and L. R. Scott. 1991. Individual and group feedback and performance: An attributional perspective. *Basic and Applied Social Psychology*, 12(1):41–62.

Wood, J. V. 1989. Theory and research concerning social comparisons of personal attributes. *Psychological Bulletin*, 106(2):231–248.

Zohar, D., A. Cohen, and N. Azar. 1980. Promoting increased use of ear protectors in noise through information feedback. *Human Factors,* 22(1):69–79.

Employees Forgot Their Hard Hats Again? Seven Lessons from Behavior-Based Safety for Increasing PPE Use[4]

Steve Roberts

The issue of using appropriate personal protective equipment (PPE) is very simple: It is provided to employees who either use it or face disciplinary actions. Although it may be tempting to subscribe to this viewpoint, the principles of behavior-based safety can offer better alternatives for long-term maintenance of PPE use, as well as other safe behaviors.

Basically, behavior-based safety (BBS) is a process by which employees identify at-risk behaviors, analyze the reasons they are performed, and implement and evaluate interventions to increase safe behavior and reduce at-risk behavior. One misconception is that BBS encourages the use of low-level hazard controls such as PPE over more effective, higher-level ones such as engineering controls. Therefore, let me first acknowledge the acceptance of the following intervention hierarchy:

1st - eliminate hazards (substitution of materials, automation)

2nd - introduce engineering controls (guarding, interlocks)

3rd - provide warnings (signs, alarms)

4th - include administrative controls and training (job rotation, lockout/tagout training, equipment inspections)

5th - furnish PPE (eye protection, respiratory protection, fall protection)

This hierarchy clearly regards PPE as the least effective intervention. However, it is not always feasible to engineer out or administratively control all risk. In the short term, PPE is often the only or last line of defense against certain hazards. Using behavior-based safety can increase appropriate PPE use.

Involve employees in an observation and feedback process

There are many instances in which using PPE is simply less comfortable, less convenient and less efficient than employing an engineering or administrative control. Increasing the quality and quantity of interpersonal feedback regarding safety is one of the best methods of increasing safety behaviors. Knowing your co-workers are looking out for you—not for the purpose of "writing you up" or other disciplinary action but because they care about you—can be a powerful motivator. Also, individuals who give feedback to others regarding safety-related behaviors are more likely to act appropriately themselves, even when no one is watching.

Reinforce targeted PPE behaviors

People do the things they are reinforced for doing. Reinforcement can take many forms. It is optimal to create situations in which reinforcement is intrinsic or "built in" to what we do. For example, we may use safety glasses not only because we know they protect our eyes but also because the new styles look "cool," or we may use certain gloves because they actually make handling materials easier. However, it is often not possible to build reinforcement into performing safety behaviors.

It can be useful to add recognition or rewards to reinforce safe behavior. An organization could systematically increase supervisory recognition and praise for respirator use. Lanyards or "Chums" for safety glasses could be given to those using safety glasses, gift certificates for soda or ice cream could be given for using a hard hat, or a thank-you card could be given for using a safety harness as opposed to a "STOP" card for not using one. Be careful, however, not to imply that small items such these are the payoff or the reason for performing safe work practices but are simply a token of appreciation for a job well done.

Consider the various factors influencing PPE use

Many factors contribute to performing safe or at-risk behavior. Simply labeling a behavior as inconvenient ignores the many actions that should be taken to encourage safe behavior.

In one company, employees were inspecting a pipe rack 25 feet off the ground. Workers were wearing safety harnesses but many would not attach the lanyard. After observation and discussion, the cause was identified: the horizontal lifeline had multiple breaks in it, forcing workers to hook and unhook every few feet. A simple intervention replaced the multiple lifelines with one that had no breaks or anchors, allowing workers to hook up only once.

In another company, it was a 20-minute walk to the one locker containing fall protection. And chances were, only old dirty harnesses in poor condition would be there. The intervention was to provide additional lockers placed closer to various work areas and to supply them with a variety of harnesses in good condition.

In a third situation, an employee was about to work nine feet off the ground and asked for fall protection. His co-workers called him a wimp and other derogatory names. This required a meeting with the people in the area and a frank discussion about the value of safe behaviors.

Two of the above examples involve administrative or engineering controls and may not seem to demonstrate BBS. However, the process by which the interventions were implemented was BBS. That is, employees identified an at-risk behavior, analyzed the reasons for the at-risk behavior, and suggested intervention strategies to reduce it.

Involve workers in determining appropriate interventions

This might seem obvious, but it is often time consuming and difficult to bring a group together for an honest discussion regarding safe and at-risk behaviors. Most often a key person decides which interventions will be implemented. The at-risk workers have little if any impact on the process.

We should never assume a reason for at-risk behaviors without first having a discussion with the people involved. This discussion should be structured according to the ABC model: first consider the events or conditions occurring before the behavior (A = activators), then consider the behavior (B = behavior), and finally, look at the events occurring during and after the behavior (C = consequences). Examples of activators are time pressure, environmental conditions, and co-worker behavior. Examples of consequences include comfort, time effectiveness, and rewards or recognition.

In one company, several people regularly failed to use a face shield while grinding, even though the face shield was hanging beside the grinder. The convenient location of this face shield made it seem that those who failed to use it had a "bad attitude" regarding safety. After much discussion, however, someone finally admitted they were not using the shield because one individual with allergies often sneezed into it and no one wanted to use it afterwards. The solution was to provide an alcohol spray bottle and wipes beside the face shield. Another intervention could have been to provide face shields to everyone who used the grinder.

Focus on the potential for injury, not safety rules

Interpersonal feedback citing rules can seem impersonal and cold. People often tune out. When focusing on a worker's potential for getting hurt instead of compliance with the rules, it is easier to convey a genuine concern for a person's welfare. Also, there is extensive overlap between compliance with safety rules and the behaviors to best reduce risk, but the two are not always the same.

At a large auto manufacturing facility in an area where auto bodies were being welded together, I was told I did not need safety glasses as long as I stayed within the warning lines on the floor. As I was walking through this area, large red sparks from robotic welders began to shoot in front of my face, ricochet off the wall beside me, and fly back in front of my face again. Although a change in shielding around the welding process may be the most appropriate long-term solution, the short-term solution was to immediately turn around, don safety glasses with side shields, and take an alternate path away from the welding area. While the potential for getting hurt is not always so apparent, communicating risk is key to motivating safety-related behaviors.

Use examples, images and case studies

There are multiple ways of focusing on injury potential rather than rules and regulations. Do not rely on statistics to get the job done either. Statistics such as injury rates are useful for providing context, identifying the scope of the problem, and prioritizing behaviors to target. But communicating with statistics is not the best way to motivate safe behaviors.

Years ago, many vendors selling safety eyewear used a poster-size photograph of an eyeball to motivate the use of safety eyewear. The photo was very close-up and clear; it showed a nail lodged dead center in the eye. While I have long ago forgotten statistics regarding eye injuries, the image of that eye is still as clear in my mind as the first time I saw it. Whenever I need extra motivation to put on safety glasses, I bring up that image and find putting on the glasses much easier.

Consider issues beyond PPE

The use of PPE is often seen as a behavior under the direct control of the individual. When discussing safety responsibilities, management will sometimes focus on those behaviors under direct personal control, such as PPE. There are multiple causes for injury, however, which are not under the control of the individual. Inadequate training, rewards for safety short cuts, inadequate tools and equipment, and failure to follow procedures are just a few examples. As explained above, PPE is a low-level control, yet it is often treated as the only necessary health and safety activity. On the other hand, many employees will focus on activities under the control of management such as facilities/equipment and management systems.

When one group focuses primarily on one issue without acknowledging the relevance of others, the resulting defensiveness and reactionary behaviors can damage credibility. Management's willingness to acknowledge and address the factors influencing those at-risk behaviors under their control will often increase feelings of teamwork. It will also lend more credibility to efforts targeting behaviors under the control of the individual worker, such as PPE compliance, when higher level controls are not feasible. This increased sense of teamwork and caring by the organization will not only influence employee participation in safety efforts, but it will also increase cooperation and involvement in other important areas such as production and quality.

There are often varied and complicated reasons people do not consistently use appropriate PPE. Treating all PPE use as simply a matter of personal choice often proves unsuccessful. Behavior-based safety provides a framework for identifying problem behaviors, analyzing the factors influencing their occurrence, and implementing, evaluating and increasing appropriate intervention activities.

Addressing Ergonomic Hazards through Behavior-Based Observation and Feedback[5]

Sherry R. Perdue

Abstract

A behavior-based observation and feedback process should be an integral component of a comprehensive approach to identifying and managing ergonomic hazards. Systematic behavioral observation and analysis can supplement other hazard identification and analysis tools. Routine behavioral observations conducted throughout the work environment identify situations in which ergonomically unsound work practices occur. Job-specific behavioral observations may then be conducted as part of a detailed analysis. Subsequently, the observation process serves as a hazard reduction tool by providing the peer feedback needed to encourage and reinforce the use of ergonomically sound work practices.

This paper:

- reviews the key elements of a comprehensive ergonomics program,
- presents a brief overview of a behavioral observation and feedback process,
- shows how a behavioral observation process supports an ergonomics program, and
- describes a training strategy for preparing behavioral safety observers to become effective at identifying ergonomic-related safe and at-risk situations.

Traditional Approach to Managing Ergonomic Risks

Traditional approaches to managing ergonomic-related illnesses include strategies to identify and then reduce exposure of employees to ergonomic hazards. An ergonomics training program is conducted to allow individuals to be active participants throughout the identification and elimination of ergonomic hazards. Instructional programs typically teach employees to recognize and report the early warning signs of ergonomic-related injuries, to understand the basic ergonomic risk factors (e.g., repetitive motion, forceful exertions, sustained or awkward positioning) and to use simple strategies to prevent ergonomic injury (e.g., proper seated posture, appropriate lifting techniques).

An ergonomics hazards analysis of the work site is performed to identify existing and potential hazards. A thorough analysis includes:

- a review of injury and illness records to identify patterns and trends indicative of ergonomic hazards,
- the administration and analysis of a "symptom survey" to identify employees experiencing early symptoms of an ergonomics-related injury,
- the administration and analysis of a "job/task survey" to pinpoint situations where tasks are performed for an extended duration and/or with sufficient frequency to raise concern,
- a self/peer workplace evaluation that allows employees to assess and adjust their own and/or others' work stations,
- a global workplace audit by an ergonomics specialist, including interviews with representative samples of workers, and
- a detailed analysis of those situations identified in the steps above as having a high risk potential.

Once ergonomic hazards are identified and prioritized through a systematic work-site analysis, corrective actions can be planned and taken. Risk-reduction strategies typically include administrative controls, personal protective equipment, and engineering and work practice controls. Administrative controls include rotating employees on and off certain jobs, modifying task schedules, and enlarging or enriching assignments. Traditional work practice controls involve modifying work methods and operating procedures followed by specific ergonomics training for the recommended changes. Engineering controls include designing or modifying the work station, tools, and equipment used in performing the work.

When the initial efforts are complete, program maintenance includes periodic refresher training and re-administration of the symptom survey and the peer/self evaluation instrument. Follow-up analysis and work and/or work-station redesign and adjustment are performed as needed. In addition, steps should be taken to ensure that principles of ergonomics are used in the selection and design of new equipment and work stations and that ergonomic considerations are incorporated into the identification and modification of job tasks to accommodate injured employees who are returning to work.

While the hazard identification and reduction techniques described are necessary components of a comprehensive ergonomics program, they are often insufficient. Even the best designed tools and work stations are frequently misused, either intentionally or unintentionally, even after well-designed and delivered ergonomics training. A behavioral observation and feedback process can be used to complement the traditional techniques, forming a more comprehensive approach for identifying and controlling ergonomic risks.

An Overview of A Behavioral Observation and Feedback Process

A behavioral observation and feedback process is a tool designed to help increase visibility and understanding of safe and at-risk behaviors. Using a checklist containing predetermined critical safety-related behaviors specific to the organization, employees periodically observe the work practices of one another while on the job, noting occurrences of both safe and at-risk behaviors. Following the observation period, which typically ranges from ten to 20 minutes, the observer and observee discuss the observer's results. The observer reviews the work behaviors he or she noted as having been performed safely and—using appropriate coaching techniques—reviews any behaviors which appeared to put the individual at risk of injury.

In addition, the observer/observee pair informally assesses the particular work context to identify any specific barriers to working safely. Possibilities might include environmental hazards, training inadequacies, communication failures, time constraints, and equipment or tool design deficiencies. With the results of the objective observation, they collectively problem-solve to reduce or eliminate these barriers to allow and/or encourage the safe alternative behavior.

The completed checklists are periodically collected and compiled. The compiled data are charted and shared with employee groups, providing a second form of constructive feedback. The data charts illustrate strengths and weaknesses of individual groups and/or the organization as a whole. Frequently occurring at-risk behaviors and/or those most likely to result in severe injury are studied in further detail. Teams assess the work situation to identify any environmental or design hindrances to performing a target behavior safely and to derive ways to reduce or eliminate these barriers and thereby enable or facilitate safe behavior. After careful analysis, teams develop, implement, and measure the impact of the behavior-improvement interventions. These interventions might include redesigning a workstation; obtaining more comfortable, more effective, or more available PPE; streamlining operating procedures, or training and supporting particular work practices.

Often, there are no systems barriers discouraging or preventing the safe behavior, or the conditions which make at-risk behavior uncomfortable or inconvenient cannot be removed. In these situations, constructive peer feedback can help reduce at-risk work practices in other ways. Individuals often perform at-risk behaviors without willful intent, either because the proper way to perform the behavior is not understood or a lapse in attention or mindfulness enables an at-risk activity. We can sometimes be "unconsciously incompetent" (Geller 1999, 2001). In such cases, feedback provides information and makes us aware of the risk we are exposing ourselves to so that we can work to change that behavior in the future.

Feedback can also be motivational, providing social support to encourage individuals to take the time and extra effort to perform behaviors in a safe manner. Sometimes people purposefully choose to work at risk, usually because the at-risk behavior is easier, faster, and/or more convenient than the safe alternative. Heavy emphasis on meeting work schedules and production goals often further encourage at-risk behaviors, as does the impression that many others also take the particular calculated risk. Although a chance of injury exists, it is perceived as highly unlikely. At the time, a decision to take the seemingly "minor" risk seems reasonable. In these cases, peer observation and feedback provide an accountability system to encourage and motivate people to perform safe rather than at-risk behaviors.

Finally, when peers provide positive feedback following the occurrence of safe behaviors, this timely and positive consequence will likely reinforce the worker's behavior, making it more likely to occur again. In this way, "conscious competence" is supported and increased, eventually becoming second nature, or "unconscious competence."

Integrating Behavioral Observation into an Ergonomics Program

A behavioral observation and feedback process can serve two distinct and critical functions in an ergonomics program. First, the observation process can supplement other assessment tools used to identify ergonomically-hazardous situations. As illustrated in the sample checklist shown in Figure 3.5.1, an organization's general observation checklist typically includes one or more ergonomics-related behavioral categories such as "Body Positioning and Protecting," "Pacing of Work," and "Manual Materials Handling." This checklist guides observers to identify safe and at-risk

Observer: ____________ Date: ____________

Department: ____________ Time: ____________

Operating Procedures	Safe	At-Risk	Comments
PPE Using personal protective equipment (e.g., eye glasses, hearing protection, gloves, hard hats).			
Housekeeping Work area maintained safely e.g., trash and scrap picked up, no spills, walk-ways clear, materials and tools organized).			
Using Tools and Equipment Correct tools for the job; proper use; tools in good condition.			
Body Positioning/Protecting Positioning/protecting body parts (e.g., proper posture, avoiding line of fire, avoiding pinch points).			

Figure 3.5.1 Checklist Used to Guide the Observation

occurrences of ergonomics-related behaviors. In addition, job- or task-specific checklists can be designed and used to provide more guidance to the observer interested in examining more specific ergonomically-sound work practices. An example is shown in Figure 3.5.2. Because a behavioral checklist is used on a frequent basis (perhaps as often as once per week by each employee on a work team), the process can help uncover ergonomically-unsound situations which may be overlooked by a more traditional ergonomics assessment protocol, which is usually performed infrequently, if at all. As the data pinpoint work areas or tasks where at-

Figure 3.5.2 Checklist Focusing on Specific Types of Ergonomic Hazards

Tool Use	Safe	At Risk	Comments
Using the proper tool for the job Using the tool as it was designed to be used Tool is in good working condition (sharp)			
Using proper PPE (& proper fit) for use with the tool			
Maintains a neutral wrist position			
Maintains a neutral arm and shoulder position			
Use does not require excessive repetitive movement of finger(s), thumb, wrist, elbow or shoulder			
Use does not require frequent forceful exertion (static or dynamic)			
Use does not require prolonged static position			
Use does not expose user to excessive vibration			
Pinch point safe guarded			
Does not put excessive pressure on any part of hand (sharp edges; handle gloves; end of handle resting on palm) Heavy tools (>1lb) are properly balanced Heavy tools (>6lbs) used with two hands and/or are supported			
Potential for "kick back" is minimized (e.g., launch brake mechanism and reaction bars)			
Designed and adjusted to minimize torque transmitted to wrist (e.g., torque reaction bar)			
Tool handle is neither too hot or cold; Exhaustion from power tools is directed away from operator			
Tool handle is not prone to slipping (moisture, smooth surface)			
Use of pinch grip is minimal			
For 2 handle tools: - does tool open easily (e.g., spring-loaded) - can hand or fingers be pinched between handles when closed - is grip span 25 inches – 35 inches For circular handle tools: - if tool used for precision work, diameter is 0.3 – 0.6 in - if full hand power is required, diameter is 1.25 in. – 2 in.			
For tools with trigger mechanisms: - can different fingers operate trigger - does it require static force - does it present sharp edges - does it require repetitive finger trigger reactions			

risk behaviors occur more frequently, the skilled ergonomist can assist the work team in conducting an even more thorough and detailed analysis.

In addition, the observation process can serve as a hazard reduction tool by providing individual and group feedback to encourage and support the use of ergonomically-sound work practices. For example, behaviors such as those involved in safe lifting or maintaining correct seated posture are seldom intentionally performed in an at-risk manner. Rather, individuals simply fail to attend to their posture and form because they are attending to other parts of the task at hand. As a result, even those who can recite the elements of safe lifting in their sleep are often guilty of not using these well-learned procedures. Likewise, well-designed and adjusted office furniture is often misused as weary computer users gradually slide further and further down into their chairs. A behavioral observation and feedback process which provides a means for frequent feedback can help individuals become aware of these unintentional at-risk behaviors so that they can work to become more fluent at the safe behavior and eventually develop a safe habit (Geller 1999, 2001).

Of course, not all ergonomically-risky behaviors are unintentional. Sometimes people knowingly lift loads heavier than they should, use a tool that is handy but inappropriate for the task, or work at a pace that puts them at risk for a cumulative trauma disorder (CTD). Here, the observation process can help reinforce the need to follow ergonomically sound work practices by providing a systematic mechanism for peers to remind one another of safe work practices.

Training Observers to Identify Ergonomic Risks

The potential contribution of a behavioral observation and feedback process to a comprehensive ergonomics program, however, is limited unless the individuals performing the observations know what to observe. An introductory ergonomics course can be effective in teaching observers to identify the more apparent ergonomically-hazardous situations. To prepare observers more fully to identify ergonomically at-risk behaviors, a more comprehensive training approach is usually needed. While the issue could be avoided by having only the ergonomist or safety professional conduct observations, much of the value of a peer-to-peer observation process—which uses the strength of peer influence, increases informal feedback between co-workers, and facilitates employee involvement and ownership—would be lost. One training approach for increasing the skills to identify ergonomically-hazardous situations tailored specifically for behavioral observation and feedback process participants is described below.

In introducing its behavioral observation and feedback process, a large manufacturing facility recognized the need to use the new system to help address ergonomic issues. After all, over one-third of their recordable cases during the past five years had been attributable to musculoskeletal disorders of one kind or another. So, as they rolled out their behavioral observation process, they customized it to focus on ergonomic issues. The plant-wide observation checklist contained several categories which targeted ergonomic concerns, including "Body Positioning and Protecting," "Manual Materials Handing," and "Pacing of Work." They already had an ergonomics program in place for two years, so their employees were aware of basic ergonomic principles. This introductory training, however, proved to be insufficient in providing observers the skills needed to identify the less obvious ergonomic hazards.

A subcommittee of hourly employees was formed to serve as the organization's leaders in ergonomic observation and analysis. These individuals coordinate their efforts through both the behavior-based safety committee and the ergonomics committee. Members are currently participating in an extensive training program customized to help develop observation skills for ergonomic-related conditions. The program consists of a series of 12 mini-workshops, each of which lasts 45 to 60 minutes. One workshop is held each week. After an introductory session, the curriculum focuses on a different ergonomics issue each week, addressing a variety of ergonomic risk factors one at a time (see Figure 3.5.3).

During the second week, for example, the target issue was posture. The course content contains a general description of the issue, supported by still photographs and video clips illustrating the concern taken within the facility. Video clips are also used to conduct "mock observations" in which participants practice observing for the target issue and then compare their results with those of the class and the instructor. During the week

Week	Topic
1	Overview
2	Posture
3	Force
4	Repetition
5	Soft-tissue Compression
6	Cold and Vibration
7	Duration and Recovery Time
8	Hands and Wrists
9	Shoulders and Elbows
10	Head and Neck
11	Back and Lower Body
12+	Special Issues

Figure 3.5.3 Training Curriculum Tailored to Prepare Observers to Identify Ergonomically-Hazardous Situations

following the workshop, participants conduct periodic observations in their work areas, concentrating only on the targeted issue for the week.

At the end of six weeks, the content of the workshops shifts to focus on different body parts. For example, the topic during the eighth week is "Hands and Wrists." During this week, participants focus on all six ergonomic risk factors, but with specific attention to the hand and wrist areas. After the eleventh week, participants will have had the opportunity to focus on each ergonomic issue and each major body part independently. Finally, any specific issues of concern, such as lifting practices or keyboard entry, will be covered using the same approach.

Conclusion

A behavioral observation and feedback process can be a meaningful addition to the traditional tools and techniques used to manage ergonomic risks and can serve as a hazard identification and analysis tool. Unlike traditional methods, the observations conducted as part of the process occur frequently, with several brief observations taking place within a work area each day. In addition, the observation process serves as a hazard reduction tool by providing a structured mechanism to help employees use ergonomically-sound work practices—a need not met by the components of a traditional ergonomics program. To be most effective, however, employees conducting the observations must have a fairly extensive working knowledge of ergonomic risk factors in order to be able to identify them in their work environment. A "hands-on" training approach designed specifically for this purpose can prepare them well.

References

Geller, E. S. 1999. Behavior-based safety: Confusion, controversy, and clarification. *Occupational Health & Safety*, 68(1):40–49.

———. 2001. The psychology of safety handbook. Boca Raton, FL: CRC Press.

Notes

[1] The original version of this paper appeared in *Proceedings of Light Up Safety in the New Millennium: A Behavioral Safety Symposium* (Orlando, FL: American Society of Safety Engineers, 1998):123–142.

[2] Originally appeared in *Proceedings of the 39th ASSE Annual Professional Development Conference and Exposition* (Des Plaines, IL: American Society of Safety Engineers, 2000).

[3] Originally appeared in *Journal of Safety Research*, 31, 3(2000): 135–142.

[4] The original version of this paper appeared in *Safety and Health* (October 2000).

[5] The original version of this paper appeared in *Proceedings of the 38th Annual Professional Development Conference and Exposition* (Des Plaines, IL: American Society of Safety Engineers, 1999):45-52.

4

Communication Skills for Safety

One of the most important keys to achieving a Total Safety Culture is effective interpersonal communication. Organizations with world-class safety programs also have world-class safety communication that is open, sincere, respectful and frequent. The way we talk to each other can be dramatically shaped and improved with effective instruction and coaching.

This chapter presents guidelines for improving communication to enhance organizational safety as well as to improve all other functions of a company. First we learn about the power of interpersonal conversation and how to lead such communication in productive directions. Then in the second paper, basic guidelines are given for giving and receiving recognition well.

These are key processes for sustaining safe work practices throughout a work culture. The best way to conduct a safety celebration is also explained in this article. When done right, celebrations are key to developing interpersonal trust and building a sense of belonging.

The third paper shows us how to liven up our communications with humor while the fourth paper explains the critical (but often overlooked) difference between "education" and "training." People need to be educated about principles and rationale before being trained on how to carry out a certain set of procedures. Effective training requires different techniques from education. These are reviewed in this paper.

Interpersonal and Intrapersonal Conversation: The Power to Increase Competence and Commitment for Safety[1]

E. Scott Geller

This article is about our safety conversations with ourselves and with others. How we talk with others (interpersonal communication) influences their current attitude and ongoing behavior, and how we talk to ourselves (intrapersonal communication) influences our own attitude and behavior. It is fair to say that the nature of our safety-related conversations influences the degree of involvement we have in safety and our self-accountability to achieve a Total Safety Culture. In other words, I believe the success of our efforts to prevent injuries in the workplace, in the home, and on the road is determined by conversation.

The Power of Conversation

This article offers guidelines and techniques for getting more beneficial impact from our inter- and intrapersonal communication. Four types of safety management are presented, each defined by the nature of interpersonal conversation. Before addressing ways to improve our safety conversations, however, I would like you to consider the power of conversation. What does effective conversation, or the lack thereof, do for us?

I am convinced the dramatic success companies are experiencing with behavior-based safety is essentially due to an increase in the quantity and quality of interpersonal conversation about safety. In turn, this change improves people's self-talk about safety, which results in an increase in their sense of personal control and optimism regarding the prevention of occupational injuries.

Lack of Conversation Builds Barriers

You have probably seen a minor incident escalate into major conflict because of a lack of communication. For example, did you ever acknowledge the presence of another person with a smile, wave, or friendly "hello" and get no reaction? What did you then say to yourself? Even though there may be numerous reasons for not getting a reciprocal gesture, including the possibility the person did not notice you, did you ever assume the person was unfriendly or did not like you? If so, you might now actively avoid acknowledging this individual. You might even report this person's "unfriendliness" to others—an interpersonal conversation that builds a stronger barrier between you and the other individual.

This is only one of many situations that can stifle interpersonal communication and lead to negative consequences and the avoidance of further conversation with certain people. The result can be the perception of interpersonal conflict, an unpleasant relationship, lowered work output, and reduced willingness to actively care for another person's safety.

Conversation Resolves Conflict

While the lack of conversation can initiate or fuel conflict, the occurrence of conversation can prevent or eliminate conflict. Of course, the quality of our interpersonal conversation determines whether perceived conflict is decreased or increased. This issue of conversation quality is covered later in this article. Here, I want you to consider the power of interpersonal talk. It can make or break interpersonal conflict, which enables constructive or destructive relationships. The nature of relationships determines the willingness of individuals to actively care for another person's safety and health (Geller 1996).

Conversation Makes Intangibles Real

What are the definitions of love, friendship, courage, loyalty, happiness, and forgiveness? Sure, you can describe behaviors that reflect these concepts, but where is the true meaning? I suggest the true meanings of these common words come from our conversations. Our conversations with ourselves and with others enable us to "fall in or out of love." Likewise, we can convince ourselves we are happy by our self-talk, and this conversation is obviously influenced by what we hear others say about us.

We define another person's friendship, courage, or loyalty by talking about that individual in certain ways, both to ourselves and to others. In other words, our mental scripts and verbal behavior with others gives useful meaning to concepts that define the very essence of human existence. And when this essence is combined in a group, organization, or community, we get a "culture." It is fair to say that conversation reflects culture, and vice versa.

Conversation Defines Culture

Does the term "unspoken conversation" make sense? These are the customs or unwritten rules we heed without mention. We might realize, for example, the "teacher's pets" sit in the front row, the boss does not want to hear about a "near-miss," and the real purpose of the safety incentive program is to stifle the reporting of OSHA recordables. We might also know certain characteristics that bias certain managers' performance appraisals, from gender and seniority to ability on the golf course.

It might be understood that a male with high seniority and a low golf handicap is more likely to get the special training assignment, but such prejudice is certainly not written or spoken. If it were, a productive conversation would be possible—one that could reduce the conflict and bias that hinder the optimization of a work system and the achievement of a Total Safety Culture.

The bottom line: Spoken and unspoken words define cultures and subcultures, which change as a function of interpersonal and intrapersonal conversation.

Conversation Defines Public Self-Image

Public conversation defines public image, whether referring to an individual or an organization. Sometimes, however, different populations talk about an individual or organization in different ways. As a result, we have mixed messages and an inconsistent public image. The public image of our

President, for example, depends on whether Democrats or Republicans are conversing. And the safety image of some organizations varies as a function of the people doing the talking. Safety professionals might question a company's touting of "zero injuries," whereas the conversations among the general public, unaware of other safety-related indices, would not doubt this company's elite ranking.

How we talk to ourselves both influences and reflects our self-esteem. In fact, it is probably fair to say that our mental script about ourselves **is** our self-esteem. We can focus our self-talk on the good things people say about us or on other people's critical statements about us. The result is a certain kind of self-talk we call "interpretation," which influences how we feel about ourselves—our self-esteem. Our self-esteem fluctuates to a certain extent according to how we talk to ourselves about the way others talk about us. For example, negative feedback can be rationalized away by saying, "He does not understand the situation, so his opinion does not impact how I feel about myself."

Conversation Makes a Breakthrough Possible

In his provocative book, *Leadership & the Art of Conversation*, Krisco (1997) defines a breakthrough as going beyond business-as-usual and getting more than expected. This requires an interpersonal realization of new possibilities, a strong commitment to go for more, and then a concerted effort to overcome barriers. So how can people visualize possibilities, show commitment to go for a breakthrough, and identify barriers to overcome? You guessed it—conversation.

Krisco warns us to expect barriers and resistance to change. The greater the change, the greater the resistance. Note also that most barriers to change are interpretations or self-talk about perceived reality. Conversation, both interpersonal and intrapersonal, enables us to overcome the barriers that hold back the accomplishment of breakthroughs.

The Art of Improving Conversation

Given the power of conversation to resolve interpersonal conflict, enable a breakthrough, and define public image, self-esteem, and culture, we need to address safety improvement through our interpersonal and intrapersonal conversations. But how should we do this? What kinds of conversations are more likely to provoke and maintain beneficial improvement in occupational safety?

The focus here is on interpersonal conversation, but implications for intrapersonal conversation will be evident. Some techniques for improving how we talk to others about safety are directly relevant to self-talk. Plus, when we recognize personal accomplishment from our other-directed conversations, our self-directed conversations are more positive and in turn boost our confidence and perception of personal control.

Get Beyond the Past

Has this ever happened to you? You ask for more safety involvement from a particular individual and you get a reaction like "I offered a safety suggestion three years ago and no one responded." Or, you attend a safety meeting where people spend more time going over past accomplishments or failures than discussing future possibilities and deriving action plans.

These are examples of conversations stuck in the past. The discussion might be enjoyable but little or no progress is made. Conversations about past events help us connect with others and recognize similar experiences, opinions, and motives. But such communication does not enable progress toward problem solving or continuous improvement. For this to happen, the conversation must leave the past and move on.

Krisco (1997) maintains that leaders need to help people shift their interpersonal conversations from the past to the future and then back to the present. If you want conversation to lead to improvement-focused behavior, possibilities need to be entertained (future talk) and then practical action plans need to be developed (present talk). This is the case for group conversation at a team meeting as well as for one-on-one advising, counseling, or safety coaching.

For optimal impact, a behavior-based feedback session with a co-worker should start with recognition and appreciation of correct behavior (Geller 1996). If called for, corrective feedback is then given to point out room for improvement. The most important aspect of this conversation occurs when possibilities for safety improvement are considered and specific ways to start improving today are discussed.

Seek Commitment

You know your interpersonal conversation is especially productive when someone makes a commitment to improve in a certain way. This reflects success in moving conversation from the past to the future and then to a specific action plan. A verbal commitment also indicates a shift from interpersonal to intrapersonal conversation. This implies self-motivation and increased probability that the target behavior will improve (Cialdini 2001). You can now have a conversation of mutual support or accountability. For example, one person might offer to help a co-worker meet an obligation through verbal reminders. Or, an individual might agree to honor a commitment by showing a coach behavioral records that indicate improvement. This is, of course, the kind of follow-up conversation that facilitates personal achievement.

Listen Before Giving Directives

In their eagerness to prevent injury, safety advocates often give corrective feedback in a top-down, controlling manner. Passion for safety sometimes leads to an overly directive approach to get others to change their behavior. However, clinical psychologists have shown—and you know from personal experience—that a nondirective approach to communicating advice is often more effective, especially over the long term (Bandura 1982, 1997; Deci and Ryan 1985).

How do you respond when someone candidly tells you what to do? It probably depends on who is giving the instruction, but I bet your reaction is not entirely positive. You might change your behavior, especially if the directive comes from an individual who controls a relevant consequence (such as reporting your at-risk behavior to a supervisor). But how will you feel? Will you take personal ownership of the recommended change and make a permanent adjustment? You probably will if you asked for the direction; if you did not request feedback, you could feel insulted or embarrassed.

Corrective feedback that can be interpreted as an "adult-child" confrontation will probably not work. Sure, you mean well, but other people might not see it that way. Their interpretation can be very different from yours. Play it safe, and try being more nondirective when using interpersonal conversation to affect behavior change.

The theme of nondirective psychotherapy is active listening (Rogers 1951). The objective is to get clients to reveal their concerns, problems, and solutions on their own terms. The therapist's role is passive catalyst, enabling and facilitating a conversation that is directed and owned by the client. I am not suggesting safety leaders be this nondirective in their corrective feedback conversations all of the time, but we can take some useful lessons from this nondirective approach.

Ask Questions First

Instead of telling people what safe behavior is called for, attempt to get them to tell you what they ought to be doing in order to be safer. This can be accomplished by asking questions with a sincere and caring demeanor. Avoid a sarcastic or demeaning tone at all costs. After pointing out certain safe behaviors you noticed, you could focus attention on an apparent at-risk behavior by asking, "Is there a safer way to perform that task?"

I recommend approaching a corrective feedback conversation as if you do not know the safest operating procedures, even though you think you do. You might find your presumptions to be imperfect. The "expert" on the job might know something you don't know.

Learn about the safety of specific jobs in order to understand the rationale behind at-risk behaviors. You might discover a barrier to safe behavior that you can help to overcome. A conversation that entertains ways to remove obstacles that hinder safe behavior is especially valuable if it translates possibilities into feasible and relevant action plans.

You will know your nondirective approach to correction worked if your colleague owns up to his or her mistake even under a cloud of excuses. It is only natural to offer a rationale for knowingly taking a risk. Allow for this; it is only an attempt to protect self-esteem. Tell yourself that when someone admits a mistake **before** you point it out, there is a greater chance for both the acceptance of responsibility to change as well as an actual behavior change.

Transition from Nondirective to Directive

What if the person does not give a satisfactory answer to your question about safer alternatives? What if the individual does not seem to know the safest operating procedure? Then you do need to transition the conversation from nondirective to directive. You need to give behavior-focused advice.

If your initial nondirective approach worked, your directive instruction will likely be accepted. In fact, the person you are coaching might turn your initial question around and ask you for safety-related advice. Then you know you have had an effective safety conversation. When this happens, give yourself some intrapersonal words to build your own self-efficacy and sense of personal control at helping to achieve a Total Safety Culture.

Overcome Prejudice Filters

Every interpersonal conversation you have is biased by prejudice or prejudgment filters—in yourself and within the other person. You cannot get around them. From personal experience people develop opinions and attitudes, and these in turn influence subsequent experience. With regard to interpersonal conversation, we have subjective prejudgment filters that influence what words we hear, how we interpret those words, and what we say in response to those words (Langer 1989). Every conversation influences how we process and interpret the next conversation.

It is probably impossible to completely escape the impact of this premature bias in our conversations. But we can exert some control. Actually, each of the conversation strategies discussed here is helpful. For example, the nondirective approach attempts to overcome this bias by listening actively and asking questions before giving instruction. With this approach, an employee's biasing filters can be identified and considered in customizing a plan for corrective action.

Certain words or phrases in a conversation can be helpful in stifling the prejudice filter. When you say "as you know" before giving behavior-based advice, you are limiting the perception of a personal insult and the possibility of a "tune-out" filter. By asking people for their input first, you are using the principle of reciprocity (Cialdini 2001) to reduce the likelihood they will later tune you out. By listening first, you increase the possibility the other person will listen to you without a "tune-out" filter.

If you think an audience might tune you out because it has heard your message before, you could use opening words to diminish the power of tune-out filters; you might start the conversation with something like, "Now I realize you might have heard something like this before, but...." In this way you are anticipating the kind of intrapersonal conversation (or mental script) that triggers a tune-out reaction, thereby reducing such filtering.

In the same vein, you can talk to yourself in certain ways to reduce the influence of your prejudice filters. Do you ever listen less attentively to certain individuals, perhaps because the person seldom has anything useful to say or because you think you know what the person will say in a particular situation? If so, you have let your past conversations with this individual bias future conversation. Becoming aware of this "stuck-in-the-past" prejudice can enable more active—even proactive—listening.

Do not let the speaker limit what you hear. Tell yourself you are not listening *to someone*, but you are listening *for something* (Krisco 1997). You are not listening reactively to confirm what you already know; you are listening proactively for possibilities.

Pay close attention to the body language accompanying the words. I am sure you have heard many times that the method of delivery can hold as much or more information as the words themselves. Listen for passion, commitment, or caring. If you do not learn a new message, you might learn a new way to deliver an old message. Or you could learn whether the messenger understands and believes the message. At any rate, our intrapersonal conversations can either facilitate or hinder what we learn from interpersonal conversation.

Plant Words to Improve Self-Image

Earlier in this article I discussed how conversation influences both public and self-image. How we talk about others influences interpersonal perceptions. And when individuals learn through conversation how others view them, their self-images are affected. Through self-talk these viewpoints can be strengthened and become a permanent feature of one's self-image or self-esteem.

If you want to change how others perceive you and subsequently improve your self-image and self-esteem, change the conversations people are having about you. Through proactive listening, you can become aware of negative interpersonal conversations about you. Then you can interject new statements about yourself into conversations, especially with people who have numerous contacts with others.

If you suspect, for example, that colleagues consider you to be forgetful and disorganized, you could mention certain self-management strategies you have been using lately to improve memory and organization. Of course, you need to actually practice these techniques so you will also change your self-dialogue. If you focus on new positive qualities rather than past inadequacies in your conversations with others and with yourself, you will surely improve your self-image and self-esteem. Plant key messages about your commitment to become a more effective safety leader, and you will eventually see yourself that way and behave accordingly.

Conversation for Safety Management

Safety is managed through conversation, and the success of safety management is determined in large part by the effectiveness of interpersonal communication. This starts with listening proactively to understand the other person's situation before giving direction, advice, or support. Then one of four types of interpersonal conversation should occur depending on what kind of management is called for. As depicted in Figure 4.1.1, the conversation can reflect coaching, supporting, instructing, or delegating (Blanchard, Zigarmi, and Zigarmi 1985), depending on the amount of direction and motivation given.

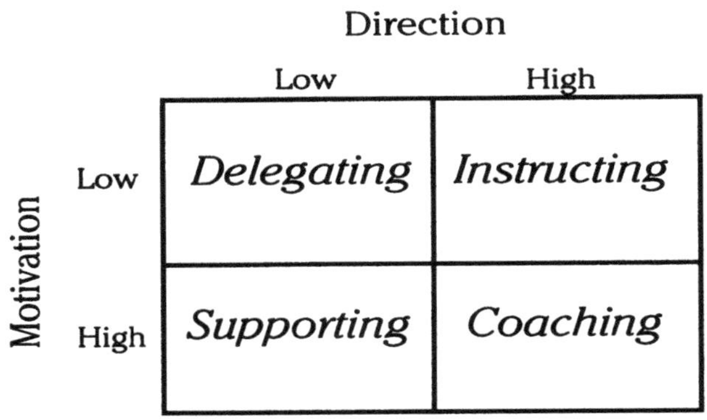

Figure 4.1.1 Four Types of Management Conversation Are Determined by Degree of Direction and Motivation Given

Coaching Conversation

Coaching typically involves a discussion of specific behaviors needed for a certain task, as well as specific feedback to pinpoint effective and ineffective behaviors. In other words, coaches give direction with a certain action plan and then follow up with appropriate support and correction. Periodic reminders keep people on the right track while intermittent recognition provides motivation to keep people going.

Delegating Conversation

There are times when it is appropriate for safety managers to give an assignment in general terms (without specific direction) and to limit interpersonal behavior-focused feedback. The expectation is that team members are already motivated to do their best and will give each other direction, support, and feedback when needed. These individuals are self-accountable (or responsible) and are expected to use self-management techniques (activators and consequences) to keep themselves motivated and on the right track (Geller 2001).

Instructive Conversation

Some people are already highly motivated to perform well but do not know exactly what to do. This is often the case with new hires. They want to make a good first impression, and the newness of the job is naturally motivating. They are nervous, however, because of response uncertainty. They are not sure what to do in the relatively novel situation. In this case, managers need to focus on giving behavior-focused instruction.

Supportive Conversation

Some people and situations require more support than direction. Take, for example, the experienced worker who does the same set of tasks day after day. This individual does not need direction but could benefit from periodic expression of sincere appreciation for a job well done. This provides the kind of interpersonal support that can motivate continued performance of a regular routine.

There are times when experienced workers know what to do but do not perform up to par all of the time. This is not a training problem but rather one of execution or motivation (Geller 2000). Through proactive listening, a manager can recognize this and provide the kind of support that increases motivation. This could involve broadening a job assignment, varying the task components, or assigning leadership responsibilities. Proactive listening includes the delivery of one-to-one recognition in ways that increase a person's sense of importance and self-worth.

The Critical Role of Proactive Listening

So how can we know what type of safety management conversation to use? That is where proactive listening is critical. You need to assess the situation carefully to determine whether a primarily coaching, instructing, supporting, or delegating conversation is needed. Given the dynamic characteristics of most work settings and the changing nature of people, you need to make this assessment periodically for each situation and worker.

Consider, for example, the new employee who needs specific direction at first. Then, as he or she becomes familiar with the routine, more support than instruction is called for. Later, you decide to expand this individual's work assignment. This situation will likely benefit most with a coaching conversation whereby both direction and support are given, at least at first. Eventually, a delegating approach might be most appropriate, whereby varying assignments are given with only expectations for certain results. These workers are able to manage themselves with self-direction and self-motivation. But these individuals still benefit from genuine words of appreciation when expectations are met.

The proactive managers of work teams change their interpersonal conversations quite dramatically as groups get more familiar with team members and their mission. In the beginning, during the forming and storming stages of team progress (Tuckman 1965), work groups need structure, including specific direction and support. This implies a coaching or directing format. Later, when the group members become familiar with each other's interests and talents and progress to the norming and performing stages of team development (Tuckman 1965), supporting and delegating conversations are needed.

The Role of Competence and Commitment

Figure 4.1.2 illustrates how a person's levels of competence and commitment determine the type of management conversation needed (adapted from Blanchard 1999). When competence is high, people know what to do and do not need a directive conversation. They need supportive conversations, however, when their motivation or commitment is low. This is particularly evident when employees perform irregularly or

Figure 4.1.2 Four Types of Management Conversation Are Determined by Degree of the Receiver's Competence and Commitment.

Commitment \ Competence	Low	High
Low	*Coaching*	*Supporting*
High	*Instructing*	*Delegating*

inconsistently. Their good days indicate they know what to do, while the occurrence of bad days suggests a motivational problem.

Causes of low commitment vary dramatically from interpersonal conflict on the job to emotional upheaval at home. Such causes can only be discovered through proactive listening. At times the diagnosis and subsequent treatment of a motivational problem requires special assistance. In this case, the best a manager can do is to recognize a need for professional help and offer advice and support.

Coaching conversations are needed when both competence and commitment are relatively low. A manager's conversation improves competence through specific direction and feedback, and it increases commitment through statements of sincere appreciation and support. Anything that increases a person's perception of importance or self-worth on the job can enhance commitment. What a particular individual needs to enhance commitment is often revealed through interpersonal conversation.

Finally, delegation is appropriate when competence and commitment are high. You can often know when an individual or work team advances to this level by observing successive progress. But it is often useful to ask people whether they are ready for this level of conversation. If they say "no," then ask them what they need in order to reach this stage. Do they need more competence through direction or more commitment through some kind of support the organization could make available?

Summary and Conclusions

I hope I have convinced you that the safety status of your organization is greatly determined by the nature of your employees' conversations about safety, from the managers' boardroom to the workers' break room. The same is true for your potential to achieve a noticeable improvement in injury prevention. Our interpretation or mental script about such conversation determines whether we feel responsible for safety and are committed to go for a breakthrough.

People generally focus their interpersonal and intrapersonal conversation on the past. This helps us connect with others, but it also feeds our prejudice filters and limits the potential for conversation to facilitate beneficial change. We enable progress when we move conversations with ourselves and with others from past to future possibilities and then to the development of an action plan.

Expect people to protect their self-esteem with excuses for past mistakes. Listen proactively for barriers to safe behavior reflected in these excuses. Then help the conversation shift to a discussion of possibilities for improvement and a personal commitment to apply a practical action plan. This is often more likely to occur with a nondirective rather than a directive approach. In a nondirective approach, more questions are asked than directives given, and opening words are used to protect self-esteem and limit the impact of reactive bias.

Proactive listening enables one to determine whether a coaching, instructing, supporting, or delegating conversation is most appropriate. Coaching involves both direction and support and is needed when a person's competence and commitment in a particular setting are relatively low. In contrast, delegating is relevant when people know what to do and are motivated to do it. They are competent, committed and can direct and motivate themselves. In this scenario, delegating conversations provide clear expectations and shows sincere appreciation for worthwhile work.

When people are internally motivated to perform well but do not know how to maximize their efforts for optimal performance, an instructive conversation is called for. In other situations, people know what is needed for optimal performance but do not always work at optimal levels. This reflects an execution problem that cannot be solved with directive conversation. Instead, supportive conversation is needed. What kind of support is called for? This crucial question can only be answered through more conversation, including proactive listening, observing, and questioning.

References

Bandura, A. 1982. Self-efficacy mechanism in human agency. *American Psychologist*, 37:747–755.

———. 1997. *Self-efficacy: The exercise of control*. New York: W. H. Freeman and Company.

Blanchard, K. 1999. *Building gung ho teams: How to turn people power into profits*. Workshop presented at the Hotel Roanoke, Roanoke, VA. November 3.

Blanchard, K., P. Zigarmi, and D. Zigarmi. 1985. *Leadership and the one minute manager.* New York: William Morrow and Company, Inc.

Cialdini, R. B. 2001. *Influence: Science and practice.* 4th ed. New York: Harper Collins College Publishers.

Deci, E. L., and R. M. Ryan. 1985. *Intrinsic motivation and self-determination in human behavior.* New York: Plenum.

Geller, E. S. 1996. *The psychology of safety: How to improve behaviors and attitudes on the job.* Radnor, PA: Chilton Book Company.

———. 2000. Behavioral safety analysis: A necessary precursor to corrective action. *Professional Safety*, 45(3):29–32.

———. 2001. *Beyond safety accountability*. Rockville, MD: Government Institutes.

Krisco, K. H. 1997. *Leadership and the art of conversation*. Rocklin, CA: Prima Publishing.

Langer, E. 1989. *Mindfulness*. Reading, MA: Addison-Wesley.

Rogers, C. R. 1951. *Client-centered therapy*. Boston: Houghton-Mifflin.

Tuckman, B. W. 1965. Developmental sequence in small groups. *Psychological Bulletin*, 63:384–399.

Behavior-Based Recognition and Celebration: Key Processes for Continuous Safety Improvement[2]

E. Scott Geller

Safety professionals intuitively realize the value of recognition and rewards for improving people's behaviors and attitudes. They understand that consequences motivate behavior and that positive consequences can create a positive attitude and boost morale. Thus, recognition and reward programs have become an integral part of most safety and health promotion programs. Unfortunately, many of these motivational efforts do not produce optimal results, and some actually do more harm than good.

While common sense might tell us that recognition and rewards are beneficial, common sense alone is usually not sufficient for optimal implementation of a reward and recognition process. Instead, we need to inform our common sense with the results of rigorous behavioral science research. A *Professional Safety* article in 1996 pointed out common-sense mistakes safety professionals make when administering incentive/ reward programs and offered research-based guidelines for using safety incentives more effectively (Geller 1996). Here, I offer rationale and guidelines for gaining more beneficial impact from interpersonal safety recognition, including group celebrations.

William James, the first great American psychologist, wrote, "The deepest principle in human nature is the craving to be appreciated" (cited in Carnegie 1981, p. 19). A little later John Dewey, the famous American educator who developed the field of school psychology, claimed, "The deepest urge in human nature is the desire to be important" (cited in Carnegie 1981, p. 18). Then in 1936, Dale Carnegie advocated that the key to winning friends and influencing people is to, "always make the other person feel important" (Carnegie 1981, p. 130). How can we readily fulfill the human need to feel appreciated and important? The answer, of course, is to give and receive recognition well.

Several pop-psychology books and audiotapes have addressed the dos and don'ts of recognizing individuals and celebrating group success, but not all of these are founded on rigorous research. And some of these presentations actually give inaccurate advice. Consider, for example, the popular and frequently followed guideline: "Praise publicly and reprimand privately." Does your common sense give credence to such advice? Perhaps you have heard this slogan so many times you have come to believe it.

Reading this article will show you why this and several other common-sense notions about recognizing and celebrating are wrong. To derive the guidelines presented in this article, I consulted research literature as well as books that discussed interpersonal recognition and celebration with a research foundation (Allen 1990; Carnegie 1981; Daniels 2000; Latham 1994; Sulzer-Azaroff and Mayer 1991). These research-based sources are consistent with regard to their advice, but some recommendations are inconsistent with common approaches in industrial health and safety. I hope this article will help to narrow the gap between ideal and actual applications of safety recognition and celebration processes and thus help readers gain greater impact from their attempts to motivate and sustain desired safety performance through interpersonal appreciation.

We Need to Give More Quality Recognition

To make us feel good and keep buying their audiotapes and self-help manuals, pop psychologists tell us we need to fail in order to learn. "Don't worry about your errors," they tell us, "We can't learn unless we make mistakes." Some support this assertion with reference to the greatest home-run hitter in baseball. "Do you know who struck out more times than any other professional baseball player?" they ask. The answer they give is Babe Ruth, and the implication is that he learned his fantastic skill by making errors.

This principle and anecdote from the annals of pop psychology might make us feel better about the errors of our ways and keep us listening to a motivational speaker, but such insubstantial verbiage does more harm than good if we believe it and act accordingly. Indeed, this could be an excuse for focusing more on people's failures than on their successes. Some people may even believe we need to push other people until they make a mistake and then point out the error in order for them to learn from the experience. Nothing could be further from the truth. Although "The Babe" might be the best-known baseball player, he did not hit the most home runs, nor did he strike out the most. Pop psychologists have fooled us again. To date, Hank Aaron hit the most home runs, and he struck out far less than Babe Ruth or Reggie Jackson—the major league player who struck out the most.

Behavioral scientists have shown quite convincingly that success—not failure—produces the most learning (Chance 1999). Edward Lee Thorndike (1911, 1931) studied intelligence in the early 1900s by putting chickens, cats, dogs, fish, monkeys, and humans in situations that called for problem-solving behavior. Then he systematically observed how these organisms learned. He coined the "Law of Effect" to refer to the fact that learning depends upon the consequences of behavior. When a behavior is followed by a "satisfying state of affairs," the probability of that behavior occurring again is increased. But, if an "annoying state of affairs" follows a behavior, that behavior (considered an error) is less likely to occur again.

Which kind of consequence, positive or negative, leads to the most learning? Must an error occur in order for the organism to solve a problem? We can reflect on our own experiences to answer the first question. A pleasant consequence gives us direction and motivation to continue the behavior. We know what we did to receive the reward and are thus motivated to earn another. In contrast, a negative consequence following a mistake only tells us what not to do. It provides no specific direction for problem solution. And overemphasis on a mistake can be frustrating, discouraging, and can de-motivate us to continue the learning process.

Errors are not necessary for learning to occur. In fact, when training results in no errors—made possible with certain presentation techniques (Terrace 1963, 1964)—learning occurs most smoothly and is most enjoyable. Errors disrupt the flow of a teaching/learning process and can lead to a negative attitude, especially if negative social consequences accentuate the mistake. Even subtle reactions to an error, such as disappointment displayed in facial expression or verbal intonation, can increase feelings of helplessness or despair and turn a person off to the entire learning process.

Before leaving this topic of learning from success versus failure, it is noteworthy that Thorndike referred to the type of learning discovered in his problem-solving situations as "trial and accidental success" (Thorndike 1911, p. 174). Many textbook authors have used the term "trial-and-error learning" when describing Thorndike's research, even though Thorndike himself opposed the term because of its inaccurate implications. This is a potential source of the pop psychology error referred to above. This mistake also alerts us to read original research and not rely completely on secondhand interpretation. But let's not focus on this error; rather let's consider the need to support safety success with quality recognition.

How to Give Quality Recognition

The antidote to depressed learning from the negative consequences of incorrect behavior is the provision of positive consequences for correct behavior. And the most powerful positive consequence readily available to motivate correct behavior and support a learning process is social recognition—the theme of this paper. I offer the following seven guidelines for maximizing the beneficial impact of interpersonal recognition.

1. Deliver Recognition During or Immediately After Safe Behavior

In order for recognition to provide optimal direction and support, it needs to be associated directly with the desired behavior. People need to know what they did to earn the appreciation. Then they are motivated to continue that behavior. If it is necessary to delay the recognition, then it is important to relive the behavior or activities that deserve recognition. Reliving the behavior means talking specifically about the performance warranting special recognition. Do not hesitate to ask the recipient to describe aspects of the situation and the desirable behavior. This assures direction and motivation to continue the behavior. Connecting a person's behavior with recognition also makes the recognition special and personal for the recipient.

2. Make It Personal for Both Parties

Recognition is most meaningful when it is perceived as personal. Recognition should not be general appreciation that could fit anyone in any situation. Rather, it should be customized to fit the particular individual receiving it. This happens naturally when recognition is linked to the individual's performance under designated circumstances.

The person giving quality recognition is expressing personal appreciation. It is tempting to say "*we* appreciate" rather than "*I* appreciate," and to refer to company gratitude rather than personal acknowledgment. However, speaking for the company can come across as impersonal and insincere. Of course it is appropriate to reflect value to the organization when giving recognition, but the focus should be personal. "I saw what you did to support our safety process and I really appreciate it. Your example illustrates actively caring and demonstrates the kind of leadership we need around here to achieve a Total Safety Culture." This second statement illustrates the next guideline for quality recognition.

3. Connect Specific Behavior with General Higher-Level Praise

Recognition is most memorable and self-esteem boosting when it reflects a higher-order characteristic. Adding a universal attitude like leadership, integrity, trustworthiness, or actively caring behavior to the recognition statement obviously makes the recognition more rewarding. But it is important to state the specific behavior first, and then make an obvious linkage between the behavior and the positive attribute it reflects.

4. Deliver the Recognition Privately and One-on-One

Because quality recognition is personal and indicative of higher-order attributes, it needs to be delivered in private. After all, the recognition is special and only relevant to one person. So it will mean more and seem more genuine if given from one individual to another.

It seems conventional to recognize individuals in front of a group. This approach is typified in athletic contests and reflected in the pop psychology slogan "Praise publicly and reprimand privately." Many managers take the lead from this common-sense statement and give their individual recognition in group settings. Isn't it maximally rewarding to be held up as an exemplar in front of one's peers? Not necessarily.

We need to realize that many people feel embarrassed when identified in a group setting. Part of this embarrassment could be due to fear of subsequent harassment by peers. Some peers might call the recognized individual an "apple-polisher" or "brown-noser," or accuse him or her of "sucking up to management."

In athletic events, the participants' performance is measured fairly and the winner is objectively determined. While behavior-based safety recognition is also objective, it is usually impossible to assess everyone's safety-related behaviors and obtain a fair ranking for individual recognition. Plus, such ranking sets up a win/lose atmosphere. This may be appropriate for sporting events, but it is certainly inappropriate for a work setting in which the elimination of injuries is dependent upon everyone actively caring for the safety of everyone else.

It is beneficial, of course, to recognize teams of workers for their accomplishments, and this can be done in a group setting. Usually group accomplishment worthy of recognition can be documented for public review. Since individual responsibility is diffused or dispersed across the group, there is minimal risk of individual embarrassment or later peer harassment. However, it is important to realize that group achievement is rarely the result of equivalent performance from all group members. Some individuals typically take the lead and work harder while others do less and count on the group effort to make them look good. Behavioral scientists call this phenomenon "social loafing" (Latané, Williams, and Harkins 1979). Thus, it is important to deliver personal and private recognition to those individuals who went beyond the call of duty for the sake of their team.

5. Let It Stand Alone and Soak In

I have heard pop psychologists recommend a "sandwich method" for enhancing the impact of interpersonal communication. "First say something nice, then give corrective feedback, and then say something nice again." This approach might sound good, but it is not supported by communication research. In fact, this mixed-message approach can cause confusion and actually reduce credibility. The impact of initial recognition is canceled by the subsequent correction, and then the corrective feedback is neutralized by the closing recognition. Keep recognition simple and to the point, and give your behavior-based praise a chance to soak in.

In this fast-track age of trying to do more with less, we all try to communicate as much as possible when we finally get in touch with a busy person. After recognizing a person's special safety effort, we are tempted to tag on a bunch of unrelated statements, even a request for additional behavior. This comes across as "I appreciate what you've done for safety, but I need more." Resist the temptation to do more than praise desired behavior. If you have additional points to discuss, it is usually best to reconnect later after the rewarding recognition has had a chance to be internalized and become a part of the individual's self-recognition system.

By giving quality recognition, we give people a script they can use to reward their own behavior. In other words, our quality recognition strengthens the other person's self-reward system. And self-reward (or self-recognition) is critical for long-term maintenance of safe behavior. Thus, by allowing our recognition communication to stand alone and soak in, we enable the internalization of rewarding words which can be used later for self-motivation of desired behavior.

6. Use Tangibles for Symbolic Value Only

Tangibles can detract from the self-recognition aspect of quality recognition. If the focus of a recognition process is placed on a material reward accompanying the social approval, the words of appreciation can become less significant. In turn, the impact on one's self-reinforcement system is lessened.

Tangibles can add to the quality of interpersonal recognition if they are delivered as tokens of appreciation. If they include a safety slogan, tangibles can help to promote safety. But how you deliver a trinket will determine whether it adds to or subtracts from the long-term benefit of your praise. The tangible must not be viewed as a payoff for the safety-related behavior, but only as symbolic of going beyond the call of duty for safety.

Even in a behavior-based safety incentive program, the tangibles should not be considered fair compensation for extra effort on behalf of safety (Geller 1996). The tangible is not enough to compensate the safety effort; it is only a token of gratitude. The real payoff, of course, is injury prevention.

Note that in an incentive program, people know beforehand what they need to do to earn a certain tangible reward. That is the incentive. In contrast, recognition is a reward without an incentive. An individual is caught doing right and is recognized for that behavior. If a tangible is presented along with verbal praise, it should be delivered with words that give it symbolic value.

7. Secondhand Recognition Has Special Advantages

Up to this point, I have been discussing one-on-one verbal communication in which one person recognizes another for a particular safety-related behavior. It is also possible to recognize a person's outstanding efforts indirectly, and such an approach can have special benefits. Suppose, for example, you overhear me talk to another person about your outstanding safety presentation. How will this secondhand recognition affect you? Will you believe my words of praise were genuine?

Sometimes people are suspicious of the genuineness of praise when it is delivered face-to-face. The recipient of praise might feel there is an ulterior motive to the recognition. Perhaps the deliverer of praise is expecting a favor in return for the special recognition. Perhaps both individuals had recently attended the same behavior-based safety course, and the verbal exchange is recognized as an extension of a communication exercise and thus devalued as sincere appreciation. Secondhand recognition, however, is not as easily tainted with these potential biases, and thus genuineness is less suspect.

Suppose I tell you that someone else in your workgroup told me about the superb job that you did leading a certain safety meeting. What will be the impact of this type of secondhand recognition? Chances are you will consider the recognition genuine because I was only reporting what someone else said. That person reported your success to me rather than you and therefore had no ulterior motive for the indirect praise. Note also that this secondhand recognition can build a sense of belonging or group cohesion among individuals. When you learn that a particular individual was bragging about your behavior, your sense of closeness (or friendship) to that individual will likely increase.

My main point here is that gossip can be beneficial—*if it is positive*. When we talk about the achievement of others in behavior-specific terms, we begin a cycle of positive communication that can support desired behavior as well as build an internal script for self-recognition. We also set an example for the kind of interpersonal communication that builds self-esteem, empowerment, and group cohesion. These are the very person states that increase actively caring behaviors and cultivate the achievement of a Total Safety Culture (Geller 1994, 1995).

We Need to Receive Recognition Well

The list of guidelines for giving quality recognition is not exhaustive, but it does cover the basics. Following these guidelines will certainly increase the beneficial effects of interpersonal recognition. Of course, the most important point is that more recognition for safe behavior is needed in every organization, whether given firsthand or indirectly through "positive gossip" (Allen 1990). It does not take but a few seconds to deliver quality recognition.

Perhaps realizing the beneficial consequences we can have on people's behaviors and attitudes with relatively little effort will be self-motivating enough for us to do more recognizing. Even more important than this awareness in increasing our recognizing behaviors are the social consequences we receive when we attempt to perform quality recognition. In other words, the reaction of the people who are recognized can have dramatic impact on whether recognition increases or decreases throughout a work culture. We need to know how to respond to recognition in order to assure that quality recognition continues.

Many people get so little recognition from others that they are caught completely off-guard when acknowledged for their actions. They do not know how to accept such appreciation when it finally comes. They might claim they do not deserve the special recognition or actually accuse the person giving recognition as being insincere or wanting something from them. This can be quite embarrassing to the person doing the recognizing and could dissuade that individual from giving more recognition.

Remember the basic behavioral science principle that consequences influence the behaviors they follow (Skinner 1969). This is true for both the person giving and the person receiving recognition. While quality recognition increases the probability the recognized behavior will continue, the reaction to the recognition influences whether the recognizing behavior will be attempted again. Thus, it is crucial to react appropriately when we receive recognition from others. Here are seven basic guidelines for receiving recognition.

1. Avoid Denial and Disclaimer Statements

Whenever I attempt to give quality recognition, whether to a colleague, student, waitress, hotel clerk, or a member of the baseball team I coach, the most common inappropriate reaction is awkward denial. Sometimes they act as if they did not hear me and keep doing whatever they are doing, or they give a disclaimer statement like "It really was nothing special," "Just doing my job," "I really could not have done it without your support," or "Other members of our team deserve more credit than I do."

Obviously, we need to accept recognition without denial and disclaimer statements and without deflecting the credit to others. It is okay to show pride in our small-win accomplishments (Weick 1984), even if others contributed to the successful outcome. After all, the vision of a Total Safety Culture includes everyone going beyond the call of duty for their own safety and that of others. In this context, numerous people deserve recognition daily. It is not "employee of the month," it is "employee of the moment."

Accept the fact that recognition will be intermittent at best for everyone. When your turn comes, accept the recognition for your current behavior and for the many safety-related behaviors you performed that went unnoticed. And keep in mind that your genuine appreciation of the recognition will increase the chance that more recognition will be given to others.

2. Listen Attentively with Genuine Appreciation

To accept recognition, first listen actively to the person giving it. Obviously you want to learn what you did right. Plus, you can evaluate whether the recognition is given well. If the recognition does not pinpoint a particular behavior, you might ask the person for that information. This will be useful in improving that person's method of giving recognition. It is important, however, not to seem critical but rather to show genuine appreciation for the special attention. Consider how difficult it is for many people to go out of their way to recognize others. Then revel in the fact that you are receiving some recognition, even if its quality could be improved. Note also that a person who recognizes you is appreciating what you do and will come to like you more if you accept the recognition well.

3. Take It In and Relive It Later for Self Motivation

As I have already indicated, most of your safety-related behaviors go unnoticed. You perform many of these when no one else is around to observe you. And even when other people are available, they will likely be so busy with their own routines they will not notice your extra effort. So when you finally do receive some recognition, take it in as well deserved (especially considering the many times you have gone the extra mile for safety but did not get noticed).

You need to listen intently to every word of praise, not only to show you care but also because you want to remember this special occasion. Do not hesitate to relive this moment later by talking to yourself. This enables self-recognition and motivates you to continue going beyond the call of duty for safety.

4. Show Sincere Appreciation

After listening actively with humble acceptance, you need to show sincere appreciation with a smile, a "thank you," and perhaps special words like "You've made my day." As I have emphasized above, your reaction to being recognized can determine whether similar recognition is apt to occur again. So be prepared to offer a sincere "thank you" and words that reflect your pleasure in the memorable interaction. I find it natural to add, "You've made my day" to the thank you, because it is the truth. When people go out of their way to offer me quality recognition, they *have* made my day. And, I will often relive such situations in order to improve a later day.

5. Recognize the Person for Recognizing You

When you accept recognition with heartfelt responsiveness, you reward the recognizing person for their extra effort and thereby motivate that individual to do more recognizing. Sometimes it is possible to do even more to assure the occurrence of more quality recognition. Specifically, often you can recognize the person for recognizing you. In this case, you can apply quality recognition principles to maximize the beneficial impact of your reaction to being recognized. You might say, for example, that you really appreciate the pinpointing of a particular behavior and the linking of it to higher-order praise. Such rewarding feedback provides direction and motivation for those aspects of the recognition process that are especially worthwhile and need to become habitual.

6. Embrace the Reciprocity Principle

Some people resist receiving recognition because they do not want to feel obligated to give recognition to others. This is the reciprocity norm at work (see the first chapter of this book). Indeed, if we want to achieve a Total Safety Culture, we need to embrace this norm. Research has shown that when you are nice to others—when you provide them with special praise—you increase the likelihood that they will reciprocate by showing similar behavior (Cialdini 2001; Schroeder et al. 1995). You might not receive the returned favor, but someone will (Berkowitz and Daniels 1963).

It is important to realize that your genuine acceptance of quality recognition will activate the reciprocity norm. And the more this norm is activated from positive interpersonal communication, the greater the frequency of interpersonal recognition and actively caring behavior. So accept recognition well, and embrace the reciprocity norm. The result will be more interpersonal involvement consistent with the vision of a Total Safety Culture.

7. Ask for Recognition When It is Deserved but Not Forthcoming

There is one final strategy I would like to recommend for increasing safety recognition throughout a culture. If you feel you deserve recognition, why not ask for it? This might result in recognition perceived as less genuine than if it were spontaneous, but the outcome from such a request can be quite beneficial. You may receive some words worth reliving later for self-motivation, but perhaps most importantly, you will remind the other individual in a nice way that he or she missed a prime opportunity to offer quality recognition. This could be a valuable learning experience for that person.

Consider the potential beneficial impact, for example, from your statement to another person that you are pleased with a certain outcome that resulted from your extra effort, including your performance of particular behaviors. With the right tone and affect, such verbal behavior will not seem like bragging but rather a declaration of personal pride in a small-win accomplishment—something more people should feel and relive for self-motivation. It is likely the other person will support your personal praise with individual testimony, and this will bolster your later self-recognition. Plus, you will teach the other person how to support the safety-related behaviors of others.

Many years ago I instituted a self-recognition process among my research students that increased our awareness of the value of receiving praise, even when it is self-initiated. I told my students that during class or group meetings they could request a standing ovation at any time. All they had to do was state specifically the behavior they felt deserved recognition and then ask for a standing ovation. Obviously, such recognition is not private, personal, and one-to-one, and therefore it is not optimal. The public aspect of the process inhibited many personal requests for a standing ovation. However, a number of students have made such a request over the years, and the experience has always been positive for everyone. The request has always included a solid rationale (from earning an exemplary grade on a project to receiving an acceptance letter from a graduate school or an editorial board for a professional journal). The actual ovation is fun and feels good, whether on the giving or receiving end. Plus, we all learn the motivating process of behavior-based recognition, even when it does not follow all of the quality principles.

Quality Safety Celebrations Are Important

The current atmosphere of "lean and mean" downsizing has reduced morale in many organizations. This can lead to apathy and helplessness with regard to safety and a decreased willingness to look out for the safety and health of others. In other words, lower worker morale means less actively caring behavior for safety and eventually more workplace injuries. What can we do to counter this unfortunate situation?

Giving and receiving quality recognition, as discussed above, are prime ways to boost the morale of individuals. Celebrations, when done correctly, can boost the morale of groups of individuals as well as motivate teamwork and build a sense of belonging. Of course the key phrase in the preceding sentence is "when done correctly." In this section, I offer guidelines for celebrating safety accomplishments thereby motivating teamwork for safety-related activities and increasing people's willingness to actively care for the safety of their co-workers.

1. Celebrations of Injury Reduction Should be Unannounced

Most safety celebrations occur when an organization reaches a certain reduced level of injuries. It is quite common, for example, for organizations to give groups of employees a celebration dinner after a particular number of weeks or months pass with no recordable injury. Achieving a record number of injury-free workdays is certainly worth celebrating, but you had better be sure the record was reached fairly. If people cheat to win by not reporting their injuries, the celebration will not mean much.

If a celebration for lower injuries is announced as an incentive, the motivation to cheat is increased. In other words, if employees are promised a reward when they work a certain number of days without an injury, it will be quite tempting to avoid reporting a personal injury so as not to spoil the fun for everyone. Under these circumstances, some people will actually feel pressure from peers to avoid reporting an injury, if they can get away with it. This is, of course, peer pressure to cheat—a situation that reduces interpersonal trust and a belief that improved levels of organizational safety cannot be reached fairly.

If the accomplishment of process activities are celebrated, then it is okay to establish a behavior-consequence contingency. That is, the behaviors needed from the group can be specified in order to warrant a celebration. This incentive-based goal setting motivates teamwork for goal accomplishment. For example, a group might decide to celebrate after carrying out a designated number of safety audits, investigating a given number of "near-miss" reports, finishing a particular training series, or completing a certain number of one-on-one safety coaching sessions. In these cases, an achievable goal is set and progress is monitored. When the goal is reached, a celebration is clearly warranted. It was earned for a successful journey destined to achieve injury prevention.

2. Celebrate Accomplishment But Focus on the Journey

Most of the safety celebrations I have witnessed give far too little attention to the journey or processes that enabled the recognized milestone in injury reduction. The focus is typically on the end result, like the achievement of zero injuries for a certain period of time, with little discussion about how the outcome was achieved. It is natural to toast the bottom line, but there is more to be gained from taking the opportunity to recognize the process.

Pinpointing aspects of the journey that were instrumental in reaching a safety milestone gives valuable direction and motivation. Participants learn what they need to do in order to continue a successful process. Those responsible for the behaviors identified as special contributors to the safety outcome receive a special boost in optimism, personal control, and self-effectiveness, and they add information to their internal recognition scripts for later self-motivation. But perhaps the most important reason for acknowledging journey activities leading to a noteworthy reduction in injuries is that it gives credit where credit is due. Focusing on the process endorses the people and their actions that made the difference. This leads to the next guideline.

3. Show Top-Down Support But Facilitate Bottom-Up Involvement

It is common for a safety celebration to start off with speeches by representation from top management who state their extreme pleasure in the accomplishment of a reduced injury rate. Sometimes charts are displayed to compare the past with the improved present. Often a sincere request for continuous improvement is made and a manager points out the amount of money the company saved with the reduction in injury rates. Sometimes promises for a bigger celebration are made if injury rates continue to decrease. Occasionally, a motivational speaker or humorist gives an uplifting and fun presentation. Often special reward placards are given to individuals or team captains along with a handshake from a top-management official. At times a certificate and a trinket with a safety slogan are given to all participants, along with a steak dinner. But rarely do the participants discuss the processes they implemented in order to reach the celebrated goal.

In the typical safety celebration, management gives and the operators receive. This certainly shows impressive top-down support. However, the ceremony would be more memorable and beneficial as a learning and motivational experience if the employees were more participant than recipient. In other words, the impact of a safety celebration will be more beneficial and last longer if management does more listening than speaking and if operators do more talking about their experiences along the journey than listening to managers' pleasure with the bottom line.

4. Facilitate Discussion to Relive the Journey

Management's primary role in a safety celebration should be to facilitate discussions of activities that led to the celebrated accomplishment. By reliving the procedures that made the journey successful, people strengthen their internal scripts that direct and motivate continued support of the successful process. When managers listen to such discussions with genuine interest and concern, they reward the participation that enabled the success and empower the employees to continue their journey toward higher-level achievement.

The most successful safety celebration I ever observed featured a series of brief presentations by teams of hourly workers. These employees shared numerous safety ideas they had implemented to prevent workplace injuries. Some workers showed off new personal protective equipment, some displayed graphs of data obtained from environmental or behavioral audits, some discussed their procedures for encouraging near-miss reports and implementing corrective action, and another group presented their ergonomic analysis and redesign of a work station.

Even the after-dinner entertainment was employee-driven. There was an employee skit related to safety issues and a talent show with entrants from all levels of the organization, including top management. There was no need to hire a band for live music. It was possible to find a number of talented musicians from the workforce of 1200. I was glad, however, they did not find a rock drummer, because that gave me a chance to participate and relive my extensive rock-band experiences of the 1960s. Incidentally, I was the only outside consultant involved in this daylong safety celebration. The program was planned and presented by the people whose daily involvement in various safety processes enabled the dramatic reduction in injury rate that was celebrated.

5. Enable Discussions of Successes and Failures

When work teams discussed their safety projects at the celebration discussed above, they related both their successes and their failures. They displayed not only the positive results of their projects, but they also relived their disappointments, their dead ends, and their frustrations. They pointed out the high points and the low points. This approach made their presentations realistic, and it made clear the amount of dedicated work needed to complete their projects and contribute to the celebrated reduction in workplace injuries.

Presentations that point out hardships along the journey to success justify the celebration. They reflect the fact that the celebrated bottom line was not the result of luck. It was achieved by synergy, hard work, win-win collaboration, interdependence, and by many people going beyond the call of duty to make a small-win contribution.

6. Use Tangible Rewards to Establish a Memory

When people discuss the difficulties in reaching a goal, the goal attainment is meaningful and memorable. When managers listen to these presentations with sincere interest and appreciation, the event becomes even more significant and momentous. When a tangible reward is distributed appropriately at such an occasion, a mechanism is established to reinforce the memory of this occasion and promote its value.

Such tangibles should include words, perhaps a worker-designed safety slogan signifying the particular safety celebration. Ideally, the tangible reward should also be something readily displayable or used in the workplace, from coffee mugs, placards, and pencil holders to caps, shirts, and umbrellas. When the tangible rewards are delivered, it should be noted that they were selected "to remind us how we achieved our real reward—fewer people getting hurt."

One week after the notable safety celebration I described above, every participant received a framed group photograph of everyone who attended the celebration. That picture hangs in my office today, and every time I look at it I am reminded of the time several years ago when management did more listening than talking in a most memorable and educational safety celebration.

7. Solicit Employee Input

When I mentioned to my graduate students that I was writing about how to celebrate, one of them quickly responded, "That's easy, a $100 bottle of cognac, a $6 cigar, and a special friend." I had to tell him, of course, I was not writing about that kind of celebration. But, it did occur to me that everyone has his or her own way of celebrating success, and when it comes to group celebration, we often inadvertently impose our prejudices on others. We often do not take the time to ask potential participants what kind of celebration party they would like.

When you ask people how they would like to celebrate, challenge them to think beyond tangible rewards. A discussion about material rewards puts the celebration in a payoff-for-behavior mode. This is not the purpose of a safety celebration. Remember that how you celebrate a safety milestone determines whether the event is meaningful and memorable or just another misguided but well-intentioned attempt to demonstrate management approval.

To motivate teamwork and build a sense of belonging among group members, a safety celebration requires top-down support but bottom-up involvement. Ideally, the event should be planned by representation from the work group whose efforts warrant the celebration. The focus should center on the projects and processes involved in achieving the bottom line. Thus, managers need to do more listening than talking. They need to show genuine interest in and appreciation for the challenges addressed and the difficulties handled in achieving the group's goals. Such discussion recognizes that it was not easy and helps to write internal scripts for continued motivation to achieve more for the next safety celebration.

Summary

Group celebrations and interpersonal recognition go hand-in-hand in cultivating a culture of actively caring people working interdependently to keep each other injury free. Group celebrations are invaluable in developing the spirit of belonging needed in a Total Safety Culture. Please note, however, that group safety celebrations do not provide sufficient recognition for those special people who make it happen. In every group project, some individuals take charge and champion the effort while others sit back and "go with the flow." In fact, some people actually exert less effort when working with a group than when working alone.

Because some participants deserve more credit than others do, it is important to follow-up a group recognition with one-on-one quality recognition. By recognizing the champions of a group effort, you are letting those individuals know you realize the importance of individual leadership in team accomplishment, and you appreciate their special contributions in this regard. This adds substantially to the internal motivational script that the person had begun to develop from the earlier group celebration. As a result, you have increased the likelihood of continued leadership directed toward the accomplishment of further group goals.

This article presented seven basic guidelines for increasing the quality of interpersonal recognition and seven ways to receive recognition so as to maximize its beneficial impact. In order to feel important and appreciated following quality recognition, we need to receive it well. We need to listen intently to every word of praise with genuine positive regard. We must not deny the commendation nor deflect the credit to others. Rather, we should take in every word and prepare to relive the situation later for continued feelings of importance and self-motivation. Then, verbalize delight in the recognition with a sincere "thank you."

If you compliment the other person for taking the time to give you quality recognition, you will increase the likelihood that person will give more interpersonal recognition. Given the reciprocity norm, your *quality* reception of *quality* recognition will likely lead to an increase in people making others feel good about their safety-related behavior. Is there a better way to keep people involved in achieving the many small wins needed to prevent workplace injuries? I don't think so.

References

Allen, J. 1990. *I saw what you did & I know who you are: Bloopers, blunders and success stories on giving and receiving recognition*, Tucker, GA: Performance Management Publications.

Berkowitz, L., and L. R. Daniels. 1963. Responsibility and dependency. *Journal of Abnormal and Social Psychology,* 66:429–436.

Carnegie, D. 1981. *How to win friends and influence people.* New York: Simon & Schuster, Inc. Originally published in 1936.

Chance, P. 1999. *Learning and behavior.* 4th ed. Belmont, CA: Wadsworth, Inc.

Cialdini, R. B. 2001. *Influence: Science and practice.* 4th ed. New York: Harper Collins.

Daniels, A. C. 2000. *Bringing out the best in people: How to apply the astonishing power of positive reinforcement.* 2nd ed. New York: McGraw Hill, Inc.

Geller, E. S. 1994. Ten principles for achieving a Total Safety Culture. *Professional Safety*, 39(9):18–24.

———. 1995. Safety coaching: Key to achieving a Total Safety Culture. *Professional Safety*, 40(7):16–22.

———. 1996. The truth about safety incentives. *Professional Safety*, 41(10):34–39.

———. 1997. The social dynamics of occupational safety. *Proceedings of the 37th Annual ASSE Professional Development Conference & Exposition*, New Orleans, LA.

Latané, B., K. Williams, and S. Harkins. 1979. Many heads make light the work: The causes and consequences of social loafing. *Journal of Personality and Social Psychology*, 37:823–832.

Latham, G. I. 1994. *The power of positive parenting,* Salt Lake City, UT: Northwest Publishing, Inc.

Schroeder, D. A., L. A. Penner, J. F. Dovidio, and J. A. Piliavin. 1995. *The psychology of helping and altruism: Problems and puzzles.* New York: McGraw-Hill, Inc.

Skinner, B. F. 1969. *Contingencies of reinforcement: A theoretical analysis*. New York: Appleton-Century-Crofts.

Sulzer-Azaroff, B., and G. R. Mayer. 1991. *Behavior analysis for lasting change*. Chicago, IL: Holt, Rinehart and Winston, Inc.

Terrace, H. S. 1963. Discrimination learning with and without "errors." *Journal of the Experimental Analysis of Behavior,* 6:1–27.

———. 1964. Wavelength generalization after discrimination learning with and without errors. *Science,* 144:78–80.

Thorndike, E. L. 1911. *Animal intelligence: Experimental studies*. New York: Hafner.

———. 1931. *Human learning*. Cambridge, MA: MIT Press.

Weick, K. E. 1984. Small wins: Redefining the scale of social problems. *American Psychologist*, 39(1):40–49.

The Value of Humor[3]

E. Scott Geller

> *There ain't much fun in medicine, but there's a heck of a lot of medicine in fun" — Josh Billings (1818–1885).*

In 1985, an article in *National Safety News* caught my attention. It was entitled "Laughter could really be the best medicine." The article quoted psychiatrist Donald W. Black of the Iowa State Research Institute as concluding from research that laughter could be the simplest and surest way to reduce stress and improve our physical and mental health. Around that time I had read similar statements in "Anatomy of an Illness" by the late Norman Cousins, former editor of the *Saturday Review*. Cousins reported the value of humor in enabling his recovery from a rare disease of the connective tissue. Physicians told him his condition was incurable and possibly fatal, but Dr. Cousins found a way to handle his painful disease. He rented videotapes of his favorite comedies (including numerous clips of *Candid Camera*) and laughed as much as possible for several months. As a result, Dr. Cousins laughed his way to better health.

Now in those days I did not personally believe in the healing power of humor as portrayed in these documents, but these reports did influence me to do something very unusual. I rarely take the elevator to the fifth floor of the university building where my office and research laboratories are, but one memorable day I boarded the first-floor elevator in a special way. I got on the elevator with one of my Ph.D. students and asked him to copy my behavior. Then I turned and faced the back of the elevator. My student, Mike Kalsher, did the same.

When the elevator stopped at the second floor, a young woman got on, and without saying a word, stood right next to us, facing the back of the elevator. Immediately I fought to hold back intense laughter, and Mike did the same. Neither of us wanted to laugh out loud and embarrass this young lady, who presumably was influenced by social conformity (a social influence principle I reviewed in the second paper of Chapter 1 of this book).

The elevator stopped at the third floor. When the elevator doors opened, two men were apparently very surprised to see three butts facing them. Both of them shouted, "What the heck is going on?" Well, Mike and I could not hold back any longer and we proceeded to laugh hysterically. Soon the girl and the new elevator occupants joined in, and we all laughed uncontrollably until the elevator doors opened at the fifth floor.

I will never forget my special mood that morning. Instead of my usual serious work-focused demeanor, I felt playful and euphoric. I said "good morning" with a smile to everyone I passed in the hall. I even gave a student genuine recognition for excelling on a class research assignment. At the time, these interpersonal behaviors were unusual for me and occurred because my mood was improved by the humorous elevator experience. These extra interpersonal communications did not reduce my accomplishments one bit. Actually, they improved my attitude toward work throughout the entire morning.

That day in 1985 showed me the value of humor in improving attitude and mood states and in turn improving interpersonal behavior. I have learned there is much humor around us everyday for our appreciation and personal benefit, but if we are in the wrong mood or when we take ourselves too seriously, we miss these opportunities. If you are in a good mood, you will appreciate the humor in these messages:

a sign at a local fast-food restaurant "Drive-thru window";

a sign at a shopping mall "Ears pierced while you wait"; and

a sign along the road "Slow men at work."

If you are not in a good mood, you will miss an opportunity to give your attitude a boost.

Over the years, I have also realized the value of adding humor to group presentations. I periodically scan magazines and newspapers for comics to illustrate a principle or lesson I teach. And often I take ideas to my artist friend, George Wills, and we develop humorous sketches to spice up a presentation for university students or safety professionals. My recent safety management books (*The Psychology of Safety Handbook* and

Working Safe) include numerous original cartoons George and I developed to increase both reading pleasure and comprehension.

Five Types of Humor

Adding humor to written and oral communication in ways that benefit attention, learning, and mood is easier said than done. We have all had the experience of telling a joke that did not work the way we wanted. It is not enough to have a good, humorous story to illustrate a point. The story needs to be told well and at the right time. Audience reaction (or lack thereof) tells me when my humor has worked or flopped. My students and professional colleagues often give me direct feedback about the appropriateness, frequency, or timing of humor-driven stories or illustrations. While it might not be fun to get corrective feedback, it is certainly useful in improving the integration of quality humor with other material.

Understanding what makes a story or illustration funny can also develop our sense of humor. In a 1996 audiotape series *The secrets of power negotiating,* Roger Dawson explains that there are only five different jokes. Although stories told by comedians, teachers, and safety coaches vary widely, each story gets its humor in one of five different ways. Understanding these five approaches to making us laugh will increase your appreciation of humor. Plus, it could facilitate your ability to add good humor to individual and group communication.

Pun

A pun is a word that has more than one meaning depending on the context. Humor occurs when the context is changed to give the word an alternative meaning. Consider, for example, the OSHA inspector who begins his safety presentation with, "I suppose you all want the latest dope from Washington; well, here I am." And then there is the boy who asks his father to make a noise like a frog. When the father asks why, he says, "Because mom said we could all go to Disneyland after you croaked."

These examples obviously get their humor from the double meaning of the words "dope" and "croak." Similarly, the double meanings of words make the following statements humorous:

"More hay, Trigger?" "No thanks Roy, I'm stuffed."

"I used to have a handle on life, then it broke."

"Where there's a will, I want to be in it."

Exaggeration

Many comedians use exaggeration to add humor to commonplace examples. Bill Cosby, for example, is a master at using this type of humor. Whether talking about raising his kids, boarding an airplane, or performing in grade school, Cosby has used exaggeration to stretch a point and create a visual image we can relate to and laugh at. For example, he talked about his experiences in kindergarten while writing with a pencil "as big as a horse's hoof" and on paper "with pieces of wood still in it."

Even an infrequent viewer of the *Tonight Show* with Johnny Carson can recall laughing at Johnny's use of exaggeration in an opening monologue. In fact, his use of exaggeration to describe everyday events became so well known his audiences would prompt him to use this type of humor. He would say, for example, that it was hot in Burbank, and the audience would shout, "How hot was it?" and then Johnny would say something like, "It was so hot that a flock of birds burst into flames from spontaneous combustion." Can you think of a humorous exaggeration to safety-related questions like, "How little involvement was demonstrated at the last safety meeting?" or "How safety-aware were the associates?" or "How fast was the forklift traveling when it rounded the corner?"

Surprise

A third type of humor is based on the element of surprise. When we are surprised by something, the fight or flight reaction of our autonomic system is activated. Our pulse quickens, yet when we see there is no danger, laughter often occurs. The surprise joke works well when you present your audience with a simple story description that stimulates thought and imagery. Then you wreck the image by adding surprise to the end. The more you catch the audience off guard, the greater the laughter.

I have often told the following story to my students at Virginia Tech to explain "modeling" or how we learn by watching others. When a father tucked his little six-year old daughter in bed, she asked, "Daddy, would you tuck me in tonight like you do Mommy every night?" "Sure," he said, as he pulled the covers up and underneath his daughter's chin. As he began to walk away, his daughter said, "Wait, Daddy! Would you give me a goodnight kiss like you do Mommy every night." "Sure, honey," said dad, and he leaned down and kissed his daughter on the cheek. As he turned to leave once again, his daughter called after him again with one more request. "Daddy, would you now whisper in my ear like you do Mommy every night?" He replied, "Sure honey," leaned down and went "buzz, buzz, buzz" in his daughter's ear. She immediately popped up her little head and said, "Not tonight, Daddy, I have a terrible headache."

I once witnessed a group facilitator use surprise to break tension and slow down an over-paced safety meeting. He announced, "We must pause a moment for a spot announcement." Immediately, a co-worker barked, "Arf, arf," and the facilitator responded with, "Thank you, Spot." This is not only surprise, but is also a pun as explained earlier, and silliness which is described next.

Silliness

Most of the humor in British sitcoms (as in "Mr. Bean") is based on silliness. Of course, there has probably been no American comedians sillier than The Three Stooges—Curly, Moe, and Larry. Perhaps readers have experienced the silliness of Gallager. He would take a large watermelon, put it on a large slab of wood, and then proceed to smash the melon so that pieces of fruit rained over the audience. And the audience would laugh hysterically, even those people in the audience whose clothes were soiled by flying food.

The following statements get much of their humor from silliness. "Suicidal twin kills sister by mistake," "He who laughs last thinks slowest," "Criminal lawyer is a redundancy," "Women who seek to be equal to men lack ambition," "Save a tree: Eat a beaver," "I don't suffer from insanity, I enjoy every minute of it." This last line derives some humor from the personal put-down contained in the statement, which is the fifth type of humor.

The Put-down

The put-down is usually used to insult other people for a laugh. Many comedy teams derived their humor from one person putting down the other. For example, the put-downs between Abbot and Costello, Laurel and Hardy, Dean Martin and Jerry Lewis, and Burns and Allen are well known. More recently, comedians derive their humor by putting themselves down instead of other people. For example, consider these following statements I found on the Internet:

> "I went to a freak show and they let me in for nothing";
>
> "My psychiatrist told me I'm going crazy. I told him, 'If you don't mind I'd like a second opinion.' He said, 'Alright, you're ugly too!'"; and,
>
> "When I played in the sandbox, the cat kept covering me up."

This latter put-down statement is also funny because it seems ridiculous or silly. As I described above, that is another form of humor.

I once heard a humorous put-down excuse given when hardly anyone showed up for a special safety meeting. The invited speaker for the poorly attended meeting asked at the start, "Did you tell them I was coming?" "No," was the reply from the safety coordinator. The speaker responded, "It must have leaked out." To those readers wondering if I was that invited speaker, let me just say I have frequently been disappointed by low turnout for a safety meeting. And, a smaller-than-expected audience has often created a mood for a less-than-optimal meeting. But, the laughter that resulted from the put-down humor at the start of this particular meeting set the stage for a very enjoyable and memorable session.

Conclusion

I hope you have enjoyed reading this article half as much as I have enjoyed writing it. Perhaps some of the stories or punch lines influenced a smile, a giggle, or even complete laughter. If so, I bet your mood improved, at least briefly. And when people are in a good mood, they are more apt to actively care for the safety or health of another person. This illustrates the power of humor to benefit both attitude and behavior and to help cultivate a Total Safety Culture.

If you have heard many or most of the humorous punch lines in this article, you have obviously experienced the advantages of humor. Understanding the five types of humor—pun, exaggeration, surprise, silliness, and put-down—can facilitate the creative application of humor for group presentations or one-on-one interaction. It is not easy to select and use the right amount of humor at the right time. But the reciprocal payoff of humor for both the giver and the receiver makes the development of a good sense of humor well worth the effort.

Are You "Training" or "Educating"? It's Smart to Do Both[4]

E. Scott Geller

When I teach a university class, it is considered "education," even "higher education," but when I teach in industry, I am considered to be "training" employees. Colleges and universities have "Centers for Educational Excellence," and industries have "Training Centers." So it seems that colleges and universities educate, and industries train. What is the difference? Actually, we use these terms interchangeably, as if they have the same meaning. But if you think about it, education and training are not the same. Each should have a different teaching style and purpose. Some situations require education, others call for training, and some need both education and training. As safety teachers we need to know when to educate, when to train, and when to do both.

First, let's dissect the difference between education and training. Actually, you already know the distinction. Do you want your teenager to receive sex education or sex training? Are you satisfied if your teenager receives only "driver education," or do you prefer some "training" with that education?

Perception problems

Our intuitive understanding of the difference between education and training can actually lead to problems. We might perceive safety training as a step-by-step procedure or program with no room for individual creativity, ownership, or empowerment. This is how safety can come to be seen as a top-down "flavor-of-the-month." If we do not educate people about the principles or rationale behind a particular safety policy, program or process, they might participate only minimally. They will perceive the program as a requirement rather than an opportunity to make a difference. They might even see themselves as animals in a circus, well trained to jump through hoops, rather than as members of a safety community, empowered to go beyond the call of duty for safety.

By the same token, safety education without follow-up training will not reap optimal benefits. Learning the theory or principles behind an intervention approach is crucial for customizing procedures for a particular work situation. But after the procedures are developed—hopefully with input from an educated work team—training is necessary. People need to know precisely what to do.

Teaching styles are not the same for education and training. When I lecture to large groups of university students or to safety professionals and hourly workers, I use various techniques—perhaps brightly colored overheads or statements written out on a blackboard or flipchart—to maintain attention and get participants involved in the learning process. I might say something extreme to get a contrary reaction or ask pointed questions and solicit answers from the audience. I am trying to influence participants' cognitive or thinking processes. This kind of education might increase profound knowledge or critical thinking skills, which could lead to behavior change. If so, I will have "thought a person into behaving differently." In other words, education targets thought processes directly and might indirectly influence what people do.

Training targets behavior directly and might indirectly influence thought processes. This typically calls for more than a lecture format. Training might start by specifying steps needed to accomplish a particular task, but more than this is needed to ensure that certain skills or procedures are learned. Participants in a training course should practice the desired behavior and receive pertinent feedback to support what is right and to correct what is wrong. If feedback is given genuinely in a trusting and caring atmosphere, behavior might not only be directly improved, but also one's thinking or attitude associated with the behavior might be positive.

My colleagues at Safety Performance Solutions use both education and training to teach safety coaching skills. They start with education, teaching the basic principles behind a behavior-based approach to coaching. Then they use group exercises to implement a training process. In one small-group exercise, participants develop a brief skit to demonstrate the coaching principles they have learned. For example, one person sets the stage, another person demonstrates safe or at-risk behavior, and a third person gives rewarding or corrective feedback. When performed in front of the group, everyone can give feedback on how principles translate into practice. If done right, the feedback from the audience and the educator/trainer improves the performance.

We have also asked groups to first show us the wrong way to coach and then to demonstrate the right way. After the group acts out appropriate safety coaching, the audience can offer supportive and corrective feedback. Usually, the educator/trainer finds opportunities to add to an observer's feedback and points out how that feedback could have been more constructive. This teaching frequently includes restating the underlying principle or rationale. In this way, education and training go hand in glove to maximize real benefits from the learning process.

That is the bottom line here: You will need a strategic combination of both education and training to improve both behavior and attitude.

Notes

[1] The original version of this paper appeared in *Occupational Health and Safety* (January 2001).

[2] Originally appeared in *Professional Safety* (1997):40-44.

[3] The original version of this paper appeared in *Industrial Safety & Hygiene News* (October 1997).

[4] The original version of this paper appeared in *Industrial Safety & Hygiene News* (November 1996).

5

Measuring Safety Performance

If you want to know what an organization values, look at what they track and measure. Most companies do a wonderful job tracking productivity and quality but a mediocre job measuring safety.

Despite a variety of proactive behavioral safety measures available to companies, most simply track outcome numbers like OSHA Recordables, First Aid Cases, and Lost Time Incidents. This is insufficient. Outcome numbers are important to assess, but they can be misleading and manipulated. A company reaches a certain milestone for man-hours worked without injury, but discussions with employees may later reveal that injuries are being hidden because of financial incentives or a fear of reprimand associated with injury reporting.

The first and third papers present convincing rationale for focusing on process numbers rather than injury statistics when evaluating safety performance and offer practical ways to measure proactive progress toward reducing injuries in the workplace. The second paper details the various factors that need to be assessed when evaluating safety-related behavior and developing a corrective action plan. A measurement system is not useful if it cannot provide information applicable to making improvement. By offering a plan for analyzing discrepancies between ideal (or safe) versus real (or at-risk) behavior, this paper defines a variety of factors that need to be assessed and evaluated in order to increase safe behavior and decrease at-risk behavior throughout a work culture.

The last paper in this chapter presents four industry-based studies that systematically measured employees perceptions, opinions, and behaviors in order to answer specific empirical questions related to the effectiveness of behavior-based safety. The methodology and results of these studies have been published in peer-reviewed research journals and therefore provide examples of reliable and valid research that could be conducted by safety leaders as part of a comprehensive measurement system.

Measuring Safety Performance for the Next Millennium[1]

Molly A. McClintock

Safety has not always been measured. In fact, measuring safety performance is a relatively new phenomenon. The 1907 Pittsburgh Survey was an early attempt to assess safety in industry by gathering incident data from one county in Pennsylvania. The study found an alarmingly high fatality rate of nearly two deaths per day over a one-year period. In recent years, the Occupational Safety and Health Administration (OSHA) has continued in the tradition of measuring safety through outcome indicators such as recordable injury rates and lost workday rates. OSHA's authority extends to requiring employers to keep records of safety and health data, and it has strict definitions of how to categorize and tally incidents.

What Gets Measured Gets Done

Measurement is critical to any process. The purpose of workplace safety and health processes is to reduce and eliminate injuries and illnesses. Without outcome indicators, no professional or organization can know with confidence that efforts will be successful at stopping incidents before they happen. The problem with measurement in safety is that we have stopped at counting injuries and illnesses and, in some cases, other non-injury incidents. We have not gone on to measure the more difficult process indicators.

For safety professionals, measurement has meant increased attention and clout within organizations. While perhaps not always on a par with other metrics, safety has at least been part of the conversation. Unfortunately, these traditional methods of measuring safety performance focus improvement efforts not on safety processes but on outcome measures. Counting and classifying injuries as a means of measuring safety performance provides inadequate—and frequently inaccurate—information. Because they emphasize outcomes or injuries, traditional measures tend to have a chilling effect on the reporting of incidents.

Additionally, outcome measures provide little insight into real safety problems and thus hamper development of genuine improvement initiatives. The result is an ineffective management of numbers without real impact on employee safety. Adding process measures provides not only better information for safety improvement but also a means of measuring employee involvement and commitment.

The Need for New Measures

Traditional measures focusing on outcome numbers such as lost time and recordable injury rates are badly polluted. That is, too many factors outside of safety impact these numbers. The previous disciplinary action taken by a supervisor suppresses reporting in his department. The desire of an operator to earn his quarterly incentive prize keeps him from discussing his burn. The burden of having her site hold the corporate record for days without a lost-time injury encourages the company nurse to re-evaluate her recommendation. As a result, a mechanic takes a few vacation days rather than seeking medical help for his chronic back pain.

Each of these influences occurs daily in facilities everywhere. While the result may be an ever-lower incident rate, it is almost certainly not improved safety. The success of one location to artificially lower incident rates often results in pressure among sister facilities to meet the same or lower targets. In effect, the pressure to keep pace forces other individuals or businesses to practice the same tactics as their "competitors." This spiraling competition leads to more unrealistic goals.

The effect on individuals can be disastrous. In fact, pressure to meet outwardly imposed and often unrealistic outcome indicators causes frustration and perhaps diminished respect for safety. Take, for example, the poem in Figure 5.1.1 that was given to a safety consultant (and a colleague of mine) by an employee of a chemical processing plant following a celebration at his facility recognizing one million work hours without a lost-time injury. The celebration included speeches by the organization's CEO and other dignitaries.

Sure the employees received kudos and the facility probably has a trophy or plaque somewhere, but what really changed? If the sentiment expressed by this poet is any indication, safety was actually harmed in the process of winning the award. Do you think it is likely he or others will take the next safety initiative seriously? Will he offer suggestions to improve safety or volunteer to serve on a safety committee? It is doubtful. The credibility of leaders with regard to safety has been terribly damaged because they focused on the wrong things.

> With crippled limbs and mangled feet
> A million man-hours we did meet.
> With record-keeping such as these
> We'll reach a zillion, it'll be a breeze.
>
> Rewards are for achievements met
> But we ain't reached a million yet.
> Their safety program is a sham
> As for you and me, they don't give a damn.

Figure 5.1.1 Poem Written by an Hourly Employee after a Safety Celebration

Consider the plight of a fictional but typical manufacturing plant of 250 employees. With a static recordable rate of 3.6, the plant's management team is stymied as to what to do next. Stressing the importance of reaching the corporate-established goal of 1.5 in the company newsletter and during safety meetings did not help. Neither did the new safety raffle program. The safety manager analyzed available injury data and discovered the most frequent injuries were contact dermatitis and lower back strain. Despite implementing lifting training and a new policy on glove use, injuries persist. "What else can we do?" he pleads.

Without other performance-related information, it is difficult to do anything but repeat more of the same stale programs and hope for a change. Process-based performance measures provide a more accurate assessment of the actual safety process as well as guiding development of safety improvement efforts.

What Else Should Be Measured?

Outcome measures will surely be with us for the foreseeable future. Safety managers and other organizational leaders will likely be held accountable for these numbers into the next millennium. But additional measures could provide better, more accurate, upstream indicators of safety performance. Additionally, employees at lower levels in the organization should be held accountable by different measures, ones they can control. A lab technician does not have control over her plant's safety outcome numbers or those of her department. Yet, it is reasonable and desirable to hold her accountable for her own safety performance. So, what do we measure?

Process measures differ from outcome numbers in three distinct ways. First, they are activity-based so they measure what employees are actually doing, not just what they failed to do (e.g., get injured and report it). Second, process measures are within the control of the individual or team being measured. Finally, they indicate levels of employee involvement and buy-in to the safety process itself. Figure 5.1.2 depicts sample quantitative and qualitative measures.

Sample Quantitative Measures	Sample Qualitative Measures
• Number of Job Safety Analyses (JSAs) performed • Time-to-closure on safety action items • Participation of employees on audit teams • Number of safety suggestions submitted • Participation of leaders on team safety meetings • Increase in percent safe behavior observed	• Effectiveness of safety training (for both new and existing employees) • Employee perceptions of incident investigation process • Type of interventions developed through observation process • Value of safety suggestions • Usability of procedures • Effectiveness of safety committee(s) • Reliability of peer observations as measured by a second observer

Figure 5.1.2 Sample Safety Process Metrics

By identifying, gathering, and analyzing process-type measures, safety professionals and others will gain greater insight into the existing barriers to safety achievement. The reduction or elimination of these barriers can be validated by further measurement. Overall success will be evident in less frequent and less severe incidents as well as by greater involvement on the part of all employees in safety improvement initiatives.

Let's return to the fictional manufacturing facility and apply process metric analyses. Instead of more raffles, training, and policies, what if the safety manager could implement new and targeted interventions? He already knows his injuries are coming generally from two areas: contact dermatitis in the pack-out room and lower back strain in the warehouse. Peer observations conducted in these areas identify safety process concerns, including insufficient glove use and several ergonomic issues. Discussions with employees working in each area pointed to potential solutions or interventions.

While the contact dermatitis problem is widely known by employees in the pack-out room, they have been unsuccessful at getting their glove supply stabilized. Sometimes gloves are plentiful and other times not. Some products do not require glove use, so workers may go several weeks without a need for gloves. Employees admit to hoarding gloves when they can get them, so the supply dwindles quickly. Employees in the storeroom express frustration with pack room workers. "We order gloves by the truckload and they all disappear. They have to give us more notice. We only have one approved supplier and he can't always meet our need in time," they say.

Increased communication and coordination between these two departments along with change in the expectations of suppliers can go a long way toward resolving the glove use issue. More training and stronger policies would do nothing but frustrate the already frustrated employees. Having the observation data and involving employees in the development of behavioral interventions can successfully reduce contact dermatitis. Relations between employees in the pack room and the storeroom have also been improved as evidenced by the recent decision to form a formal liaison between the areas. Line employees will meet regularly to discuss supply concerns. Peer observation data demonstrate an increase in glove use. The safety manager can realistically predict a decrease in the contact dermatitis because he has upstream indicators telling him the processes affecting this illness have improved.

What about the lower back strains occurring in the warehouse? An ergonomics assessment of the warehouse identifies a need for more accessible hoists to alleviate the perceived need for employees to manually lift loads in excess of 75 pounds. A policy exists regarding this lifting limit, but it is routinely disregarded because so few hoists are located in the area they are needed. This relatively inexpensive equipment fix along with observation and coaching by peers and supervisors can greatly enhance hoist use and reduce back strains.

Process-Based Accountabilities Measure Individual Performance

One use of outcome numbers is particularly troublesome for organizational leaders: accountabilities. Most managers, many supervisors, and some line employees are held accountable for outcome numbers. These measures are meant to assess the performance of the individual at managing safety within his area of responsibility. The problems with this type of accountability are many.

First, as mentioned earlier, pressure on outcome numbers, either intentional or unintentional, leads to under-reporting. When a manager's pay, promotion or bonus is tied to hitting a particular target, she is extremely motivated to make that goal. Now, that *could* motivate her to survey workers on the effectiveness of safety audit recommendations, improve the quality of safety meetings, and encourage additional employee observations. Those actions *may* have a positive impact on outcome measures. (I say may because we have all watched baseball games in which the team with the best process measures—hits, stolen bases, double plays—ended up losing in the outcome measure—runs.)

On the other hand, the manager may apply pressure to her supervisors by deploying individual outcome accountabilities to them, introducing a hefty incentive for those who don't report an injury all year, or cracking down on injured employees who violate safety procedures. If managers and supervisors are not aware of viable methods of improving the process of safety, they will return to improving the numbers.

Second, outcome-based accountabilities are influenced by a great many issues outside the realm of safety, including decisions by medical personnel and labor-management disputes. A doctor who chooses to write a prescription rather than recommend an over-the-counter medication may cause the injury to be classified differently. Employees who are angry about new work schedules target the supervisor for high incident reporting. The number is so polluted by other factors, its credibility is questionable.

Finally, those held accountable for these measures often experience distress because they do not feel in control of the numbers. A supervisor is frustrated because he is doing all he knows to improve safety but has had a number of employees report injuries. His counterpart in another department has no reported injuries but has accomplished very little in the way of process improvement. Who will receive the promotion at year's end?

One organization drafted the following process-based accountabilities for supervisors:

- Support and reward employee participation in safety activities,
- Set process-based safety goals and expectations with employees,
- Provide regular formal and informal safety performance feedback,
- Model appropriate safety-related behaviors,
- Solicit and encourage employee input on safety-related matters,
- Communicate safety information to employees regularly, and
- Demonstrate visible support for safety policies, rules, and regulations.

Clearly, additional detail including frequency rates would be helpful in tailoring these accountabilities to individual settings. In the case of this company, supervisors developed specific examples for each of the above activities and then invited superiors, peers, and subordinates to assess the success of each supervisor on a quarterly basis.

The Reluctance to Change Measures

Traditional outcome measures often are embedded not only in accountabilities but also in reward and recognition processes. If employees have grown accustomed to receiving a hefty check at the end of each quarter or year that they do not report an injury during that time, the company may have an even greater reluctance to change what is measured and reinforced for safety. Some organizations have successfully cleared this hurdle by first combining outcome and process measures and then gradually shifting the balance to more and more process measures.

An example of a purely process-based employee incentive program is presented in Figure 5.1.3. A joint safety team established targets, and rewards were distributed based on overall plant performance. This change was one of many that helped this facility reduce its number of recordables from 17 to seven over two years.

Figure 5.1.3 Sample Process Measures Used for a Behavior-Based Safety Incentive Program

Quarterly Safety Incentive Metrics
• Safety meetings/training per employee • Safety inspections/audits/drills • Behavioral observations • Safety concerns identified • Incidents reported/investigated • Behavioral process participation • Incident-to-injury ratio • DO IT* processes completed • Safety work orders completed • "Ten Most Wanted"** completion rate * A safety problem-solving process developed by E. Scott Geller, Ph.D. ** Facility/equipment repair

In general, measurement is an important component of any process, and safety is no exception. Managers may be reluctant to halt the deployment of injury numbers to their supervisors in the absence of a legitimate alternative. Adding process measures to assess not only individual performance but also the overall safety system provides organizations with better, more accurate, and best of all, upstream indicators of safety performance.

Behavioral Safety Analysis: A Necessary Precursor to Corrective Action[2]

E. Scott Geller

A prescription without diagnosis is malpractice ~~ Socrates

As I have addressed in other publications, there appears to be substantial confusion about the benefits of behavior-based safety (Geller 1997b, 1998a, 1999). Part of the problem is lack of agreement regarding the operational definition of this particular approach to injury prevention. Even leading consultant firms seem to disagree about the key components of a behavior-based safety process. See, for example, the multiple perspectives on behavioral safety presented by Manuele (1998) and La Bar (1998).

I attempted to promote some clarity and consensus-building regarding behavioral safety by documenting the basic principles of this approach as derived from the academic and research literature (Geller 1997b, 1998a, b, c). I have seen an increase, however, rather than a decrease in confusion and misunderstanding as behavior-based safety becomes increasingly popular and as more consultants claim to cover this approach in their safety services.

Some of the harshest criticism of behavior-based safety has come from representatives of the United Auto Workers (UAW) Health and Safety Department (Howe 1998; Mirer 1999). In a widely disseminated document, the Assistant Director of the UAW Health and Safety Department claims that behavior-based safety programs "are completely reactive, blame the workers for almost all health and safety problems, and drive problems underground, inject fear into the workplace and discourage workers from reporting injuries and illness" (Howe 1998, p. 21). Similar critical statements were expressed by the Director of the UAW Health and Safety Department in a general session panel discussion at the 1999 *ASSE's Best Practices in Safety Management Symposium* held in Las Vegas, Nevada.

Some concern and confusion regarding behavioral safety is based on the misperception that this approach is only for corrective action. In fact, many believe behavior-based safety is essentially observation and feedback. This usually translates into an interpersonal coaching process whereby observers fill out a behavioral checklist while watching other people work and then use this checklist to deliver behavior-focused feedback in one-on-one and/or group settings (Geller 1995; Geller et al. 1998).

Interpersonal observation and feedback is certainly critical for behavior improvement in many work situations, but it is not the only injury prevention approach reflective of behavior-based safety. In articles for Professional Safety, I described applications of behavioral principles for designing safety incentive programs (Geller 1996), delivering individual recognition and group celebrations (Geller 1997a), and implementing safety self-management procedures for the lone worker (Geller and Clarke 1999). All of these publications address ways to improve the behavior of workers, and that's the root of most criticism directed at behavior-based safety.

It seems those most critical of behavior-based safety are concerned that intervention or corrective action focuses on the worker rather than management or the system in general (Howe 1998; Hoyle 1998; Lessin 1997; Mirer 1999). In fact, it is true that observation and feedback, recognition, incentive/reward, and self-management programs do focus on work behavior. But this does not make engineering, management systems, or environmental fixes any less important. Factors in all of the domains contribute significantly to occurrences of at-risk behavior, which eventually lead to injury.

This article illustrates how the principles of behavior-based safety can help determine what factors contribute to at-risk behavior and should be changed to reduce such behavior. In other words, the theme of this article is not corrective action but behavior *analysis*. It offers basic guidelines for analyzing the behavioral aspects of a near-hit or injury in order to determine the most cost-effective corrective action. The reader will understand that the principles and methods of behavior-based safety do not involve fault-finding or victim-blaming and do focus on factors beyond the individual worker. Plus, you will see how the principles and methods of behavior-based safety can be used for more than corrective action. They are invaluable for conducting the kind of safety diagnosis needed to select the most appropriate approach for safety improvement.

Too often "retraining" or "discipline" (meaning punishment) are selected impulsively as a corrective action for behavior change when another less costly and more effective approach is called for. This can be revealed through a proper behavioral safety analysis. Of course, this kind of analysis is only relevant if the problem involves behavior, or more specifically, a "behavioral discrepancy."

What Is a Behavioral Discrepancy?

It is important to consider human performance problems as a discrepancy rather than a deficiency (Mager and Pipe 1997). This places the focus on the behavior, not the individual. In other words, a difference exists between the behavior demonstrated and the behavior desired. When evaluating safety problems, this discrepancy is between behavior considered to be at-risk versus behavior considered to be safe.

The behavioral discrepancy could be a "sin of omission" or a "sin of substitution." The worker might have failed to perform a particular safe behavior, as when taking a short cut. Or the individual could have performed a certain behavior that puts someone at risk for injury. After determining what is safe and what is at-risk for a particular individual and work situation, an action plan can be designed to reduce the discrepancy between what is and what should be.

Should the Work Context Be Changed?

From their more than 60 combined years of analyzing and solving human performance problems, Mager and Pipe (1997) conclude that many discrepancies between ideal and real behavior can be eliminated with relatively little effort. More specifically, behavior might be more at-risk than desired because expectations are unclear, resources are inadequate, or feedback is unavailable.

In these cases, solutions to reducing a behavioral discrepancy should include changing the context in which the behavior is performed, typically referred to as environmental system factors or interpersonal dynamics. Behavior-based instruction or demonstration can overcome invisible expectations, and behavior-based feedback can enable continuous improvement. A work team could decide what resources are needed to make a safe behavior more convenient, comfortable, or efficient. Explore answers to these questions:

- Does the worker know what safety precautions are expected?
- Are there obvious barriers to safe work practices? Is the equipment as safe as possible under the circumstances? Is protective equipment readily available and as comfortable as possible?
- Do workers receive behavior-based feedback related to their safety?

Is Safe Behavior Punished?

A key principle of behavior-based safety is that behavior is motivated by its consequences. In other words, our behavior results in favorable or unfavorable consequences, and these consequences determine our future behavior. Sometimes the natural consequences work against us. This is especially true in safety because safe behavior is usually less comfortable, convenient, or efficient than the at-risk alternative.

Safety leaders need to try to see the situation through the eyes of the performer. Some consequences might actually seem positive to an observer but are viewed as negative by the performer. The plant manager, for example, might consider a person's public safety award as reinforcing, but for the individual it could be punishing because of expected harassment from co-workers. Or how about those "rate busters" in school and industry who excel in work quantity or quality. Often the negative consequences from peers lead to subsequent reductions in individual output.

In some work cultures the interpersonal consequences for reporting an environmental hazard or near-hit are more negative than positive. After all, these situations imply that someone was irresponsible or careless. It is not unusual for people to be ridiculed for wearing protective gear or using an equipment guard. It might even be considered "cool" or "macho" to work unprotected and take risky short cuts. The hidden agenda might be that "only a 'sissy' would wear that fall protection."

Mager and Pipe (1997) refer to these situations as "upside-down consequences" and insist that whenever a behavioral discrepancy exists, part of the problem is due to desired behavior being punished. Are people put down when they should be lifted up? Explore answers to these questions:

- ⋇ What are the consequences for safe behavior?
- ⋇ Are there more negative than positive consequences for safe behavior?
- ⋇ What negative consequences for safe behavior can be reduced or removed?

Is At-risk Behavior Rewarded?

As mentioned above, at-risk behavior is often followed by natural positive consequences. Short cuts are usually taken to save time and can lead to a faster rate of output. So taking on at-risk short cuts can be labeled "efficient" behavior. I have analyzed several work environments, for example, where avoiding or overriding the power lockout switches was acceptable because it benefited production—the bottom line. In these cultures, the worker who could fix or adjust equipment without locking out the power was a hero—a "macho" hero. He could handle equipment problems without slowing down production.

Behavior does not occur in a vacuum. Most people perform the way they do because they expect to achieve soon, certain, and positive consequences. Or they expect to avoid soon, certain, and negative consequences. People take calculated risks because they expect to gain something pleasant and/or avoid something unpleasant. Explore answers to these questions:

- ⋇ What are the soon, certain, and positive consequences for at-risk behavior?
- ⋇ Does a worker receive more attention, prestige, or status from co-workers for at-risk than safe behavior?
- ⋇ What rewarding consequences for at-risk behavior can be reduced or removed?

Are Extra Consequences Used Effectively?

Because the natural consequences of comfort, convenience, or efficiency are often available to support at-risk over safe behavior, it is sometimes necessary to add extra consequences to motivate the occurrence of safe over at-risk behavior. These frequently take the form of incentive/reward or disincentive/penalty programs. Unfortunately, many of these programs do more harm than good because they are implemented ineffectively.

Disincentives are often inadequate because they are used inconsistently and motivate avoidance behavior rather than achievement. Safety incentive programs based on outcomes usually discourage injury reporting (Geller 1996; Hislop 1993; Krause 1998). More effective alternatives include behavior-based incentive/reward programs and interpersonal recognition (Geller 1996, 1997a; Loafman 1998). Explore answers to these questions:

- Can safe behavior lead to soon, certain, and positive consequences?
- Does a safety incentive program discourage the reporting of personal injuries?
- Are workers recognized individually and as teams for completing process activities related to safety improvement?

Figure 5.2.1 depicts a variety of extra and extrinsic consequences that can influence occurrences of safe or at-risk behavior. Three categories of behavior are shown (safe, at-risk, and production-related) as potentially influenced by four different types of behavior-based consequences. None of the consequence examples are natural or intrinsic to the task but are added to the situation in an attempt to sustain desired behavior or change undesired behavior. Therefore, each consequence manipulation in Figure 5.2.1 can be considered a corrective action. They are not presented, however, as recommended interventions but are only intended to illustrate the variety of potential individual and group consequences that can change the context of a work situation and occurrences of safety-related and production-related behaviors.

Obviously, a careful behavior analysis of the work situation is needed (as outlined above) to determine what kinds of motivating consequences should be removed from or added to the work context. Sometimes an analysis of natural behavioral consequences reveals the need for a more fundamental intervention. Specifically, it might be possible to implement an environmental or engineering intervention to simplify the task.

Figure 5.2.1 Consequences for Behaviors

	Individual Consequences		Group Consequences	
	Reward	Penalty	Reward	Penalty
Safe Behavior	"Thank You Card" for cleaning up spill	"Sissy" or other non-macho comment for using PPE	Group celebration after 100 coaching sessions	Team ranked at bottom for attendance at safety meetings
At-Risk Behavior	Praise for adjusting equipment without locking out power	Verbal reprimand for walking outside yellow line	High-fives for team lifting without a hoist	Group reprimand for unreported property damage
Production Behavior	Praise for working 12 hours overtime	Written warning for omitting a quality check	Group efficiency plaque for fastest work process	Work team ranked last on "Resource Management" chart

Can the Task Be Simplified?

Before developing a training program to increase on-the-job safety, make sure all possible engineering "fixes" have been implemented. For example, explore the many ways the environment could be changed to reduce physical effort, reach, and repetition. Entertain ways to make the job more user-friendly before deciding what behaviors are needed to prevent injury. This is, of course, the rationale behind ergonomics and the search for engineering solutions to occupational safety and health risks.

Sometimes behavior facilitators can be added, such as designing controls with different shapes so they can be discriminated by touch as well as sight, placing clear instructions at the point of application, using color coding to aid memory and tasks, differentiation (Norman 1988), and applying machine lifts or conveyor rollers to help with physical tasks. Plus, it is possible that complex tasks can be redesigned to involve fewer steps or more people. To reduce boredom or repetition, simple tasks might allow for job swapping.

Explore answers to these questions to determine whether a task can be simplified:

- Can an engineering intervention make the task more user-friendly?
- Can the task be redesigned to reduce the physical demands?
- Can a behavior facilitator be added to improve response differentiation, reduce memory load, or increase reliability?
- Can the challenges of a complex task be shared?
- Can boring, repetitive tasks be swapped?

Is There a Skill Discrepancy?

But what about those times when the individual really does not know how to do the prescribed safe behavior? The person is "unconsciously incompetent." This situation might call for training, which is usually a more expensive intervention approach. Mager and Pipe (1997) claim that most of the time a behavioral discrepancy is not caused by a genuine lack of skill. Usually people can perform the safe behavior if the conditions and the consequences are right. Training should really be the least used approach for corrective action.

Ask these questions to determine whether the behavioral discrepancy is caused by a lack of skill:

- Could the person perform the task safely if his or her life depended on it?
- Are the person's current skills adequate for the task?
- Did the person ever know how to perform the task safely?
- Has the person forgotten the safe way to perform the task?

What Kind of Training is Needed?

Answers to the latter two questions can help pinpoint the kind of intervention needed to reduce a skill discrepancy. A "yes" answer to these questions implies the need for a skill maintenance program. Skill maintenance might be needed to help a person stay skilled, just as police officers practice regularly on a pistol range to stay ready to use their guns effectively in the rare situation when they need it. This is, of course, the rationale behind periodic emergency training. People need to practice the behaviors that could prevent injury or save a life during an emergency. Fortunately, emergencies do not happen very often, but because they don't, people need to go through the motions just to "stay in practice." Then if the infrequent event does occur, they will be ready to do the right thing.

A very different kind of situation also calls for skill maintenance training. This is when certain behaviors occur regularly, but discrepancies still exist. Contrary to circumstances requiring emergency training, this problem is not lack of practice. Rather, the person gets plenty of practice doing the behavior ineffectively or at risk. In this case, practice does not make perfect but rather serves to entrench a bad (or at-risk) habit.

Vehicle driving behavior is perhaps the most common and relevant example of this second kind of situation in need of skill maintenance training. Most drivers know how to drive a vehicle safely and at one time showed little at-risk driving. But for many drivers, safe driving has deteriorated over time, with some safe driving practices dropping out of a person's driving repertoire completely.

Practice with appropriate behavior-based feedback is critical for solving both types of skill discrepancies. However, if the skill is already used frequently but has deteriorated (as in the driving example), it is often necessary to add an extra feedback intervention to overpower the natural consequences which have caused the behavior to drift from the ideal. This is the basic rationale for the observation and feedback process in behavior-based safety.

While the police officer gets task-inherent feedback to improve his performance on the pistol range, at-risk drivers might need behavior-based coaching to improve. The coach needs to systematically complete a critical behavior checklist (CBC) while watching for safe versus at-risk driving and then use this CBC to give supportive and corrective feedback (Geller 1995). Explore answers to these questions to determine whether the cause of the apparent skill discrepancy is due to lack of practice or lack of appropriate feedback:

- How often is the desired skill performed?

- Does the performer receive regular feedback relevant to skill maintenance?
- How do performers find out how well they are doing?

Is the Person Right for the Job?

It should be obvious from this discussion that a skill discrepancy can be handled in one of two ways: change the job or change the person. The first approach is exemplified by simplifying the task while the latter approach is reflected in practice and behavior-based feedback. This is behavior-based training. But what if a person's interests, skills, or prior experiences are incompatible for the job?

The person might be like me, for example, and not be "mechanically inclined." Sure, I could learn how to fix a computer if my life depended on it, but the training process would be relatively long and I would not like it. Before investing in skill training for a particular individual, it is a good idea to assess whether the person is right for the job. If the person does not have the motivation or the physical and mental capabilities for a particular job, the cost-effective solution is to replace the performer. If you do not, you will not only sub-optimize work output, but you will also increase the risk for personal injury.

Ask these questions to determine whether the individual has the potential to handle the job safely and effectively:

- Does the person have the physical capability to perform as desired?
- Does the person have the mental capability to handle the complexities of the task?
- Is the person overqualified for the job and thus prone to boredom or dissatisfaction?
- Can the person learn how to do the job as desired?

Conclusion

This article offered some basic guidelines for diagnosing the human behavior aspects of safety-related problems. Numerous factors have been entertained that can contribute to a behavioral discrepancy—a distinction between the behavior performed and the behavior desired. In safety terms, this is the difference between at-risk and safe behavior.

Most of the factors contributing to behavioral discrepancies are due to the context in which the task is performed or the task itself. Contextual variables include unclear or misunderstood expectancies, upside-down contingencies which reward at-risk behavior or punish safe behavior, and the lack of behavior-based feedback to help people improve. Often a job can be simplified or re-engineered to reduce physical or mental effort and thus decrease the probability of personal injury.

Training should be considered only after analyzing and correcting critical contextual and task variables. Some training is needed to keep people in practice for handling a relatively rare event while other training is needed to help people change frequently occurring at-risk behavior to safe behavior. There is also the training needed to introduce a new procedure or process. Each of these training situations requires behavior-based feedback, but obviously the situation and the individuals involved determine the protocol for delivering the feedback. This is one more analysis challenge.

Figure 5.2.2 summarizes the main practical points of this article with a flow chart of the ten basic questions to ask when conducting a behavioral safety analysis. Task context questions are asked first, and engineering strategies are considered for task simplification before the individual worker is targeted with a training intervention.

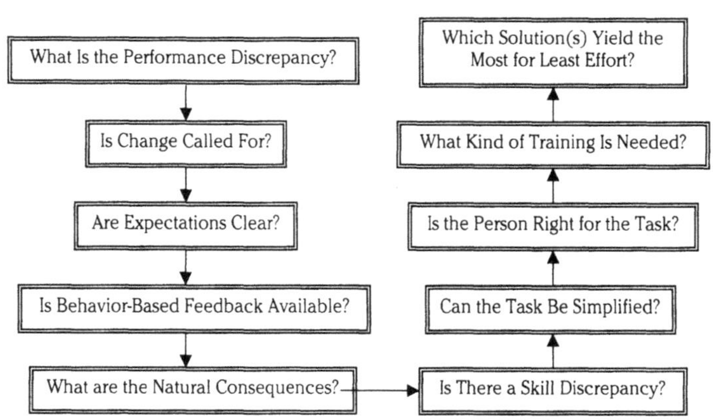

Figure 5.2.2 Ten Basic Steps in a Behavioral Safety Analysis

The bottom line: Before deciding on an intervention approach, conduct a careful analysis of the situation, the behavior, and the individuals involved in an observed discrepancy between desired and actual performance. Do not impulsively assume corrective action requires "training" or "discipline." A behavioral safety analysis will likely give priority to a number of alternative intervention approaches. Doing such an analysis before intervening will help ensure your corrective action plan will not reflect malpractice.

References

Geller, E. S. 1995. Safety coaching: The key to achieving a Total Safety Culture. *Professional Safety*, 40(7):16–22.

———. 1996. The truth about safety incentives. *Professional Safety*, 41(10):34–39.

———. 1997a. Key processes for continuous safety improvement: Behavior-based recognition and celebration. *Professional Safety*, 42(10):40–44.

———. 1997b. What is behavior-based safety, anyway? *Occupational Health and Safety*, 66(1):25–35.

———. 1998c. *What can behavior-based safety do for me?* Neenah, WI: J. J. Keller & Associates, Inc.

———. 1998a. Principles of behavior based safety. *Proceeding of The American Society of Safety Engineers Behavioral Safety Symposium, "Light Up Safety in the New Millennium."* Orlando, FL. 13–24.

———. 1998b. *Understanding behavior-based safety: Step-by-step methods to improve your workplace.* 2nd ed. Neenah, WI: J. J. Keller & Associates, Inc.

———. 1999. Behavior-based safety: Confusion, controversy, and clarification. *Occupational Health & Safety*, 68(1):40–50.

Geller, E. S., and S. W. Clarke. 1999. Safety self-management: A key behavior-based process for injury prevention. *Professional Safety*, 44(7):29–33.

Geller, E. S., T. E. Boyce, J. Williams, C. Pettinger, J. DePasquale, and S. Clarke. 1998. Researching behavior-based safety: A multi-method assessment and evaluation. *Proceedings of the 37th Annual Professional Development Conference and Exposition American*

Society of Safety Engineers/Canadian Society of Safety Engineering. Seattle, WA. 537–559.

Hislop, R. D. 1993. Developing a safety incentive program. *Professional Safety*, 38(4):20–25.

Howe, J. 1998. *A union critique of behavior safety*. Paper presented at The American Society of Safety Engineers Behavioral Safety Symposium, (February). Orlando, FL.

Hoyle, B. 1998. *Fixing the workplace, not the worker: A worker's guide to accident prevention*. Unpublished paper for the Oil, Chemical and Atomic Workers International Union, Lakewood, CO.

Krause, T. R. 1998. Safety incentives from a behavioral perspective. *Professional Safety*, 43(8):24–28.

La Bar, G. 1998. A national symposium frames the possibilities and challenges for behavior-based safety. *Occupational Hazards*, May:49–54.

Lessin, N. 10 July 1997. Testimony on the Occupational Safety and Health Act and the Occupational Safety and Health Administration before the Subcommittee on Public Health and Safety, Senate Committee on Labor and Human Resources, Washington, DC.

Loafman, B. 1998. Behavior-based safety: Power & pitfalls. *Professional Safety*, 43(8):20–23.

Mager, R. F., and P. Pipe. 1997. *Analyzing performance problems or you really oughta wanna*. 3rd ed. Atlanta, GA: The Center for Effective Performance, Inc.

Manuele, F. A. 1998. Perspectives on behavioral safety: Observations of American Society of Safety Engineers behavioral safety symposium. *Professional Safety*, 43(8):32–37.

Mirer, F. E. 1999. Safety engineering or safety management? Fix the system, don't blame the victim. *Proceedings of the American Society of Safety Engineers Best Practices in Safety Management Symposium*. Las Vegas, NV. 24–33.

Norman, D. A. 1988. *The psychology of everyday things*. New York: Basic Books.

Measuring Performance in Occupational Safety and Health

Sherry Perdue

The ultimate purpose for any organization's safety management system is to reduce and ultimately eliminate injuries in the workplace. Not only does an organization's safety performance have a direct impact on employee safety and health, it has a direct effect on employee productivity and quality of work life, and on company profits and community image. Enormous human and financial losses can result from poor safety performance; therefore, organizations must continually improve safety performance. But, how do organizations know when their safety program is effective? How can they measure safety performance, set goals for improvements, and track progress toward these goals? What are appropriate measures for safety performance?

Process vs. Outcome Performance Measures

> *Traditional measures of safety "tell management how things have been going, but they do not point the way to improvement."*
>
> ~~ Dr. Edwards Deming

Traditional measures of safety performance tend to all be injury-related: recordable rates, lost workday rates, and worker compensation costs. Achieving reductions in these outcome measures, however, requires effective safety management processes to manage the situations and events which lead to injuries. Activities such as safety training, safety meetings, behavioral observations, incident investigations, and job safety analyses provide proactive approaches that have been demonstrated to promote improvement in overall safety performance. The more effective these processes, the fewer the injuries. However, in terms of measuring safety performance, many organizations focus a disproportionate amount of attention on the outcome measures of safety like incident rates rather than on measuring and improving the processes responsible for achieving lower rates. Consider, for example, the amount of time given at corporate safety meetings to the presentation of outcome statistics compared to the time spent discussing process results.

Focusing on outcomes rather than processes is analogous to trying to play golf, tennis, or baseball by watching the scoreboard instead of the ball. Although the final score is ultimately what counts, the players' attention must be focused on the processes involved in playing the game in order to achieve a winning score. Outcome-focused measures can only indicate that there are problems; they do not identify what the specific problems are or suggest how to solve them. Similarly, managers must not merely monitor the "safety scoreboard," they must measure and improve the effectiveness of the safety management processes in order to affect the outcome injury statistics. These improvements must be made by following a structured, problem-solving process involving analysis, measurement, and evaluation of the systems. Otherwise, in absence of true *process* improvements, any changes in overall safety performance can only be attributed to random variation. Widely varying injury statistics indicate a process that is not in control. Although this principle is intuitive, managers all too often give more attention to outcome statistics than to the processes responsible for those outcomes.

As a result, when managers are measured or ranked according to their injury rates with little consideration for safety improvement processes or activities, or when employee incentives are based on injury reductions, attention is diverted from the development and maintenance of processes to reduce work injuries. Rather, such pressure to reduce the number of injuries encourages a short-term "fix-the-symptom" view rather than a long-term "fix-the-system" perspective. Consider, for example, the work team who *does* make significant improvements in their safety management processes but has an injury in spite of their efforts, or the work team who largely ignores safety improvements but is fortunate enough not to have an injury during the measured time period.

Overemphasis on the outcome measures leads managers to react by not only looking for a "quick fix" but also by inevitably "tampering" with the safety management processes. Rather than following a systematic problem-solving approach necessary in order to make truly beneficial system-level changes, they react by haphazardly making uneducated modifications. In addition, such pressure to reduce the numbers is usually more likely to influence the *reporting and classification* of injuries than the actual number of injuries incurred. Bottom line: what gets measured gets done. In order to encourage organizational focus on the processes and activities that will achieve long-lasting effect on safety performance, overall safety performance measures must balance outcome-oriented metrics with process or activity-oriented ones.

Unfortunately, process-based performance measures are much more difficult to identify and track than injury statistics. Unlike outcome measures, they are not defined and regulated by an outside organization like OSHA. Rather, metrics must be developed to measure the effectiveness of the organization's specific safety improvement activities (e.g., number of behavioral observations completed, number of inspections performed, average time-to-closure on safety work orders). Additional time and resources are required for their continued monitoring.

Quantitative vs. Qualitative Performance Measures

Process measures can be either quantitative or qualitative in nature. Quantitative measures typically focus on the number of activities performed or the percentage of personnel participating in or completing a given activity. Examples may include the number of JHA's completed, the percentage of employees participating on safety process improvement teams, or the time-to-closure on safety-related action items. However, to truly measure the effectiveness of the various safety management processes, qualitative measures are also necessary. Qualitative metrics attempt to incorporate measures of process quality and effectiveness as well as to assess the process' impact on employee perceptions and the overall safety culture. Qualitative measures are often incorporated into formal process assessments or audits in which each safety management system is judged against predetermined criteria of effectiveness. Employee perception surveys and/or employee interviews are performed in conjunction with or in addition to the systems audit.

While qualitative metrics provide rich information, collection and interpretation of the results are time consuming. Qualitative metrics are also more difficult to use for comparative purposes since they are often even more subjective than most quantitative measures. However, despite the resources required to continually track qualitative measures of performance, the effort is essential. Alternatively, emphasis on quantitative measures alone places the organization's focus on completion of the activities, rather than the quality of the effort or outcome. As a result, the safety process tends to become a bureaucratic system resulting in little or no true value to the organization.

Several examples of both quantitative and qualitative performance measures for a variety of safety management systems are shown in Table 5.3.1.

Goal Setting Based on Performance Measures

Another issue to consider when determining organizational safety and health performance measures pertains to setting goals. Metrics which are measured and monitored as a means of assessing the performance of safety systems are not all appropriate metrics to use for goal setting. Goals should be set based on metrics which people personally can control through their own actions. Further, thought should be given to any alternative, undesirable ways in which a goal might be met. That is, the goal should not inadvertently reinforce an undesirable behavior. The most obvious example is the common practice of setting goals based on injury rates. As mentioned previously, when people are reinforced for low injury rates, low injury rates are usually attained. However, the lower rates may come from under-reporting and "creative classification" rather than fewer injuries.

A second, somewhat less-obvious example of inappropriate goal setting concerns a behavioral observation and feedback process (BOFP). A common performance measure for a BOFP is the "percent safe behaviors" observed. As observations take place and employees receive feedback on their safety-related behaviors, at-risk behaviors should be replaced by the safer alternatives. Hence, over time the "percent safe" index should increase, indicating the process is successful in changing behavior. This metric, however, is not entirely objective and bias-free. The process of observing safety-related behaviors is difficult and subjective. To scrutinize the work practices of others and "see past the obvious" to identify some of the more subtle at-risk behaviors which can lead to injury takes great

Table 5.3.1 Quantitative and Qualitative Safety and Health Performance Measures

Quantitative Measures	Qualitative Measures
Management Planning and Commitment	
Safety Accountabilities	Formal assessment results which may include a measure of:
Percent of departments with comprehensive, written activity-oriented safety accountabilities for managers and employees	Quality of safety accountabilities established, etc.
Percent of accountabilities developed with involvement of employees	Employee perceptions (via survey)
Percent of performance evaluations that include performance of safety accountabilities	
Number of JSA, SJP, SOPs, etc. completed	
Number of JSA, SJP, SOPs, etc. approved	
Number of Process Safety Reviews completed	
Behavioral Observation and Feedback Process	
Number of observations completed (vs. goal)	Formal assessment results which may include a measure of:
Percent Participation (in processes where participation is voluntary)	Quality of observations, based on:
Percent safe vs. Percent at-risk behaviors	Quality of comments
Number of job- or injury-specific checklists developed & used	Quality of follow-up items identified
Number of action items identified for follow-up	Quality of feedback given
Number of action items completed	2nd observer reliability
Average time to closure on action items	
Reduction in incidents associated with <u>targeted</u> behaviors	Employee perceptions (via survey)
Safety Inspections	
Number of safety inspections completed (vs. goal)	Formal assessment results which may include a measure of:
Percent involving managers	Quality of inspections, based on:
Percent involving hourly employees	Quality of follow-up items identified
Number of action items identified	2nd observer reliability
Number of action items completed	
Average time-to-closure on action items	Employee perceptions (via survey)
Near-Miss Management System	
Number of near-misses reported	Formal assessment results which may include a measure of:
Number of safety improvement items identified	Quality of items reported
Number of action items completed	Quality of follow-up items identified
Average time-to-closure on action items	Employee perceptions (via survey)
Incident Investigations	
Percent of incident investigations initiated within 24 hours	Employee perceptions of the process (via survey)
Average time-to-completion of investigation	Formal assessment results
Average time-to-closure on action items identified	

Table 5.3.1., continued

Employee Safety Suggestions	
Number of safety suggestions submitted	Employee perceptions of the process (via survey)
Number of safety suggestions implemented	Formal assessment results which may include a measure of:
Average time-to-closure on safety suggestions	Quality of suggestions submitted
	Timeliness of feedback provided
Safety Work Orders	
Average time-to-closure on safety work orders	Employee perceptions (via survey)
	Formal assessment results
Training	
Percent employees completing safety training requirements	Employee perceptions (via student evaluations & survey)
Scores on safety quizzes	Formal assessment results
Percent employees trained in incident investigation techniques	
Percent employees trained in behavioral observation & feedback	
Percent employees trained in conducting safety inspections	
Safety Meetings	
Percent attendance at safety meetings	Employees perceptions (via survey)
Percent of meetings which include a review of <u>each</u> of the following:	Formal assessment results
Incidents, incident investigation results, status of action items	
Behavioral observation data, status of action items	
Safety inspection results, status of action items	
Near-miss data, status of action items	
Employee Participation	
Percent of employees participating in one or more of the following:	Employee perceptions (via survey)
Safety committee	Formal assessments results
Site safety inspection	
Planner and leader of group's safety meeting	
Review/update JSA, SJP, SOP, etc.	
Incident investigation	
Volunteer Emergency Response/Rescue/Fire Teams	
Improvement teams	
Injury and Illness Rates	
Total Cases (first aid and recordables)	
Recordable Cases	
Total Lost Workday Cases (restricted duty and days away)	
Days Away Cases	
Total Lost Workdays (# of days of restricted duty & days away from work)	
Workers Comp cost	

skill. The classification of a given behavior as either "safe" or "at-risk" can often be a matter of judgment.

Finally, personal biases often limit an individual's ability to make accurate assessments. Therefore, much time is spent developing skilled observers, particularly during the initial stages of process introduction. Observers are coached to mark any potential at-risk behavior as "at-risk" so that the opportunity to coach an employee to minimize injury potential is not missed (stressing to the observee, of course, the uncertainly of the noted behavior). Essentially, observers are instructed to maximize the number of at-risk behaviors they observe. If too much emphasis is placed on reducing the performance metric of "percent safe behaviors," we are likely to encourage observers to "not see" and therefore not give feedback on many at-risk work practices.

Developing a Health and Safety Measurement System

A Health and Safety Measurement System should consist of a set of both quantitative and qualitative process measures as well as safety outcome measures. The process measures should reflect the performance of the organization's key safety management processes. In determining the features to be measured and units of measurement, resources required for collection and record-keeping must be considered. The safety outcome measures should indicate not only the frequency but also the severity of injuries incurred.

The time period for reporting each metric may vary, depending on the nature of the system being assessed and the resources required for gathering the data. Employee perception surveys and formal management systems assessments may occur only once every one to two years, whereas quantitative process and outcome measures (such as the number of observations completed, the number of safe operating procedures reviewed, or the number of recordable injuries) may be collected and monitored monthly. Some Health and Safety Measurement Systems attempt to combine individual metrics into a single numeric "safety score" by weighting each of the indices and then summing the results. If such a practice is done, the individual contributing indices should always be reflected along with the cumulative score.

Using a Health and Safety Measurement System

As managers at all levels of the organization monitor these metrics and communicate the results, they must be careful to balance the amount of attention paid to the process and outcome measures. And they must take care in interpreting the measures as well as the relationship between the two types. In determining the current status of the organization's safety performance, both process and outcome measures must be jointly assessed, taking into account that outcome measures may lag process measures in time. As the outcome measures reflect performance improvements, the process measures should be examined and then emphasized to reinforce to everyone the efforts and activities responsible for the accomplishments. Likewise, when outcome measures reflect performance inadequacies, the process measures should be analyzed. The process measures may suggest areas on which to concentrate improvement efforts. Alternatively, they may suggest that the system is, in fact, in control and that the outcome measure merely represents a system "outlier," averting the temptation to tamper with a system operating effectively.

Researching Behavior-Based Safety: A Multi-Method Assessment and Evaluation[3]

E. Scott Geller, Thomas E. Boyce, Joshua H. Williams, Charles B. Pettinger, Jason P. DePasquale, and Steve Clarke

Abstract

Behavior-based (BB) approaches to injury prevention have a number of advantages: a) they can be administered by individuals with minimal professional training; b) they can reach people in the setting where a problem occurs (community, school, or workplace); and c) the leaders in these settings can be taught the BB techniques most likely to work under specific circumstances (Baer, Wolf, and Risley 1968, 1987; Daniels 1989; Geller 1998). Research has also shown this approach to be cost effective, primarily because BB techniques are straightforward and relatively easy to administer and because intervention progress can be readily assessed by indigenous personnel monitoring target behaviors (Daniels 1989; Geller 1996; Geller, Winett, and Everett 1982; Rudd and Geller 1985; Sulzer-Azaroff and De Santamaria 1980).

Introduction

Behavior-based approaches to safety focus on systematically studying the effects of various interventions on target behaviors, first by defining the target behavior in a directly observable and recordable way, and second by observing and recording it in its natural setting. When a stable baseline measure of the frequency, rate, or duration of behavior is obtained, an intervention is implemented to change the behavior in beneficial directions.

Interventions typically involve modifying or changing the antecedents and/or consequences of specified target behavior(s). To determine intervention effectiveness, the frequency, duration, or rate of the target behavior is recorded during and/or after the intervention and compared with baseline measures of behavior (Daniels 1989; Geller 1996, 1998, 2001).

Intervention Effectiveness Theory

A Multiple Intervention Level (MIL) Hierarchy.

Over two decades of behavior change research at corporate and community sites has led to our development of the *multiple intervention level* (MIL) hierarchy depicted in Figure 5.4.1. We use this model to categorize behavior change approaches and to evaluate the cost-effectiveness of successive intervention strategies to alter the behavioral patterns of large numbers of individuals (Geller, Berry, Ludwig, et al. 1990; Ludwig and Geller 1991).

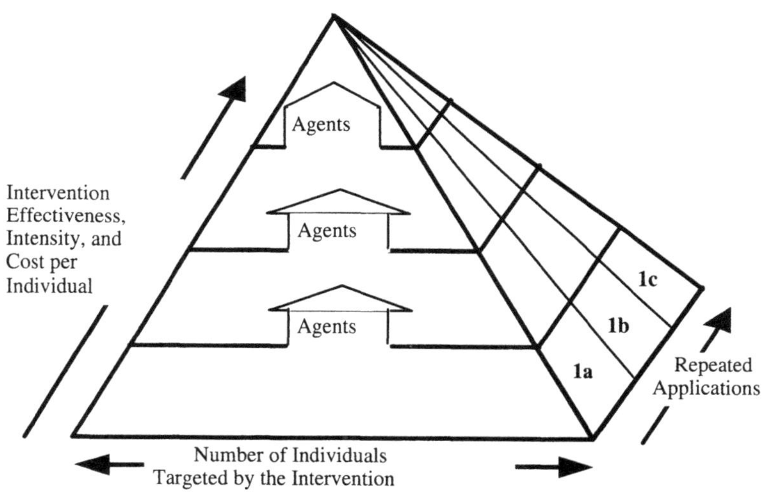

Figure 5.4.1 The MIL Hierarchy Differentiating Repeated Prevention Interventions at the Same Effectiveness Level and Higher Level Prevention Interventions that Are Progressively More Influential at Changing Behavior

The MIL hierarchy is characterized by dividing intervention strategies into multiple tiers or levels with each level defined by certain dimensions of intervention effectiveness. At the first (bottom) level, the interventions are least intrusive and target the maximum number of people. At this level, the intervention is designed to have maximum large-scale appeal and minimum individual-to-individual contact. Those individuals affected at a particular intervention level may benefit from repeated exposure to similar interventions (booster sessions), but we assume most individuals uninfluenced by the first exposure to a particular intervention will be uninfluenced by repeated exposure to interventions at the same level of intrusiveness. Thus, these individuals require a higher-level (more involving) intervention.

Successively higher intervention levels are more costly and intrusive, but they are needed for the "hard core" problem individuals. These problem persons are likely at greater risk for injury. Thus, an MIL approach to public health has critical implications for evaluating the cost-effectiveness of a BB safety program. The MIL model also suggests that individuals who are influenced at a particular level of the intervention hierarchy become potential intervention agents for the next level of intervention (Katz and Lazarfeld 1955).

The current research explored ways to involve indigenous personnel in the implementation of successful intervention strategies. We also studied personality, lifestyle, and attitudinal characteristics of those individuals who volunteer to help implement a program to promote safety at work. Additionally, evaluation of the MIL model was the catalyst for the research targeting driving behaviors.

Study 1: Nationwide Survey on Behavior-Based Safety

Safety professionals were solicited for input through a nationwide survey published in *Industrial Safety and Hygiene News* (*ISHN*), a monthly magazine for safety professionals with 62,000 company subscribers. The survey was designed to assess readers' knowledge and interest in BB safety and to explore ideas for improving the communication and implementation of BB principles and procedures for reducing industrial injuries. The survey also allowed us to begin constructing a database of organizations currently active in BB safety efforts.

A total of 162 completed surveys were returned to us by mail or fax. An appreciation of the BB approach was shown by 80 percent (n = 129) of the respondents answering "yes" to the question "Do you believe behavior-based safety is a viable approach for reducing at-risk work behaviors and activities?" (Only 3 percent responded "no" to this question; the rest said they did not know). In addition, more participants responded "no" (48 percent) than "yes" (34 percent) to the question "Do you think a safety program should put more direct focus on attitudes than on behaviors?" This is interesting because it not only shows preference for a BB approach, but it also reflects a shift from the traditional educational approach to injury prevention.

The astute reader will note, however, that the sample of surveys we analyzed was not random and was likely biased toward the BB approach. The survey was presented within the context of research aimed at discovering how to make the BB approach more effective. Thus, it is likely most people who took the time to complete and return the survey were at least interested in this particular approach to industrial safety. In fact, several respondents asked specifically to be included in our sample of organizations to visit for an on-site evaluation of factors contributing to the impact of a BB safety process. Compared to the average *ISHN* reader, those who answered the questions and returned our survey were probably more informed about BB safety and had higher confidence in the effectiveness of BB safety. Even with this positive bias, however, the survey revealed some misperceptions about BB safety that can limit its application for safety improvement.

What is behavior-based safety?

The first part of the survey asked respondents to give their impression of BB safety by checking all of the items they believe are true from a list of 16 possible characteristics. In general, the respondents' selections indicated accurate knowledge of BB safety, but there were a few notable exceptions.

Three items were most often selected as representing BB safety: 1) an intervention approach for increasing safe behavior (selected by 143 respondents), 2) an observation and feedback process (n = 130), and 3) a tool for managing safety (n = 114).

Fewer respondents considered other characteristics of BB safety to be relevant. Only 42 of the 162 respondents considered BB safety as an approach useful for investigating injuries. Only 88 respondents (54 percent) felt BB safety is useful for evaluating safety achievement, and 99 respondents (61 percent) considered BB safety an intervention approach for decreasing at-risk behavior.

It is likely that people have a rather narrow viewpoint regarding BB safety. This limited perspective is also reflected in numerous safety articles, sales pitches from safety consultants, and presentations at safety conferences. In fact, BB safety is much more than a tool for doing observation and feedback. It is actually "a general philosophy that can be applied to many aspects of safety management"—the general definition that was checked by 71 percent of the respondents.

Principle versus application

Most survey respondents were aware that BB safety focuses on positive consequences to influence behavior change; only four individuals indicated that BB safety was "an approach focusing on the use of punishment to decrease unsafe behavior." A different story emerged, however, when the survey asked respondents to check which techniques were actually used in their plant "to influence safety-related behaviors in the workplace."

Activators (or antecedent strategies) were the most popular techniques chosen, with policies (n = 149), posted safety signs (n = 124), demonstrations (n = 108), and lectures (n = 102) leading the list. Goal setting, feedback, and incentive/reward programs were selected frequently, but more companies focused on outcome (accidents or injuries) rather than process (safety-related behaviors or activities) when setting goals (n = 95 vs. 48), when giving group feedback (n = 83 vs. 60), when giving individual feedback in coaching sessions (n = 96 vs. 74), and when rewarding people for safety improvement (n = 72 vs. 56).

The absence of checks for many techniques was quite revealing and inconsistent with an appreciation for BB safety principles. For example, the most cost-effective BB approaches to improve safety are BB goal setting and feedback for individuals and groups, yet these intervention approaches were being used at less than half of the sites represented by the survey respondents. It was encouraging, however, that almost two-thirds of the sample (n = 102) use safety steering committees to manage their safety programs.

Only 15 percent (n = 24) of the respondents indicated they monitor percent safe behavior to assess the success of their safety programs. The traditional outcome measures were most popular, with 77 percent (n = 125) using OSHA recordables, 75 percent (n = 122) using lost-time accidents, 42 percent (n = 68) using total recordable injury rate, and 44 percent (n = 66) using total recordable rate, including illness. Interestingly, slightly more respondents reported they use attitude or perception surveys (17 percent) rather than percent safe behaviors (15 percent).

Implications

The responses of those who completed and returned our BB safety survey published in *ISHN* reflected appreciation for a BB approach to injury prevention, but they also demonstrated substantial misunderstanding and misapplication. A majority of respondents, for example, perceived BB safety as an observation and feedback tool rather than a general approach to improving the human dynamics of safety that is relevant for ergonomics, injury analysis, and the design of incentive/reward programs.

Even with substantial appreciation for BB observation and feedback as a way to increase safe behavior, relatively few respondents indicated use of a relevant metric for monitoring the success of a behavior-improvement process. Thus, while safety leaders are increasing their belief in the power of observation and feedback to improve behavior, companies are apparently slow to apply appropriate feedback measures to evaluate and improve their safety programs. This is likely not due to inconsistencies between people's beliefs and behaviors but rather to management system variables that prevent a paradigm shift from an outcome-based and reactive evaluation process to one focused on upstream process activities that contribute to the prevention of workplace illnesses and injuries.

Study 2: Applications of Strategies to Increase Safe Driving: A Systematic Evaluation of Intervention Effectiveness After a Decade of Culture Change

Despite mandatory belt-use laws (BULs) and engineering strategies designed to protect vehicle occupants, many motorists still refuse to use the devices provided to safeguard them. This study had three primary objectives: a) to investigate the use of safety belts among industry workers after a decade of culture change, influenced in part by a mandatory BUL; b) to test the impact of several common behavior change techniques on the safe driving behaviors of industry employees; and c) to study generalization between changes in safety-belt use and the use of turn signals.

Our recent observations of employees revealed remarkable increases in baseline safety-belt use from 1982 and a significant difference in mean belt use among wage (46 percent, n = 2,897 observations) and salaried workers (65 percent, n = 1,744 observations). More than a decade earlier, Geller, Davis, and Spicer (1983) found that during baseline, mean use of safety belts among industry workers at the same plant was only 3.4 percent for wage workers and 17.4 percent for salaried employees. At that time, an incentive program increased mean belt use more effectively among salaried (50.6 percent) than wage employees (5.5 percent).

Since the 1982 research, Virginia has instituted a mandatory BUL. The state average of safety-belt use remains near 70 percent which approximates the national average of 67 percent (Novack 1995). The current study describes a systematic effort to increase safety-belt use among workers not influenced by Virginia's BUL.

Subjects and Setting. The subjects consisted of 556 hourly and salaried employees at an engine bearing manufacturing plant in southwest Virginia. The population of employees ranged in age from 19 to 63 years, and employee tenure at the facility ranged from six months to more than 25 years. The proportion of hourly to salaried workers was approximately five to one.

Procedure. Trained research assistants sat in two distinct parking lots of the facility during the first shift arrival, second shift arrival/first shift departure times for hourly workers, and at the arrival and departure time for salaried employees during the regular work week. A minimum of three data collection times were scheduled at each lot every weekday.

Observers collected data on driver gender, safety-belt use, turn-signal use, complete stops at a parking-lot stop sign, the presence or absence of traffic, and whether the targeted vehicle was entering or leaving the parking lot. Information on weather conditions (clear or rain, etc.) and road conditions (dry or wet, etc.) and the time of day were indicated on the data sheet.

To assess inter-observer reliability, a second independent observer collected data with the primary observer on 30 percent of all observation sessions. Because it was impossible to record data on every vehicle entering or leaving the parking lots, the primary observer identified vehicles to observe by calling out their make and color as they passed an obvious stationary landmark. Inter-observer reliability exceeded 90 percent for all measures collected.

Design and intervention techniques. Over the course of two years and after four weeks of baseline observations, a series of progressively more invasive interventions was implemented at the facility to test the validity of the MIL model. Interventions in sequential order consisted of written prompts for the hourly workers, written prompts for the salaried workers, plant-wide written prompts, an employee safety-slogan contest and celebration, assigned goals, assigned goals with performance feedback, and a pledge-card commitment. Detailed descriptions of each intervention appear below.

Written prompt

This intervention consisted of a flyer displaying the logo of the industry and a message selected by a safety steering committee of wage workers which read: "*We Buckle-Up Because Safety is Not Only for the Workplace.*" These flyers were attached to paychecks, first of all wage employees and two weeks later of all salaried employees. After the initial attempts to prompt salaried and wage employees separately, the same prompt was displayed plant-wide on table tents in the workers' cafeteria, in bathrooms above hand dryers, and on doors located in strategic places throughout the plant. (Interventions B, C, and D respectively in Figures 5.4.2 and 5.4.3).

Plant-wide safety slogan and safety celebrations

A celebration announcing the winner of a plant-wide safety slogan contest was scheduled near the Christmas holidays during the first year of the research. This celebration consisted of rewarding the winner with a $50 gift certificate at a store of their choice. When the research team learned of this event, Vince and Larry (the famous Crash Test Dummies) made an appearance at the plant during the scheduled celebration to distribute posters featuring two prominent Virginia Tech football players encouraging safety-belt use. To increase involvement, the event occurred between the first and second workshifts and included refreshments and photograph sessions with the dummies. (Intervention E in Figures 5.4.2 and 5.4.3)

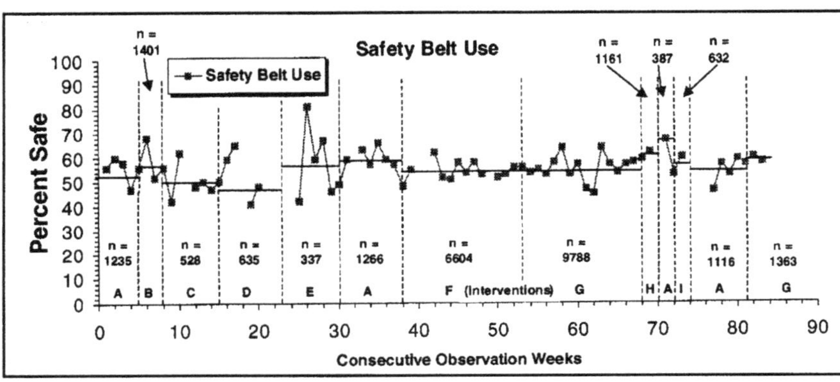

Figure 5.4.2 Percent Safety-belt Use over Duration of Study

Assigned goal setting

This intervention involved setting a specific, difficult, but attainable goal for plant-wide safety-belt use. Feedback about the current amount of safety-belt use among employees was not provided. Flyers displaying the plant logo, the winning safety slogan ("Bearings in Mind, Safety First!"), and the set goal were posted on table tents in the workers' cafeteria, in bathrooms above hand dryers, and on doors located in strategic places throughout the plant. These flyers were inspected weekly by research assistants and the plant safety manager for damage and loss. Damaged and missing flyers were replaced throughout the goal-setting intervention. The goal was set at 80 percent, approximately 25 percent above their current rate of belt use. It met the criteria for effective assigned goals in that it was specific and difficult but achievable. (Intervention F in Figures 5.4.2 and 5.4.3).

Goal setting plus feedback

This intervention was identical to the goal-setting intervention described above with one noteworthy exception—feedback in written and graphic form on the plant-wide safety-belt use from the previous week was included on the flyer. This feedback was updated weekly, and new flyers were posted during first shift each Monday. As a result of an employee suggestion and after an initial period of feedback, the flyers were printed in bright colors which were changed weekly to attract attention. (Intervention G in Figures 5.4.2 and 5.4.3).

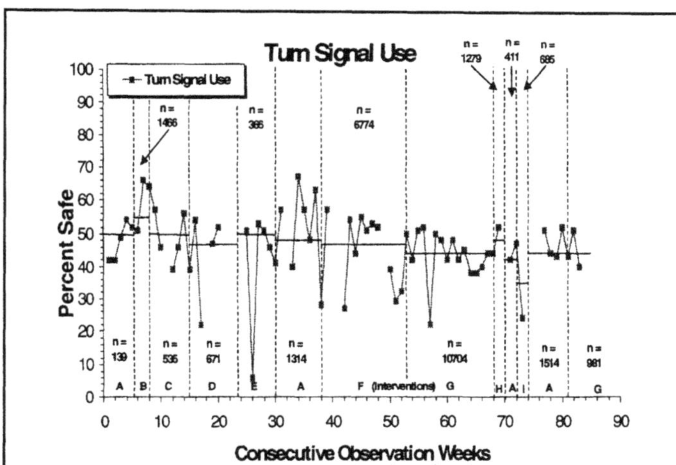

Figure 5.4.3 Percent Turn-signal Use over Duration of Study

Pledge-card commitment

This intervention consisted of written pledge cards distributed to all employees with their paychecks. The voluntary pledge encouraged safety-belt use for a two-week period. The pledge cards contained the company logo, the winning safety slogan, and a formal statement promising the use of safety belts throughout the pledge period. The pledge also contained a line for the employees to sign and a box they could check if they would allow their pledge to be posted in the plant. Pledges indicating posting (n = 200, 82 percent of the signed pledges) were laminated on a four foot by four foot poster-board and posted on the plant safety bulletin board after the second week of the pledge period. Blank pledge-cards were distributed next to the signed pledges and labeled "Second-Chance Pledges." This was done to encourage workers who had not initially pledged to do so. The initial pledge period was extended an additional two weeks. New pledges (n = 31) were posted (with approval) at the time they were signed. The flyers displaying the safety-belt use goal and performance feedback were posted as described above throughout the pledge period. (Interventions H and I respectively in Figures 5.4.2 and 5.4.3).

Return to baseline

Each return to baseline was marked by removal of all intervention materials. Trained research assistants continued their observations as usual during baseline periods. (Represented by A in Figures 5.4.2 and 5.4.3).

Safety-belt use. Figures 5.4.2 and 5.4.3 display the weekly percent of safety-belt and turn-signal use respectively for all employees throughout the research. Vertical lines indicate the introduction of a new intervention or return to baseline as described above. Horizontal lines represent the mean percent safe for each specific phase of the research. The number of observations per phase is indicated above each intervention condition.

A visual inspection of the data indicates some surprising results. Most noteworthy is the lack of significant increases in safety-belt use, the target behavior, as interventions became more intrusive. The lack of impact resulting from the addition of performance feedback to the initial goal-setting intervention is of particular interest. Previous research (Sulzer-Azaroff, Loafman, Merante, and Hlavacek 1990) suggests that timely feedback would have an additive effect with goal setting alone. The absence of this effect may have resulted from the failure of the initial goal-setting strategy to influence safety-belt use (54 percent, n = 6604 observations) significantly above the baseline mean (52 percent, n = 1235).

A closer look at the data indicates that the most powerful intervention to date was the employee selected safety-slogan contest and celebration. This intervention resulted in an immediate increase in safety-belt use for six weeks to 57 percent (n = 337) and a longer-term maintenance of 58 percent (n =1266) during a subsequent six-week return to baseline. These increases resulted after three attempts to prompt the use of safety belts resulted in a decrease of safety-belt use to 47 percent (n = 637) during the plant-wide prompt condition.

Intrinsic motivation theory would suggest that the beneficial impact of the safety-slogan contest resulted because it was developed "in-house" as opposed to the other efforts which were imposed by "outside" researchers. Similarly, in their conceptual review of the industrial safety literature, Boyce and Geller (2001) describe the powerful effects of using indigenous personnel on maintenance of safety-related behaviors. This intervention was also the only one associated with a monetary consequence for participation. Although not contingent on safety-belt use, employees participating in the contest may have felt the need to match their behavior to their thoughts about safety as a result of submitting a slogan. This result is consistent with the literature on cognitive dissonance (Festinger 1957).

It is particularly noteworthy that approximately 59 percent of the workers submitted a legitimate safety slogan. This level of participation matches the proportion of employees using their safety belts during this phase of the research. The modest increase in safety-belt use above baseline suggests that the contest may have influenced some safety-belt nonusers, but it is likely that workers who already used their safety belts submitted the majority of the slogans.

The only intervention to date that has had similar effects was the pledge-card commitment campaign. Two hundred and seventy-five (49 percent) of the workers signed the buckle-up pledge. This strategy increased safety-belt use to 61 percent (n = 1161) during the two-week pledge period. An increase to 67 percent (n = 387) was observed during a return to baseline, but upon posting the pledges publicly, use of safety belts dropped to 57 percent (n = 632), the level obtained during the safety-slogan contest.

Like the voluntary slogan contest, the voluntary pledge technique provided employees with choice and may have enhanced perceptions of internal control as predicted by intrinsic motivation theory. It is noteworthy that safety-belt use decreased when pledge cards were posted. It could be speculated that upon public posting (done by research personnel) some of the feelings of personal control and ownership established by the initial voluntary pledge was diminished, resulting in observed decreases of the target behavior. As of this writing, safety-belt use in a goal-setting plus feedback phase remains near 60 percent (n = 1363 observations), 10 percentage points higher than the initial baseline taken nearly two years ago.

Turn-signal use. Although not targeted, turn-signal use was observed to decrease throughout the course of the research. During the initial phases of the study, turn-signal use covaried positively with safety-belt use, suggesting support for response generalization (RG) theory (Bandura 1969; Carr 1988). This theory proposes that some behaviors (including safe-driving behaviors) are organized in a class of functionally-related responses. As a result, RG theory predicts that an increase in one safe behavior will be accompanied by increases in other related safe behaviors. Firm conclusions cannot be drawn, however, because of the failure of the interventions to impact the target behavior, safety-belt use.

A systematic relationship between turn-signal use and the use of safety belts was lost after the first return to baseline. Visual inspection of the data indicates that turn-signal use did not differ markedly throughout the course of research from the baseline mean of 50 percent (n = 1190) but trended downward to its current level of 46 percent (n = 1301). Thus, from the big picture, the trend upward in safety-belt use (10 percent over initial baseline) coupled with the trend downward in turn-signal use (4 percent below baseline) would support risk compensation (RC) theory (Peltzman 1975), which states that as people perceive themselves to be less at-risk in some situations, they will compensate by taking more risks in others.

Conclusions. Overall, these data support the MIL model. One could argue that the antecedent strategies used to motivate safety-belt use in the current research are all first-level interventions and are less powerful than Virginia's BUL. As such, it follows that the hard-core resistors, those who are not influenced by the BUL, would not be impacted by repeated applications of lower level interventions. As suggested by the MIL, the modest improvements seen in the use of safety belts may have been due in part to the involvement of belt users as intervention agents to get some non-users to buckle-up. Modeling appropriate behavior has been shown to increase the likelihood of others producing the desired response; when people change cultural norms regarding a behavior, make the costs versus benefits of a target behavior more salient, and demonstrate the ease at which appropriate behavior can be done, others are easily persuaded to incorporate appropriate behavior into their own actions.

Study 3: The Effects of Employee Involvement on Behavior-Based Safety

This research evaluated intervention procedures commonly taught by industrial safety trainers and consultants as part of a BB safety process. This research focused on the effects of involving employees in the development and implementation of BB safety interventions as opposed to the typical top-down approach to occupational safety.

Training BB safety. The BB safety process at a large manufacturing plant in southwest Virginia began by training volunteer safety facilitators from representative areas and then conducting facility-wide training sessions for the remaining employees (n = 550). The format of the training sessions was manipulated in order to investigate the impact of employee participation during BB safety training. The materials for all sessions were held constant; however, the safety trainers in the *Choice* condition (n = 230 on Shift 1) were instructed to ask questions and facilitate group discussion and involvement with workers. This included group exercises in which workers presented their suggestions for safety enhancement. In contrast, the same trainers in the *Assigned* conditions (n = 246 on Shifts 2 and 3) presented the safety material in a lecture format without asking questions or facilitating workers' input. Further, group exercises in this condition were "yoked" from the responses of the groups in the *Choice* condition (Shift 1).

Four-hour training sessions presented by two trainers were held for 12 *Choice* groups and 14 *Assigned* groups, which ranged in size from seven to 30 individuals. To assess the impact of the two training approaches, three variables were measured—the amount of verbal participation directed at the trainers, participants' reported satisfaction with the training, and the participants' retention of safety training information.

All verbal behaviors from the participants, including questions asked, questions answered, and reactive statements, were evaluated for each training session. As shown in Figure 5.4.4, a one-way analysis of variance (ANOVA) of verbal behavior for training format (*Choice* vs. *Assigned*) indicated that participants in the *Choice* condition emitted significantly more verbal behaviors (M = 6.48) than participants in the *Assigned* conditions (Shift 2: M = 1.09, Shift 3: M = 1.19, $p < .05$).

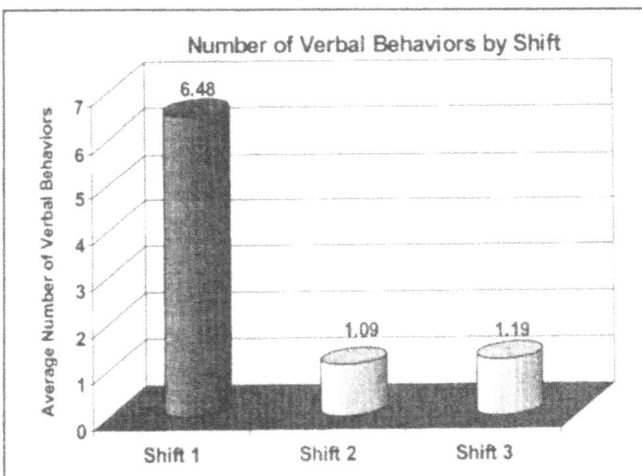

Figure 5.4.4 Number of Verbal Behaviors by Shift

Further analyses were conducted on each type of verbal response: questions answered, reactive statements, and questions asked. As shown in Figure 5.4.5, a one-way ANOVA of questions answered per shift indicated that participants in the *Choice* condition answered significantly more questions (M = 4.02) than participants in the *Assigned* condition (Shift 2: M = .35, Shift 3: M = .38, $p < .05$). As illustrated in Figure 5.4.6, a one-way ANOVA of reactive statements per Shift indicated that

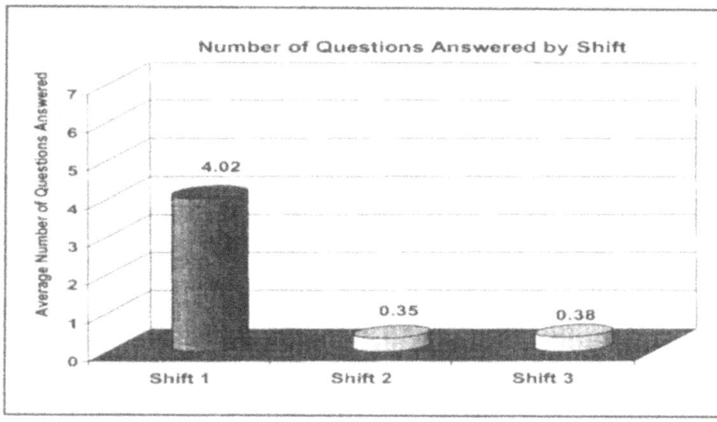

Figure 5.4.5 Number of Questions Answered by Shift

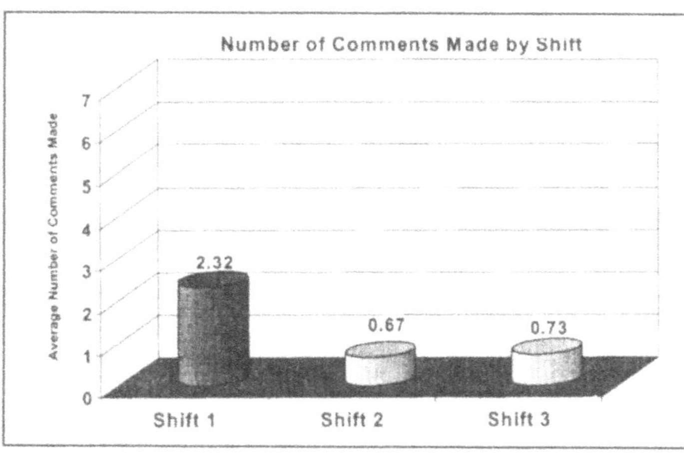

Figure 5.4.6 Number of Comments Made by Shift

participants in the *Choice* condition made significantly more reactive statements (M = 2.32) than participants in the *Assigned* conditions (Shift 2: M = .67, Shift 3: M = .73, $p < .05$).

Additionally, a one-way ANOVA of questions asked per shift indicated no significant difference in the average number of questions asked by participants in the *Choice* (M = .14) and *Assigned* conditions (Shift 2: M

= .08, Shift 3: M = .08, $p > .05$). Overall, these findings suggest participants in the *Choice* condition were significantly more involved in the training process than employees in the *Assigned* condition, although there were no differences in number of questions from participants.

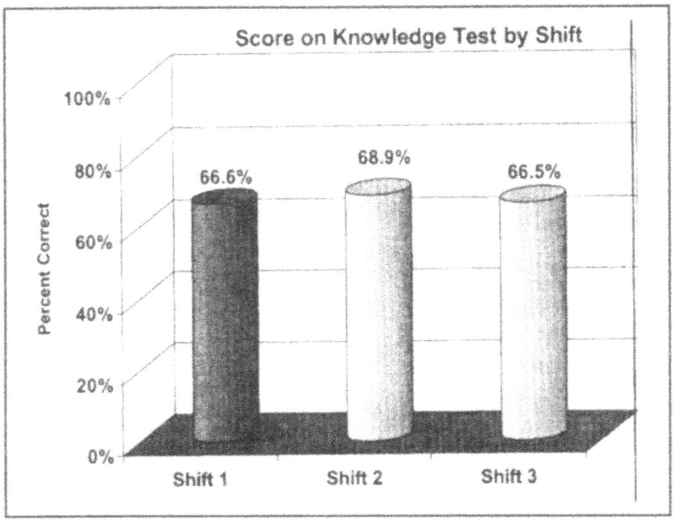

Figure 5.4.7 Score on Knowledge Test by Shift

Participants' responses to the following post-training questionnaires were also evaluated: a) an 18-item knowledge test, b) a five-item measure of involvement, and c) a one-item measure of satisfaction with the training. As illustrated in Figure 5.4.7, a one-way ANOVA by shift indicated no significant differences between the knowledge scores of participants in the *Choice* (M = 66.6) and *Assigned* conditions (Shift 2: M = 68.9, Shift 3: M = 66.5, $p > .05$).

Analyses of participants' perceptions of involvement using a one-way ANOVA by shift indicated no significant difference between the perceived involvement of participants in the *Choice* (M = 3.44) and *Assigned* conditions (Shift 2: M = 3.58, Shift 3: M = 3.70, $p > .05$). Finally, analysis of participants' self-reports of satisfaction with the training process using a one-way ANOVA by shift indicated that participants in the *Choice* condition (Shift 1) were more satisfied with the training process (M =

2.56; 1 = *strongly liked*, 7 = *strongly disliked*) than Shift 3 participants in the *Assigned* condition (M = 3.05, $p < .05$). However, the satisfaction of Shift 2 participants in the *Assigned* condition (M = 2.80) was not significantly different from the satisfaction of participants in the *Choice* condition.

These results were counter to our hypothesis that participative training would be more effective and appreciated than nonparticipative training. The lack of significant differences between conditions regarding information retention, satisfaction, and involvement suggest that group participation may not be directly measurable by verbal behavior alone, the nature of the training material itself may have involved workers cognitively, or *Choice* training simply may not be more effective than *Assigned* training in terms of information retention and personal satisfaction.

Implementing BB safety. Many involvement manipulations were made to give the Shift 1 facilitators (n = 8) as many opportunities to make key decisions in the BB safety process as possible. For example, the Shift 1 safety facilitators selected: a) the initial safety-related behavior to observe plant-wide (hearing protection was chosen), b) the design and location of group feedback prompts, c) the initial plant-wide safety intervention, d) the design of a safety slogan contest, and e) the safety-related behaviors to observe in individual work areas. The choices made by the Shift 1 safety facilitators were assigned (or yoked) to the Shift 2 safety facilitators (n = 6), thus limiting Shift 2 input into the BB safety process. This extended the *Choice* vs. *Assigned* manipulations implemented during training.

For nine weeks, the safety facilitators of Shift 1 (n = 230) and Shift 2 (n = 210) made behavioral observations on hearing protection. The data were graphed and posted on a safety bulletin board located at the entrance to the production areas of the plant. The dependent measure used to assess facilitator involvement was number of observations taken.

Results and discussion. Over the nine-week period, the Shift 1 facilitators (n = 8) made significantly more observations per facilitator per week (M = 10.57, SD = 3.39) than the Shift 2 facilitators [(n = 6, M = 6.13, SD = 3.76, t (16) = 3.1, $p < .05$]. Also, the Shift 1 facilitators observed a significantly greater [t (8) = 2.6, $p < .05$] percentage of their shift than did the Shift 2 facilitators (37 percent vs. 23 percent, respectively). Shift 1 facilitators also took significantly more observations per person each week than the Shift 2 facilitators did [t (16) = 3.05, $p < .05$].

These findings show promise concerning the effects of giving employees choices regarding the implementation of a BB safety process. In addition, a time series tabulation of this plant's injury statistics shows the bottom-line beneficial impact of the BB safety process across all three shifts (see Figure 5.4.8). Lost work days due to injuries decreased markedly after the BB safety training and observation/feedback programs were put in place (10.9 lost work days per month prior to BB safety versus 1.4 lost days per month after the initiation of BB safety).

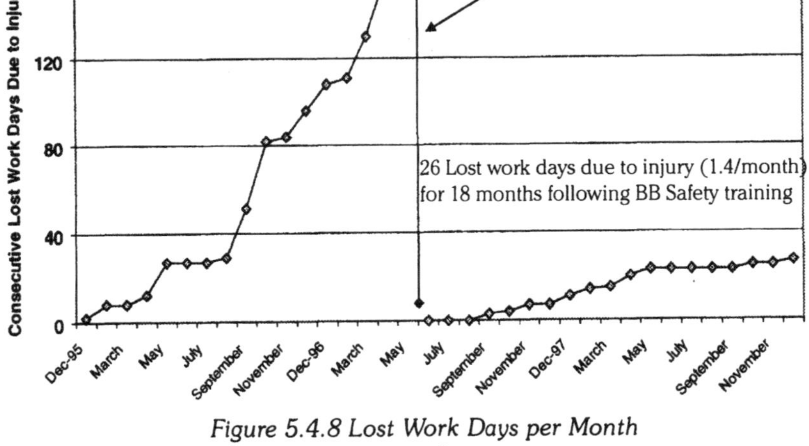

Figure 5.4.8 Lost Work Days per Month Due to Injury

Study 4: Effects of Self-Observation on Safe Driving Practices

Self-management (Mahoney 1971, 1972) is an improvement process whereby individuals change their own behavior in a goal-directed fashion by manipulating behavioral antecedents, observing and recording specific target behaviors, and self-administering rewards for personal achievements (Kazdin 1994; Watson and Tharp 1997). The practical benefits of self-management processes have been demonstrated in numerous clinical settings, including the reduction of alcohol consumption (Sobell and Sobell 1995), weight control (Baker and Kirschenbaum 1993), and smoking cessation (Curry, Marlatt, and Gordon 1987; Shiffman 1984).

Social cognitive theory (Bandura 1991) suggests that self-observation is one of the primary components of the self-regulation of behavior change. According to Bandura (1991), critical elements in the application of self-observation include specific self-set goals and a belief in the "value" of behavior change. Research also indicates that perceptions of choice and increased self-awareness of specific behaviors can focus individuals' attention on behavior improvement (Binswanger 1991). In addition, the impact and maintenance of a self-management effort is determined by an individual's ability to sustain focus on the consequences motivating behavior change (such as avoiding injury).

Need for Self-Management in Safety

While a variety of procedures have been developed to improve safety performance (reward systems, feedback, etc.), most rely on the interpersonal observation of behavior by supervisors or co-workers. While these techniques have been effective for increasing safety, there is a need for effective behavior-change techniques among solitary workers and for infrequent safety-related behaviors (such as lockout-tagout), which are not likely to be observed in most behavior-based coaching systems.

Self-management strategies. Research indicates that five self-management procedures can facilitate behavior change: a) activator management, b) social support, c) goal setting, d) self-observation and self-recording, and e) self-rewards.

Activator management involves identifying environment, behavior, and person factors that precede the occurrence of safe and at-risk behaviors. Strategies are then employed to eliminate activators that precede at-risk behaviors and add activators such as reminder messages to increase the probability of safe behaviors.

Creating a supportive environment can also improve self-management success (Stuart 1967). Sources of social support can include supervisors, co-workers, friends, and family who can encourage the targeted behaviors (Heinzelman and Bagley 1970; Lawson and Rhodes 1981; Moss and Arend 1977). Social support can include positive feedback and encouragement for personal successes (Perri and Richards 1977), particularly when self-observation data is displayed so others can see accomplishments (Taylor, Neter, and Wayment 1995). Additionally, specific behavioral goals should be set high yet be achievable, should specify expectations for improvement, and should include the tracking of progress (Geller 1996, 1998, 2001).

Regularly engaging in the self-observation and self-recording of specific behaviors is also a key component of self-management (Bagozzi 1992; Kazdin 1994). Self-observation involves observing one's own target behavior soon after it occurs as well as noting related environmental events that precede and follow the target behavior (Watson and Tharp 1997). Recording the occurrence of safe and at-risk behaviors provides an objective record of current safety performance; charting progress toward reaching specific behavior-based safety goals provides feedback on achievements and identifies areas for improvement.

Finally, using self-rewards to recognize personal achievement motivates continued self-management. While self-management techniques are new to the field of occupational safety, studies in clinical settings indicate that individuals who use self-rewards are more likely to sustain attempts to change their behavior and maintain behavior change (Watson and Tharp 1997). However, the efficacy and "social validity" of using self-rewards to improve personal safety in the workplace needs to be established.

Implementing safety self-management. Self-management is typically a six-step process:

1) establish behavioral baseline using self-observation, including the

identification of antecedents and consequences associated with the occurrence and non-occurrence of the target behaviors (Cormier and Cormier 1991);

2) identify target behaviors (Cervone and Scott 1995);
3) select a self-management strategy to promote desired behavior change and chart progress;
4) select a goal which is specific, motivational, attainable, recordable, and trackable (Geller 1996);
5) self-observe and self-record target behaviors to measure progress toward the goal (Kirschenbaum et al. 1982); and
6) administer self-rewards that are accessible, individualized, valued, varied, and follow the targeted behavior as immediately as possible (Cormier and Cormier 1991).

Identification of target behaviors can be accomplished by reviewing injury and crash data, analyzing "near-misses," and most importantly, evaluating previous behavioral observations to identify individual problem areas. Since self-management is a self-directed process, the most beneficial approach to identifying target behaviors is for individuals to survey their own work practices to determine personal at-risk behaviors.

Self-observation of behavior has been shown to be effective at identifying and changing behavior within clinical psychology. To date there has been little research on the use of self-observation to improve individual safety-related performance in work settings, and more specifically, driving behavior. This study investigated the viability of using self-observation technology to increase the safe driving of short-haul truck drivers.

Method

A sub-sample of 12 male drivers from Site A (self-observation) and 11 male drivers from Site B (control) were randomly selected to evaluate the self-observation intervention. During the week prior to implementing the self-observation intervention, trained research assistants observed the driving behaviors of participants while riding in the passenger seat of their delivery truck. All employees at Site A (n = 30) were then asked to complete a self-observation checklist each workday for a two-week period.

While on their sales routes, participants recorded the following behaviors when arriving at each of their accounts: a) use of running/head lights, b) safety-belt use, and c) walking around the vehicle before leaving account. At the end of the day, participants estimated their percent safe on the following behaviors: a) following distance, b) lane change, c) traffic lights, d) complete stops at stop signs, and e) vehicle speed.

Self-observation forms were collected and graphed on a daily basis. These graphs were posted in a break-room where employees gathered before and after their routes each day. The self-observation data that were graphed each day are presented in Figure 5.4.9.

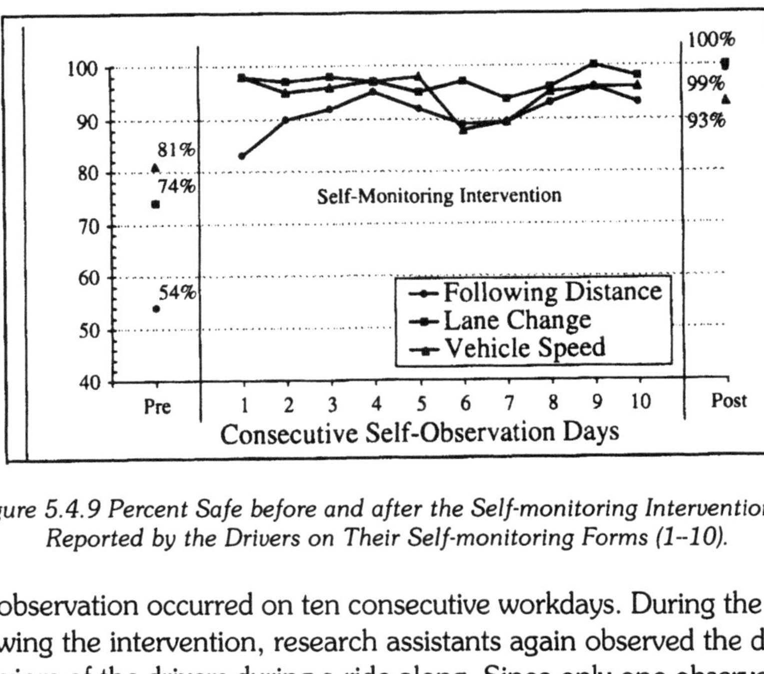

Figure 5.4.9 Percent Safe before and after the Self-monitoring Intervention as Reported by the Drivers on Their Self-monitoring Forms (1–10).

Self-observation occurred on ten consecutive workdays. During the week following the intervention, research assistants again observed the driving behaviors of the drivers during a ride-along. Since only one observer was permitted to ride with each participant, it was necessary to assess reliability (inter-observer agreement) of the driving measures in an alternative setting using the same observers. The inter-observation reliabilities for all behaviors were above .88.

Results

A percent safe score was calculated for each participant's driving behaviors by dividing the number of safe behaviors by the number of safe and at-risk behaviors. A total percent safe score was also calculated based on the total number of safe behaviors (collapsed across all driving behaviors). An ARCSIN transformation was calculated on all percent safe scores. A 2 Time (pre- vs. post-intervention) x 2 Group (self-observation vs. control) mixed model ANOVA was calculated for each dependent variable. Results indicated a significant Time x Group interaction for overall percent safe, $p < .05$. Follow-up analyses indicated that overall safety did not improve in the control group (from 71.5 percent to 75.4 percent), $p > .05$. In contrast, there was a significant increase in the overall safety of the self-observation group (from 66.8 percent to 87.8 percent), $p < .05$.

Based on this finding, each driving behavior was examined separately. Sample sizes varied for each analysis because there was no opportunity to observe some behaviors for some drivers on either the first or second ride-along. Thus, n-sizes are reported for each analysis. As shown in Figure 5.4.9, results indicated a significant Time x Group interaction for lane changes, $p < .01$, and following distance, $p < .01$. Follow-up analyses indicated that safe lane changes and following distances *did not* improve in the control group (from 94.3 percent to 81.9 percent for lane changes, $n = 5$; from 64.4 percent to 75.0 percent for following distance, $n = 5$), $ps > .05$. In contrast there was a significant increase for both lane changes and following distance in the Self-Observation group (from 0.0 percent to 100.0 percent for lane changes, $n = 5$; from 48.1 percent to 96.9 percent for following distance, $n = 7$), $ps < .05$. No other effects were significant for the remaining driving behaviors, but they were in the expected and desired direction.

Discussion. Self-observation may be one of the most effective behavior change techniques in situations where social support is not available. In fact, it is the only technique that can be applied by solitary workers without extensive support from supervisors and co-workers. Self-management is non-invasive, simple, and practical. If self-management activities can be integrated with other job activities, safety practitioners will have an effective tool for improving safety-related behaviors that occur when there is little or no opportunity for interpersonal BB observation and feedback (as with low base-rate behaviors like lockout-tagout, and with workers who work alone). The successful integration of self-management behaviors into daily work tasks is invaluable for the maintenance of safe work practices.

Summary

Our comprehensive review of the research literature turned up convincing evidence that a BB approach to occupational safety can be extremely effective at reducing at-risk behavior and workplace injuries. Key ingredients of every effective BB intervention were observation and feedback. In other words, participants need a mechanism for learning what to do differently in order to prevent the possibility of personal injury. Little is known, however, about the best way to implement a BB observation and feedback process, nor has there been any systematic study of the organizational factors needed to institutionalize a BB process for improving safety-related work practices. This defines the mission of our research reviewed here. Four experiments described here enable the following conclusions:

- Although appreciation for BB safety is on the rise, BB safety principles are rarely applied in the development of a company-wide accountability system. In other words, outcome numbers (such as total recordable injury rate) rather than process numbers (such as quality and quantity of participation) are used most often to evaluate safety performance, even on the shop floor.
- Marked reductions in workplace injuries occurred at two study sites immediately after a BB observation and feedback process was put in place. Interviews and focus groups at sites with varying success implementing BB safety indicated the importance of an employee steering committee, visible management support, one or more champions to drive the process, and personal perceptions of program ownership and interpersonal trust. The most successful organizations do it their own way, and the procedures and support systems vary dramatically across cultures.
- A systematic evaluation of successive interventions to improve safety-belt use supported a key assumption of a multiple intervention level (MIL) model: People who are uninfluenced by a BB intervention presented at one level of intrusiveness will not be influenced by

subsequent interventions with the same degree of intrusiveness. In other words, repeating the same kind of intervention process will not influence the critical resistors unless a more intense and intrusive procedure is added.

- Self-management techniques developed in clinical settings seem to be a viable behavior change tool for workers in a solitary work environment.

We realize we have only cracked the surface in determining the critical success factors of BB safety. We have only begun an important discovery process. Clearly, much more research is needed in this human factors domain of industrial safety and health.

References

Baer, D. M., M. M. Wolf, and T. R. Risley. 1968. Some current dimensions of applied behavior analysis. *Journal of Applied Behavior Analysis*, 1:91–97.

———. 1987. Some still-current dimensions of applied behavior analysis. *Journal of Applied Behavior Analysis*, 20:313–327.

Bagozzi, R. P. 1992. The self-regulation of attitudes, intentions and behaviors. *Social Psychology Quarterly*, 55(2):178–204.

Baker, R. C., and D. S. Kirschbaum. 1993. Self-monitoring may be necessary for successful weight control. *Behavior Therapy*, 24:377–394.

Bandura, A. 1969. *Principles of behavior modification*. New York: Holt, Rinehart & Winston.

———. 1977. Self-efficacy: Toward a unifying theory of behavioral change. *Psychological Review*, 84:191–215.

———. 1991. Social cognitive theory of self-regulation. *Organizational Behavior and Human Decision Processes*, 50:248–287.

———. 1997. *Self-efficacy: The exercise of control*. New York: W. H. Freeman and Company.

Binswanger, H. 1991. Volition as cognitive self-regulation. *Organizational Behavior and Human Decision Processes*, 50:154–178.

Boyce T. E., and E. S. Geller. 2001. Applied behavior analysis and occupational safety: The challenge of response maintenance. *Journal of Organizational Behavior Management.*

Carr, E. G. 1988. Functional equivalence as a mechanism of response generalization. In *Generalization and maintenance.* R. H. Horner, G. Dunlap, and R. L. Koegel, eds. Baltimore: Paul H. Brooks. 221–241.

Cervone, D., and W. D. Scott. 1995. Self-efficacy theory of behavioral change. Foundations, conceptual issues and therapeutic implications. In *Theories of behavior therapy: Exploring behavior change.* eds. W. O'Donohue and L. Krasner. Washington, D.C.: American Psychological Association.

Cormier, W. H., and L. S. Cormier. 1991. *Interviewing strategies for helpers: Fundamental skills and cognitive behavioral interventions.* Pacific Grove, CA: Brooks/Cole.

Curry, S. G., G. A. Marlatt, and J. R. Gordon. 1987. Abstinence violation effect: Validation of an attributional construct with smoking cessation. *Journal of Consulting and Clinical Psychology,* 55:145–149.

Daniels, A. C. 1989. *Performance management.* Tucker, GA: Performance Management Publications.

Festinger, L. 1957. *A theory of cognitive dissonance.* Stanford CA: Stanford University Press.

Geller, E. S. 1996. *The psychology of safety: How to improve behaviors and attitudes on the job.* Radnor, PA: Chilton Book Company.

———. 1998. *Understanding behavior-based safety: Step-by-step methods to improve your workplace.* 2nd ed. Neenah, WI: J. J. Keller & Associates, Inc.

———. 2001. The psychology of safety handbook. Boca Raton, FL: CRC Press.

Geller, E. S., L. Davis, and K. Spicer. 1983. Industry-based incentives for promoting seat belt use: Differential impact on white-collar versus blue-collar employees. *Journal of Organizational Behavior Management,* 5:17–29.

Geller, E. S., R. A. Winett, and P. B. Everett. 1982. *Preserving the environment: New strategies for behavior change*. New York: Pergamon Press.

Geller, E. S., T. D. Berry, T. D. Ludwig, R. E. Evans, M. R. Gilmore, and S. W. Clarke. 1990. A conceptual framework for developing and evaluating behavior change interventions for injury control. *Health Education Research: Theory and Practice*, 5(2):125–137.

Heinzelman, F., and R. W. Bagley. 1970. Response to physical activity programs and their effects on health behavior. *Public Health Reports*, 85:905–911.

Katz, E., and P. E. Lazarfeld. 1955. *Personal influence: The part played by people in the flow of mass communication*. Glencoe, IL: Free Press.

Kazdin, A. E. 1994. *Behavior modification in applied settings*. Pacific Grove, CA: Brooks/Cole.

Kirschenbaum, D. S., A. M. Ordman, A. J. Tomarken, and R. Holtzbauer. 1982. Effects of differential self-monitoring and level of mastery of sports performance: Brain power bowling. *Cognitive Therapy and Research*, 6:335–342.

Lawson, D. M., and E. C. Rhodes. 1981. *Behavioral self-control and maintenance of aerobic exercise: A retrospective study of self-initiated attempts to improve physical fitness*. Paper presented at the meeting of the Association for the Advancement of Behavior Therapy (November). Toronto.

Ludwig, T. D., and E. S. Geller. 1991. Improving the driving practices of Pizza deliverers: Response generalization and moderating effects of driving history. *Journal of Applied Behavior Analysis*, 24:31–44.

Mahoney, M. J. 1971. The self-management of covert behavior: A case study. *Behavior Therapy*, 2:575–578.

———. 1972. Research issues in self-management. *Behavior Therapy*, 3:45–63.

Moss, M. K., and R. A. Arend. 1977. Self-directed contact desensitization. *Journal of Consulting and Clinical Psychology*, 45:730–738.

Novak, L. 1995 (May). National Highway Traffic Safety Administration. Personal communication with E. Scott Geller.

Peltzman, S. 1975. The effects of automobile safety regulation. *Journal of Political Economics*, 83:677–725.

Perri, M. G., and C. S. Richards. 1977. An investigation of naturally occurring episodes of self-controlled behaviors. *Journal of Counseling Psychology*, 24:178–183.

Rudd, J. R., and E. S. Geller. 1985. A university-based incentive program to increase safety-belt use: Toward cost-effective institutionalization. *Journal of Applied Behavior Analysis*, 18:215–226.

Shiffman, S. 1984. Coping with temptations to smoke. *Journal of Consulting and Clinical Psychology,* 52:261–267.

Sobell, L. S., and M. B. Sobell. 1995. *Guided self-change case study: Lisa*. Toronto, Canada: Addiction Research Council.

Stuart, R. B. 1967. Behavioral control of overeating. *Behaviour Research and Therapy*, 5:357–365.

Sulzer-Azaroff, B., and M. C. De Santamaria. 1980. Industrial safety hazard reduction through performance feedback. *Journal of Applied Behavior Analysis*, 13:287–295.

Sulzer-Azaroff, B., B. Loafman, R. J. Merante, and A. C. Hlavacek. 1990. Improving occupational safety in a large industrial plant: A systematic replication. *Journal of Organizational Behavior Management*, 11(1):99–120.

Taylor, S. E., E. Neter, and H. A. Wayment. 1995. Self-evaluation processes. *Personality and Social Psychology Bulletin,* 21:1278–1287.

Watson, D. L., and R. G. Tharp. 1997. *Self-directed behavior: Self-modification for personal adjustment.* 6th ed. Monterey, CA: Brooks/Cole.

Notes

[1] Originally appeared in *Proceedings of the 38th ASSE Annual Professional Development Conference and Exposition.* (Des Plaines, IL: American Association of Safety Engineers, 1999):375–381.

[2] Originally appeared in *Professional Safety* 45:3 (2000):29–32.

[3] Originally appeared in *Proceedings of the 37th Annual Professional Development Conference and Exposition.* (Des Plaines, IL: American Society of Safety Engineers, 1998):537–559.

6

Safety Leadership

The same qualities of great athletic coaches apply to great safety leaders: vision, charisma, high standards, compassion, communication skills and much more. Safety leadership goes beyond getting employees to comply with safety rules. Great safety leaders engender a sense of responsibility beyond simple accountability so employees not only follow rules, but they also make corrections themselves when possible, coach co-workers about safe and at-risk behaviors, and generally go beyond what is required to make safety improvements for themselves and others.

Effective leaders also cultivate interpersonal trust, a key ingredient to achieving a Total Safety Culture. This chapter defines the characteristics of the most effective safety leaders and offers strategies for enhancing leadership throughout a work culture.

Managers hold people accountable, but leaders inspire people to feel responsible. The first paper distinguishes between safety management and safety leadership and gives ten guidelines for cultivating more leadership within a workforce. The second and third papers describe what leaders can do to sustain employee involvement in occupational safety.

The third paper focuses on a most critical ingredient of a successful BBS process—interpersonal trust. A simple device for assessing the trust level in a work culture is provided as well as practical ways to increase interpersonal trust throughout an organization. Safety leaders bring out the best in people when they practice the various strategies described in this chapter for cultivating interpersonal trust, personal responsibility, and actively caring for safety and health. While the focus is on occupational safety, the leadership principles provided here are relevant for every aspect of life, from the workplace to one's family and every life event in between.

Safety Management Is Not Enough: Ten Leadership Qualities for a Total Safety Culture[1]

E. Scott Geller

Safety professionals use the term "management" more than they use the term "leadership." This is reflected in the title of the ASSE Professional Development Conference held in Las Vegas last February. It was titled "Best Practices in Safety Management" (see ASSE 1999). However, in his keynote address for that conference, John Nance referred to "leadership" rather than "management." He discussed the need for safety leaders to overcome the organizational and cultural barriers that prevent open discussion about environmental hazards and at-risk behavior. In his words, "Good leaders invigorate the flow of interpersonal communication" and "orchestrate others to achieve synergy."

Later at this ASSE conference Tom Krause defined leaders as people who "inspire people to want to do something," as opposed to managers who "hold people accountable for doing something." This distinction was perfectly consistent with the topic of my address at this conference entitled "How to develop personal accountability for safety." The theme of my presentation was that a Total Safety Culture requires people to take personal accountability or responsibility for safety. I discussed specific ways to help people feel more responsible for safety (Geller 1999, 2001a).

As a follow-up to the inspirational talks delivered at the 1999 ASSE Management Conference, I would like to consider some specific differences between safety management and safety leadership. This is not to belittle management or even to suggest we need less management. But I do suggest we need more leadership in safety, which is not the same as management. The ten qualities I suggest for effective safety leadership are relevant for everyone in a work culture, including managers.

1. Leaders Focus on Process

Starting in kindergarten the focus of our schooling seems to be on the final result rather than on the process by which an outcome is achieved. Students in my university classes seem obsessed with their grades and seem to lose sight of the important purpose of their "higher" education. Rather than focusing on critical thinking and the problem-solving processes related to a particular theory or research finding, students memorize the facts and formulae needed to perform well on an exam. And rather than helping students think creatively about a set of circumstances and explore the many possibilities to a problem, many teachers just lecture the facts. Final grades are given and received. The critical outcome is attained.

Managers are typically held accountable for outcome numbers, and they in turn use these outcome numbers to motivate others. In safety, the outcome numbers are based on the relatively rare occurrence of an injury. Numbers are reactive, they reflect failure, and they are not diagnostic for prevention.

Safety leaders hold people accountable for accomplishing proactive process activities that can prevent injuries. When people see improvement in the process numbers, they are reinforced for their efforts and develop a sense of personal responsibility for continued contributions and never-ending improvement.

A process orientation asks "How did they do it?" instead of "What did they do?" It is not "What is our total recordable injury rate?" but "What are the steps needed to keep people safe every day?" It is not about failing to achieve an injury-free workplace; it is about deriving corrective action plans for ongoing reports of environmental/behavioral audits, near-misses, first-aid cases, property-damage incidents, and occurrences of recordable and lost-time injuries.

When an outcome such as a reduced injury rate is viewed as an achievement of successive steps or "small wins," people's sense of personal control is enhanced. They see the outcome as hard-won through their involvement in a process. And by continuing to increase people's attention to the process, injuries can be reduced even more. So while managers track outcomes, leaders enable and reinforce discussions of the ongoing processes needed to prevent injuries. Conversations about process help to keep people aware of what they need to continue doing in order to keep themselves and others safe.

2. Leaders Educate

In industry, "training" is a more common term than "education." This reflects the concern that employees know exactly what to do in order to complete a particular task effectively and safely. With a "training" mindset, however, managers can come across as demanding a certain activity because "I said so" rather than because "it's the best way to do it."

Education involves an explanation of the principles or the rationale behind a particular set of procedures. Education enables the listener to understand why a certain protocol needs to be followed. With the proper education, people can develop responsibility for an action plan rather than doing something a certain way because a manager is holding them accountable.

Pertinent instruction of principles can also inspire creative customization and ownership. In other words, when leaders offer rationale and examples rather than policy and directives, individuals or work teams can select procedures that best fit their situation. And in the process of refining a set of procedures, people assume ownership and follow through from a self-directed or responsibility perspective.

3. Leaders Use Conditional Statements

The style of an instructor can encourage or inhibit creativity and ownership. When facts are presented unconditionally as an absolute truth, alternative ways of thinking about something are stifled. Such single-minded teaching and learning might be more efficient and involve the least amount of discussion and interpersonal conflict, but the cost can be loss of involvement, personal responsibility, and resourcefulness.

Dr. Ellen J. Langer (1989) studied the impact of unconditional versus conditional instruction by introducing a collection of objects in an ordinary unconditional way to one group: "This is a hair dryer," "This is an extension cord," and "This is a dog's chew toy." For a second group, the objects were introduced conditionally with the extra phrase "could be." So instead of "This is a hair dryer," the subjects heard "This could be a hair dryer."

After the objects were introduced, the subjects were asked to complete some survey forms. Then, while the forms were being completed, the experimenter announced that the study could not continue because the wrong instructions had been given and no spare survey forms were available. Several subjects suggested, however, that the rubber chew toy could be used as an eraser to correct the flawed forms. Interestingly, this novel way of using the chew toy was only suggested by subjects in the conditional group, those who were told "This could be a dog's chew toy."

Consider the differential impact of discussing safety rules as unconditional mandates versus discussing safety principles that imply certain guidelines for action but can be operationalized according to relevant circumstances. It is obvious which approach will stimulate more resourcefulness and ownership among both individuals and work teams. The uncertainty of the second approach allows for inventive adaptation of a safety principle to a particular work situation. The process of customizing a safety principle for certain work circumstances develops a sense of ownership and a commitment to comply with a personalized set of operations.

4. Leaders Listen First

Under pressure to get a job done, managers often speak first and then listen to concerns or complaints. This is a reasonable strategy for efficient action. After all, the challenge of management is to make things happen according to an established plan, and this requires specific directives and a mechanism for motivating compliance. After describing an action plan and accountability system, managers answer a lot of questions from workers who want to make sure they will do the right thing.

In contrast, leaders take time to learn another person's perspective before offering direction, advice, or support. Active listening is key to diagnosing a situation before promoting change or continuous improvement. This is not the most efficient approach to getting a job done. It requires patience and a communication approach that asks many questions before giving advice. In this way an individual or work team can personalize an action plan or process for achieving a particular outcome.

5. Leaders Promote Ownership

When the development of an action plan involves the people expected to carry out that plan, ownership for both the process and the outcome is likely to develop. In other words, when leaders give a reasonable rationale for a desired outcome and then offer opportunities for others to customize methods for achieving that outcome, they facilitate a special kind of motivation. This motivation comes from inside people and is commonly referred to as internal or self-directed motivation (Geller 1996). In this state, people participate because they want to, not because they have to.

When managers direct by edict for efficient transfer of an action plan, they might get compliance, but they might also stifle self-directed motivation. Behaviors performed to comply with a prescribed standard, policy, or mandate are other-directed. They are accomplished to satisfy someone else, and they are likely to cease when compliance cannot be monitored. This happens, for example, when personal protective equipment is used at work but not at home for similar or even riskier tasks.

6. Leaders Encourage Choice

The advantages of giving people choices are well documented (Geller 2001a; Langer 1989; Steiner 1970). Having more opportunities for personal choice increases both motivation and a sense of personal control. The greater people's personal control, the more likely they are to participate in efforts to improve their own health and safety, as well as that of others. In other words, people with personal control are more likely to actively care.

Ellen Langer and her colleagues (1989) have conducted highly regarded research to support the value of giving people opportunities to make choices. In one seminal research project, Langer and Rodin (1976, 1977) gave one group of elderly residents in a Connecticut nursing home the opportunity to care for a houseplant and to make a number of minor decisions about their daily routines. A year and a half later, these residents were more cheerful, alert, and active than a similar group of residents who were not given these choices. These residents had also been given a houseplant, but the nurses had taken care of it. The most remarkable result of this choice manipulation was that after a year and a half, less than half as many of the residents in the choice group had died as had those in the other group.

Langer (1989) suggests we can become more motivated and aware of personal control by becoming more mindful of our numerous choices during ordinary activities. We choose, for example, when to get up in the morning, what to wear, what to eat for breakfast, and how to travel to work. The routine might seem very familiar and similar every morning, and therefore it is easy to become mindless about personal control. But consider that there are many alternatives for each set of behaviors, and you do willingly select an option.

Besides actually giving people more choices, leaders can help people become more aware of how they already shape their days. This can increase their perception of personal control and thus their motivation. Helping people see their options can also give them pause to consider other alternatives that could be more useful, productive, or effective. When leaders help others become more observant of their everyday choices, they not only increase people's awareness of personal control, but they also set the occasion for more effective decisions.

7. Leaders Set Expectations

Most safety-related behavior starts as other-directed, meaning it is performed because someone asked for it (Watson and Tharp 1997). The important issue is whether the behavior remains other-directed or advances to self-directed (Geller and Clarke 1999). This depends to some extent on the method of asking. A behavioral request that comes across as a mandate or an unconditional statement is likely to stay other-directed. This is often the management approach to safety, as illustrated by regulatory compliance issues and the common slogan "Safety is a condition of employment."

Leadership can facilitate a shift from other-directed to self-directed motivation by initiating a process or action plan with expectations rather than mandates. What is the difference? Both approaches specify desirable outcomes and establish the need for certain behaviors as process activities. However, expectations imply choice. While a certain outcome is anticipated, there is room for individual and group decision-making regarding procedures and methods. When people realize what is expected of them but perceive some personal control in how to reach specific goals, they are more likely to own the process and transition from an other-directed to a self-directed mindset.

8. Leaders Are Confident But Uncertain

In her provocative book *Mindfulness*, Ellen Langer (1989) proposes that a leadership quality conducive to promoting innovation and initiative is displaying a degree of uncertainty. She suggests that leaders should show confidence that a particular job will get done but without being sure of the best way to do it. This allows employees to be alert, innovative, and self-motivated. They feel their involvement is a necessary part of the process. They feel important (Carnegie 1936).

People are less likely to hide mistakes when they work for a confident but uncertain leader. It is okay to be imperfect if the boss is not sure exactly how the job should be done. In this context, workers are more willing to suggest ways to improve a process. In Langer's words, "admission of uncertainty leads to a search for more information, and with more information there may be more options" (p. 143).

To test her theory about confident but uncertain leadership, Langer assessed the general level of confidence among supervisors at a particular company and then asked them how many of their daily decisions have absolutely correct answers. In addition, the employees of the organization completed surveys that evaluated their work relationships with their supervisors. Results showed that the confident but relatively uncertain supervisors were viewed as allowing more independent judgment and innovative action.

Relating these findings to industrial safety, a leader should show confident expectation that appropriate precautions will be taken to prevent injuries. However, the safety leader should not pretend to know exactly how the injury-free job should be accomplished. After all, the employees of a work team know what hazards need to be eliminated or avoided and what safety-related behaviors need to be improved. These individuals are also most likely to be effective at encouraging safe behaviors and discouraging at-risk behaviors at their job sites. Thus, it seems a leadership quality of "confident uncertainty" can be instrumental in empowering workers to go beyond the call of duty for safety and health.

Langer also believes that certainty and familiarity contribute to fatigue and burnout. When the job is seen as a mindless routine, energy and enthusiasm wane. Workers lose interest and a sense of choice and personal control. It is easy to get lulled into a false sense of security that doing things the way they have always been done will continue to work. This "certainty" mindset can not only hinder creative innovation, but it can also contribute to feelings of fatigue and burnout which in turn puts people at risk for personal injury.

9. Leaders Look Beyond the Numbers

Managers focus on the numbers, which in safety means injury records and compensation costs. When I discuss behavior-based safety principles and procedures with managers, I inevitably get the question, "What's the return on investment?" Managers want to know how much the process will cost and how long will it take for the numbers (as in total recordable injuries) to improve. This analytical approach to safety is obviously inspired by the popular management principle "You can only manage what you can measure."

Leaders certainly appreciate the need to hold people accountable with numbers, but they also understand you cannot measure everything. There are some things you do and ask others to do because it is the right thing to do. Leaders believe, for example, that it is important to increase self-esteem, self-efficacy, personal control, optimism, and a sense of belonging throughout a work culture.

In other sources (Geller 1994, 2001 a, b), I have discussed these five intangibles as feeling states that influence people's tendency to actively care or go beyond the call of duty for another person's safety or health. Here I simply want to make the point that leaders do things on a regular basis to inspire these feeling states in others but do not worry about measuring their direct impact on these intangibles. They have faith in the research-supported theory that promoting these five states is important.

Now and then it is a good idea to assess whether certain actions are influencing people's subjective feelings in a desired direction. This can be done informally through personal interviews, unaided by a scorecard. It is a given that certain interpersonal and group activities are useful. For example, genuine one-to-one recognition increases trust and feelings of importance, behavior-based goal setting builds feelings of empowerment, and group celebrations facilitate a sense of belonging. You need to perform and support these sorts of activities without expecting to see an immediate change in the numbers of a safety accountability system.

Leaders do not need a monitoring scheme to motivate their attempts to help people feel valuable and part of an important team effort. This kind of leadership is self-directed and responsible, and it helps to inspire self-directed responsibility in others.

10. Leaders Make More Distinctions Between People

We tend to give global labels for people, such as student, patient, union representative, safety professional, or athlete. Each label prompts a particular image and a set of characteristics. Similarly, the best-seller *Men are From Mars, Women are From Venus* (Gray 1992) teaches us generic differences between men and women which we presumably need to understand if we want to be more successful at dating, loving, and sustaining a marriage. The general label we give people influences how we view them, judge them, and react to their communication with us. This is the kind of destructive bias or "premature cognitive commitment" (Langer 1989) that leads to stereotyping, prejudice, interpersonal conflict, and sometimes even hate crimes.

Educational efforts to combat prejudice focus on teaching people that everyone should be considered equal and that putting people into different categories is wrong. In other words, to decrease discrimination and its accompanying problems we are told to stop discriminating. Langer (1989) believes this is the wrong approach. Categorizing people and things according to discernable characteristics is a natural learning process. It is how we come to know and understand our surroundings.

Dr. Langer suggests the key to reducing prejudice is to make more, not fewer, distinctions between people. When people become more attentive to the numerous differences among individuals and how these differences vary according to the environmental or interpersonal context, it becomes increasingly difficult to put individuals into universal categories. It becomes impossible to view people and their behavior as black or white, normal or abnormal, masculine or feminine, safe or unsafe.

Leaders put people's attributes and skills on a continuum. A person is not good or bad, skilled or unskilled, safe or unsafe at a particular task, but rather is a particular degree of good, skilled, and safe. Plus, one's quality level for a certain attribute can fluctuate dramatically from one situation to another. Thus, leaders make more distinctions between people and fewer global stereotypes. This enables objective and fair linkage between people's talents and job descriptions, and it facilitates the kind of interpersonal trust needed for a Total Safety Culture (Geller 1999b).

Conclusion

This article described ten approaches that leaders can use to help people transition from an other-directed perspective about safety to self-directed responsibility or personal accountability. Although I contrasted most of the leadership qualities with a typical safety management perspective, my intent was not to criticize or demean management. In fact, most willful behavior is initially other-directed and motivated by an external accountability system. Some people remain other-directed for various types of behaviors and situations. Managers are needed to keep people going.

On the other hand, it is usually more desirable for people to be self-directed rather than other-directed, especially in the domain of safety and health promotion. When people are self-directed with regard to safety they do not need an external accountability system to keep them performing safely. They are prone to actively care for the safety and health of others. The leadership characteristics described here can help build self-directed responsibility for safety while command-and-control management styles can stifle its development.

The bottom line: safety management is necessary at times to motivate people to do the right things for injury prevention. But such management is not sufficient to achieve a Total Safety Culture. Safety managers need to know when to become safety leaders and build personal responsibility rather than hold people accountable. Most importantly, whether or not you hold a safety management position, you can be a safety leader and help people transition from an other-directed to a self-directed motivational state. Remember, few manage but many must lead.

References

American Society of Safety Engineers 1999. *Proceedings of best practices in safety management symposium* (February). Las Vegas, NV.

Carnegie, D. 1936. *How to win friends and influence people*. New York: Simon & Schuster Inc.

Geller, E. S. 1994. Ten principles for achieving a Total Safety Culture. *Professional Safety,* 39(9):18–24.

———. 1996. The truth about safety incentives. *Professional Safety,* 41(10):34–39.

———. 1999a. How to increase personal accountability for occupational health and safety. *Proceedings of best practices in safety management symposium*. Des Plaines, IL: American Society of Safety Engineers. 213–230.

———. 2001a. *The psychology of safety handbook*. Boca Raton, FL: CRC Press.

———. 2001b. *Beyond safety accountability*. Rockville, MD: Government Institutes.

———. 1999b. Interpersonal trust: Key to getting the best from behavior-based coaching. *Professional Safety,* 44 (4):16–19.

Geller, E. S., and S. W. Clarke. 1999. Safety self-management: A key behavior-based process for injury prevention. *Professional Safety*, 44 (7):29–33.

Gray, J. 1992 *Men are from Mars, women are from Venus*. New York: HarperCollins Publishers, Inc.

Langer, E. J. 1989. *Mindfulness*. Reading, MA: Perseus Brooks.

Langer, E. J., and J. Rodin. 1976. The effects of enhanced personal responsibility for the aged: A field experiment in an industrial setting. *Journal of Personality and Social Psychology*, 34:191–198.

———. 1977. Long-term effects of a control-relevant intervention among the institutionalized aged. *Journal of Personality and Social Psychology*, 35:897–902.

Steiner, I. D. 1970. Perceived freedom. In *Advances in experimental social psychology, Vol. 5*. ed.L. Berkowitz. New York: Academic Press.

Watson, D. L., and R. G. Tharp. 1997. *Self-directed behavior: Self-modification for personal adjustment*. 7th ed. Pacific Grove, CA: Brooks/Cole Publishing Company.

How to Sustain Employee Involvement in Occupational Safety[2]

E. Scott Geller

The key to preventing more work-related injuries is to get more people involved in programs and processes designed to improve occupational health and safety. This is not profound; it is obvious. Yet in so many situations, safety is managed in such a way that involvement is actually inhibited rather than enhanced. For example, common safety slogans, safety incentive programs, and safety metrics actually stifle interpersonal trust and the kinds of proactive conversations and learning needed to prevent injury. In fact, the language we use to talk about safety—words like "accident investigation," "loss control," and "root cause"—hinders proactive participation.

It does not take major complicated change to turn current situations around and get more employee involvement in occupational health and safety. But it does take a paradigm shift. We need to perceive the problem of workplace injuries differently and to intervene differently with regard to the people aspects of safety. Two of the three "E" words for industrial safety are still appropriate and critically important: Engineering and Education. However, to get more participation, we need to replace the third "E" word of traditional safety, Enforcement, with another: Empowerment.

This paper offers a number of basic strategies relevant to cultivating a work force that feels empowered with regard to safety improvement and does something about it on a regular basis. In other words, this paper suggests ways to get more people actively caring for their own and others' health and safety. The principles and techniques presented are not based on common sense but on research-tested theory and practical applications. Let's start with the most basic strategy, one that defines culture and therefore determines whether all my other suggestions can be accepted, implemented, and sustained.

Watch Your Language

Words shape our feelings, expectancies, attitudes, and behavior (Hayakawa 1978). How you talk about something influences both how others feel about it, but especially how you feel about it. In other words, our verbal behavior affects our attitudes and beliefs; in turn, these determine more behavior. Question: Does your safety-related language increase or decrease employee involvement?

"Accident investigation" is a common phrase in industrial safety and health. What does it mean? Or, more to the point, what does it imply? Safety pros use this phrase to define one of their basic job requirements, and they attend professional development workshops with this label to improve their skills. But, really, what is your assignment when investigating an accident? Let's look more closely at this language.

The word "accident" implies "a chance occurrence" that is outside your immediate control. When a child has an "accident" in his pants, we presume he was not in control. He could not help it.

And what about the word "investigation?" Doesn't this term imply a hunt for some one thing or person to blame for a particular incident, as in "criminal investigation"? How can we promote fact-finding over fault-finding with a term like "investigation" defining our job assignment?

To learn more about how to prevent injuries from an analysis of an incident, we need to approach the task with a different mindset. It is not "accident investigation." It is "incident analysis." This simple change in our language suggests the following shifts in perspective, leading to more participation in the process and greater preventive impact.

From One Root Cause to Many Contributing Factors

There seems to be a common myth in the safety field that injuries are caused by one critical factor—the root cause. "Ask enough questions," advises the safety consultant, "and you'll arrive at the critical factor behind an injury." Come on, do you really believe there is a single root cause of an incident, whether a near-hit, damage to property, or personal injury?

Consider the three sides of "The Safety Triad" (Geller, Lehman, and Kalsher 1989) as a framework for defining the challenges of injury prevention. One side is for environment, including tools, equipment, engineering design, climate, and housekeeping factors. Another side of this triangle stands for behavior, the actions everyone did or did not perform related to an incident. And the third side represents person factors, or the internal feeling states of the people involved in the incident, including their attitudes, perceptions, and personality characteristics.

Given the dynamic interdependency of environmental, behavioral, and personal factors in everyday events, how can anyone expect to find one root cause of an incident? Instead, take a systems approach and search for a variety of contributory factors within the environment, behavior, and person domains. Then decide which of these factors can be changed to reduce the chance of another unfortunate incident. Environmental factors are usually easiest to define and improve, followed by behavioral factors. Most difficult to define and change directly are the person factors, but many of these can benefit indirectly with proper delivery of a behavior improvement process (Daniels 2000; Geller 2001a, b; McSween 1995).

From Avoiding Failure to Achieving Success

Interpersonal conversation is key to finding and correcting the potential contributors to an incident. People need to engage in open communication about the various environment, behavior, and person factors related to a near-hit, injury, or damage to property. But, this will not happen in an atmosphere of loss or failure. If the focus is on finding a single reason for failure, people will resist admitting any personal involvement.

It is human nature to deny personal influence in a loss. As kids we blamed the other kid—"he made me do it." As adults we just keep our mouths shut. To get people to open up, we need to approach incident analysis as an opportunity for success. Let's get away from the perspective "incident-equals-failure." The focus should be on how an incident gives us the chance to learn and improve. This can lead to more reports of personal near-hits and property damage. The more we report and analyze, the more opportunity we have to correct the factors that can contribute to a major injury to a fellow co-worker.

From Top-Down Correction to Bottom-Up Involvement

You can expect more participation in the reporting and analysis of an incident if you involve workers in the actual correction phase of the process. People will contribute more if they have a say in the outcome. Of course, management needs to approve and support the corrections recommended by the workforce, but workers, more than anyone else, know what it will take to make environmental, behavioral, and personal factors more safe. Use their critical expertise and you will motivate more ownership and involvement in the entire process.

From Narrow to Broad Application of Solutions

Traditionally the corrective action following an incident is not only designed narrowly, it is also applied narrowly. The safety director presents a report to management, and then the recommended solution to eliminating the "root cause" is implemented in the work area where the incident occurred. An equipment guard might be replaced, more comfortable personal protection equipment might be ordered, or a certain employee might be "retrained" or even punished (incorrectly referred to as "discipline" in the safety literature).

You will get broader interest and involvement in the incident analysis process if corrective action plans were applied to all relevant work areas. This also promotes a systems perspective rather than the piecemeal Band-Aid® approach common in so many work cultures. Look at the bigger picture. Use the results of an incident analysis to improve relevant environmental, behavioral, and personal factors plant-wide. This sends the kind of actively caring message that not only promotes more participation but also makes that participation more constructive.

From Outcome to Process Accountability

Both the quantity and quality of participation in incident analysis depend on the numbers you use to evaluate success or failure. The bottom-line measure—the total recordable injury rate (TRIR)—provides neither instructive guidance nor motivation to continue a particular safety process. It tells us nothing about why we are succeeding or failing.

Instead, keep track of the various components of incident analysis. Monitor the amount property damage and the number of near-hits and injuries reported. Track the number of corrective actions implemented for environmental, behavioral and personal factors. Now you have an accountability system that facilitates participation.

Keep the focus on successfully completing these various steps of the process. Keep score of your process achievements rather than waiting to see a reduction in injuries. This is a more valid and instructive measure of your success than any outcome measure currently available.

Other Language Barriers

Space prohibits a discussion of the many other language barriers to employee involvement, but consider the following ten pairs of terms or phrases. Which ones are more likely to encourage participation in a safety process?

behavior modification	or	behavioral safety
occupant restraint	or	safety belt
air bag	or	safety cushion
priority	or	value
safety manager	or	safety leader
compliance	or	achievement
peer pressure	or	peer support
program	or	process
training	or	coaching
loss control	or	injury prevention

Improve Interpersonal and Intrapersonal Conversation

The first paper in Chapter 4 gives specific guidelines for improving interpersonal and intrapersonal communication. These recommendations are extremely important for increasing involvement in occupational safety. Simply put, I believe the success of your efforts to prevent injury in the workplace, in the home, and on the road is determined by conversation. Our *interpersonal* communication, how we talk to others, influences our current attitude and ongoing behavior. And our *intrapersonal* communication, how we talk to ourselves, influences our own attitude and behavior. I am convinced, in fact, that improving the quality of our interpersonal communication about safety and the quantity of these conversations will reap the most cost-effective benefits in safety performance. Indeed, one key to the success of behavior-based safety is that it teaches new objective ways to talk about the human dynamics of occupational safety and health. So I urge you to re-read the communication lessons in Chapter 4 and apply as many of the principles you can on a daily basis. And include the critical lessons about behavioral safety contained in the next section of this paper.

Make Feedback Behavior-Focused

Feedback is the mechanism by which interpersonal conversation improves human performance. Practice does not make perfect. Only with appropriate feedback can we improve. The key to improving performance through feedback is to be behavior-based, both in diagnosing a problem and in suggesting ways to improve. Behavior-focused feedback is objective and impersonal. It merely displays a specific discrepancy between desired and observed behavior (Geller 2000a). In addition, behavioral feedback can include specific directions on how to reduce the behavioral discrepancy.

When feedback points out a behavioral discrepancy it is essentially motivational and informs participants how much improvement is needed. For optimal results, this kind of behavioral feedback should come as soon after the relevant behavior as possible. On the other hand, feedback intended to be more instructional than motivational is most influential when it occurs just prior to an opportunity to perform the behavior. In this case, you need to note the corrective action needed to make a certain behavior safer and then offer behavior-focused instruction when an occasion arises for the target behavior to occur again.

Specific techniques for making behavior-based feedback more effective are given elsewhere (Geller 1995, 2001b; McSween 1995). Here I want to add a few important points pertaining to getting people more involved in giving and receiving behavior-focused feedback. The success of any industrial behavior-based safety process depends on the number of employees (from managers to line workers) giving each other quality feedback. How can we get more people involved in such a process?

Make the Context Positive

I have heard several behavior-based safety consultants discuss feedback as if it is naturally accepted and used. They imply that involving employees in the development of a behavioral checklist and the posting of behavior-related numbers are all that is needed to put an effective feedback process in place. It is as if people naturally look forward to receiving feedback about their performance.

How do you feel when someone asks, "Can I give you some feedback?" Do you really expect a positive experience? Most people do not expect to enjoy a feedback session. Based on a lifetime of experience, people more often link feedback with "reprimand" than "praise." So do not expect people to naturally accept and look forward to receiving behavioral feedback.

The context of a feedback conversation is crucial. The nature of the conversation or group discussion surrounding a feedback session will determine whether such a process will be appreciated, supported, and sustained. Therefore, the first feedback session really needs to be positive and constructive. Realize that many people will not look forward to their initial feedback meeting because they expect to be corrected, perhaps even criticized.

Give Feedback Privately

Have you ever heard the pop psychology slogan "Praise publicly and reprimand privately"? This statement might sound good, but it can lead to ineffective or even detrimental delivery of *positive* feedback. Consider situations when a manager or supervisor gives a person positive feedback at a group meeting. Is it always helpful to be held up as an exemplar in front of one's peers? Is this delayed and public feedback a positive learning experience?

Many people feel embarrassed when receiving special attention in a group context. Part of this discomfort is due to fear of subsequent harassment by peers. They might expect to hear something like "Why did you deserve that special recognition about your safe work practices? Have you been kissing up to management again?" I advise you never to give individual feedback or recognition publicly without the person's prior permission.

It can be useful to give *group* feedback publicly. Since individual responsibility is diffused or dispersed across the group, there is little risk of individual embarrassment or peer harassment. But it is critical for this group feedback to be more positive than negative. It should be presented in terms of achievement rather than failure. Remember, we want people to look forward to the next feedback session.

Ask Others for Feedback

Safety leaders realize improvement is only possible with feedback. So they not only give feedback, they also ask others for feedback. When you ask another person for feedback and show appreciation for that person's specific, timely, and appropriate advice, you increase the likelihood that your feedback will be accepted by that person. When you show a desire to improve your own performance by asking for specific advice from an observer, you are teaching by example the key to safety improvement. Plus, you have established a reciprocal feedback relationship with another individual.

Your request and appreciation for another person's behavioral advice opens the door for that person's receptivity of feedback from you. If you ask questions to make the feedback you request specific, timely, appropriate, and real (Geller 2001b), you are teaching the proper way to give feedback.

So, how should you ask for feedback? Merely asking "How did I do?" after performing a certain job ensures timely advice. But the reaction you get might not be specific enough to be instructive. In fact, this kind of question might come off as though you are really not asking for genuine advice but are only "fishing" for a compliment. This is especially the case when your performance was pretty good, which is usually the case when a person asks for feedback.

It is quite rare for people to ask for feedback when they know their performance was inferior or below average. Ironically, this is when behavior-based feedback is most instructive. If you do not ask for specific feedback in these situations, it is possible you will never learn how to improve. Your friends and co-workers are most reluctant to volunteer advice in those very situations where instruction is most needed. They do not want to hurt your feelings.

My advice is to ask for feedback often and in such a way that it invites correction. Begin by asking for comments after you have done something well. Then show sincere appreciation for what you hear. This will make it easier for you to ask for feedback following your below-average performance. This also allows your friends and co-workers to build up your "emotional bank account," making it easier for them to make "withdrawals" when you ask for feedback after your below-average performance.

As I already mentioned, be careful when you start out asking for feedback following a good performance, or you risk appearing insincere. One way around this is to use a ten-point rating scale. After you ask "How did I do?" follow up with "Please give my performance a rating from one to ten, with one being completely inadequate and ten being perfect." This opens the door for specific feedback. People might give you a high rating of eight or nine, even nine and a half, but probably not a perfect ten. So now you ask, "What can I do to come closer to a perfect ten?"

Since you have already received a fairly high rating, it is easier for your friend or co-worker to follow-up with some corrective feedback. And here is another benefit of using this system: Perhaps you will convince your friend to be a more careful observer of your performance next time, in order to help you achieve that "perfect ten." Nothing, in fact, can bring you closer to "perfect ten" performance, especially in situations where appropriate feedback is not naturally available.

Give More Positive than Negative Consequences

When a checklist is used to observe behavior and give instructional feedback, the "safe" column will be checked more often than the "at-risk" column. In other words, when you have a list of things to look for in a particular situation, you will usually find more things right than wrong. This allows for a feedback session to be more positive than negative, which is exactly what you need for your improvement suggestions to be appreciated, accepted, and followed.

To be sure, the same situation can be viewed as control by negative or positive consequences. Some of the students in my university classes, for example, are motivated to avoid failure (a poor grade), whereas other students are motivated to achieve success (a good grade or even increased knowledge). Which of these groups of students feel more in control of their class grade and get more involved? Of course you know the answer to this question because you can reflect on your own feelings or attitude in similar situations where you perceived your behavior as influenced by positive or negative consequences.

Giving more positive than negative consequences not only increases involvement, but it also increases learning. Contrary to the popular press, we learn more from our successes than our failures (Chance 1999). Giving a person positive recognition for safe behavior will facilitate more learning than giving a person negative consequences for at-risk behavior will.

When delivered correctly, positive recognition for safe behavior provides direction and motivation to continue that behavior and improves personal attitude toward safety in general. Details on how to give positive consequences for safety-related behavior are given elsewhere (Daniels 2000; Geller 1997, 2001b, and the second paper of Chapter 4). Here, the issue is how to get more people involved in giving positive recognition for safe behavior.

Obviously, people need to learn how to recognize others appropriately. Sessions to teach the principles of recognizing people well should include role-playing exercises whereby participants practice giving behavior-based recognition to another person and then receive behavior-based feedback on their performance. The use of small rewards or "actively caring thank-you cards" (Geller 2001b) can be helpful in "breaking the ice" and initiating a positive approach to promoting safety-related behavior. Please refer to the second paper in Chapter 4 for practical guidelines on how to deliver and receive recognition.

Teach Theory and Principles before Procedures

Many scholars have written about the need to have a guiding theory or set of principles to consult when designing and refining methods and procedures (Covey 1990; Deming 1993). In fact, by summarizing the right theory or principles into a mission statement, you have a standard or map for judging the value of your company's procedures, policies, and performance expectations.

When it comes to safety, many companies start with teaching step-by-step procedures (referred to as "training"). They do not educate people first about the principles or rationale behind a particular safety policy, program, or process; as a result, many safety programs are referred to as "flavor-of-the-month." Such hand-me-down programs do not get many people involved, and they do not last very long.

Provide Guidance for Customizing a Process

When people are educated about the principles and rationale behind a safety process, they can customize specific procedures for their particular work areas. Then the relevance of the training process is obvious, and participation is enhanced. People are more likely to accept and follow procedures they helped to develop. They see such safe operating procedures as "the best way to do it" rather than "a policy we must obey because management says so."

Too many safety efforts start as off-the-shelf programs. A videotape is shown and ready-made workbooks are followed to train step-by-step procedures. Much more involvement occurs when consultants begin a new safety effort by teaching rationale and principles and then guiding participants through the development of specific procedures. Subsequently, people want to be trained on the implementation procedures.

When effective leaders guide the customization of a process, they state expectations but they do not give mandates or directions. They show both confidence and uncertainty (Geller 2000b, and the first paper in this chapter; Langer 1989, 1997). In other words, they are confident a set of procedures will be developed but do not know the best way to do it. This allows employees room to be alert, innovative, and self-motivated. The result: Ownership and interpersonal trust increase, which in turn leads to more involvement.

Build Ownership and Interpersonal Trust

When the development of a safety process involves the people expected to carry out the process, ownership for both the process and the outcome develops. When people are given a reasonable rationale for a desired outcome and then offered opportunities to customize methods for achieving that outcome, they feel more motivated to contribute over the long haul. This motivation comes from inside people and is commonly referred to as internal or self-directed motivation (Geller 2001a; Geller and Clarke 1999).

The process of developing ownership enhances interpersonal trust. People trust leaders who give them opportunities to tailor a set of procedures. The process of customization builds trust between team members. When employees trust both the intentions and abilities of their leaders and co-workers, they are more likely to contribute willingly to whatever process is called for. Specific strategies for increasing interpersonal trust are defined by these seven words that begin with the letter "C": Communication, Caring, Candor, Consistency, Commitment, Consensus, and Character These are explained in the next paper (from Geller 1999).

Teach and Promote Interdependency

Developing interpersonal ownership and trust in a safety process contributes to cultivating the ultimate mindset in a Total Safety Culture: interdependency. People experience situations every day—from home to work and traveling in between—that reflect win/lose independence. But involvement in a process designed to keep everyone safe requires the perspective of win/win interdependence.

The attitude "it won't happen to me" is used to justify an independent decision to work at risk. But when people understand interdependency, they realize their safety-related behaviors influence the safety of others. They participate in a safety process because they do not want anyone to get hurt, and they realize their good example contributes interdependently to the vision of an injury-free workplace. Effective leaders demonstrate an interdependency mindset whenever possible, which is the best way to teach this important concept—by example.

Do More Leading Than Managing

At times safety management is necessary to hold people accountable for their participation in a safety process. But this results in behavior being other-directed rather than self-directed (Watson and Tharp 1997). Other-directed behavior requires external activators and consequences for maintenance. Continual participation in a safety process needs to be self-directed, and effective leadership facilitates this. Safety leaders inspire people to feel personally responsible for the success of a safety process as opposed to safety managers who hold people accountable for their participation.

The first paper in this chapter (Geller 2000b), contrasts effective leadership qualities with characteristics of traditional safety management. These are summarized in Figure 6.2.1. The leadership qualities reflect the points discussed here for sustaining involvement in occupational safety. Most of the points discussed here for maintaining involvement in occupational safety help to increase personal responsibility for safety, from improving interpersonal conversation and giving behavior-based feedback and recognition to customizing procedures from sound principles and cultivating ownership, teamwork, and interpersonal trust. See another recent publication from ABS Consulting (Geller 2001a) for more details on how to facilitate people's acceptance of personal responsibility for safety.

Enhance the Actively Caring Person States

Several years ago, I defined a Total Safety Culture as one in which "everyone feels responsible for safety and pursues it on a daily basis; employees go beyond 'the call of duty' to identify unsafe conditions and behaviors, and intervene to correct them.... [P]eople actively care on a continuous basis for safety" (Geller 1994, p.18). There is significant overlap between feeling responsible for safety and actively caring for the safety of others. In fact, people who accept personal responsibility for safety are usually willing to go beyond the call of duty for another person's safety and health. Whether such actively caring behavior actually occurs, however, depends in part on the individual's psychological state when an opportunity to help someone occurs.

Managers	Leaders
. Hold people accountable	. Build responsibility
. Focus on outcomes	. Focus on process
. Train	. Educate
. Speak first, then listen	. Listen first, then speak
. Answer questions	. Ask questions
. Promote compliance	. Promote ownership
. Mandate rules and policies	. Set expectations
. Manage what's measured	. Facilitate intangibles
. Enable mindlessness	. Facilitate mindfulness
. Direct	. Instruct, support, coach

Figure 6.2.1 Management Is Not the Same as Leadership

Research has shown that five person dimensions influence people's willingness to help others:

- self-esteem ("I am valuable"),
- belonging ("I belong to a team"),
- self-efficacy ("I can do it"),
- personal control ("I am in control"), and
- optimism ("I expect the best").

The latter three states influence perceptions of empowerment ("I can make a difference"). The statements in quotes define the essential meaning of each person state or dimension. These critical actively caring person states are discussed in much more detail in other publications, which also include strategies for increasing them throughout a work culture (Geller 1998, 2001a, b, see also Chapter 1 of this text). As the strategies given in this paper increase involvement in a safety-related process, they also enhance the actively caring person states.

Teach and Promote Systems Thinking

It is easy to get bogged down with handling immediate short-term demands—like production deadlines—and lose sight of the bigger picture. Systems thinkers take a broad and long-term perspective. They look beyond immediate payoffs, like the ease, speed or comfort they get by taking a risky shortcut. They consider the possibility of a bigger payoff in the future. They realize their safe behavior teaches others by example and protects them from injury. And systems thinkers are happier because they appreciate how the interdependency of equipment and people's behavior can optimize a dynamic and productive system.

Systems thinkers understand the link between behavior and attitude. A small change in behavior can result in a beneficial change in attitude, followed by more behavior change and then more attitude change, eventually resulting in total commitment. So by teaching and promoting systems thinking, you set the stage for an interdependent mindset, actively caring behavior, and continuous involvement in the achievement of a Total Safety Culture.

References

Chance, P. 1999. *Learning and behavior.* 4th ed. Belmont, CA: Wadsworth Publishing Company.

Covey, S. R. 1990. *Principle-centered leadership.* New York: Simon and Schuster.

Daniels, A. C. 2000. *Bringing out the best in people: How to apply the astonishing power of positive reinforcement.* 2nd ed. New York: McGraw-Hill, Inc.

Deming, W. E. 1993. *The new economics for industry, government, education.* Cambridge, MA: Massachusetts Institute of Technology, Center for Advanced Engineering Study.

Geller, E. S. 1994. Ten principles for achieving a Total Safety Culture. *Professional Safety,* 39(9):18–24.

———. 1995. Safety coaching: Key to achieving a Total Safety Culture. *Professional Safety,* 40(7):16–22.

———. 1997. Key processes for continuous safety improvement: Behavior-based recognition and celebration. *Professional Safety,* 42(10):40–44.

———. 1998. *Understanding behavior-based safety: Step-by-step methods to improve your workplace.* 2nd ed. Neenah, WI: J. J. Keller & Associates, Inc.

———. 1999. Interpersonal trust: Key to getting the best from behavior-based coaching. *Professional Safety*, 44(4):16–19.

———. 2000a. Behavioral safety analysis: A necessary precursor to corrective action. *Professional Safety*, 45(3):29–32.

———. 2000b. Ten leadership qualities for a Total Safety Culture: Safety management is not enough. *Professional Safety*, 45(5):38–41.

———. 2001a. *Beyond safety accountability.* Rockville, MD: Government Institutes.

———. 2001b. *The psychology of safety handbook*. Boca Raton, FL: CRC Press.

Geller, E. S. and S. W. Clarke. 1999. Safety self-management: A key behavior-based process for injury prevention. *Professional Safety,* 44(7):29–33.

Geller, E. S., G. Lehman, and M. Kalsher. 1989. *Behavior analysis training for occupational safety*. Newport, VA: Make-A-Difference, Inc.

Hayakawa, S. I. 1978. *Language in thought and action.* 4th ed. New York: Harcourt Brace Jovanovich, Publishers.

Langer, E. J. 1989. *Mindfulness*. Reading, MA: Perseus Books.

———. 1997. *Mindful learning*. Reading, MA: Perseus Books.

McSween, T. E. 1995. *The value-based safety process*. New York: Van Nostrand Reinhold.

Watson, D. L., and R. B. Tharp. 1997. *Self-directed behavior: Self-motivation for personal adjustment.* 7th ed. Pacific Grove, CA: Brooks/Cole Publishing Company.

Interpersonal Trust: Keys to Getting the Best from Behavior-Based Safety Coaching[3]

E. Scott Geller

Further prevention of occupational injuries beyond current plateaus will occur when individuals work in teams to develop checklists of safe and at-risk behavior, use these checklists to observe each other's work behaviors, and then provide constructive feedback from their observations in one-on-one and group situations (Geller 2001; Krause, Hidley, and Hodson 1996; McSween 1995). Of course, the individuals who receive the behavioral feedback must believe the feedback report. They should feel good about reports of their safe behavior, and they should be willing to correct instances of at-risk behavior.

Sometimes when I introduce these aspects of behavior-based safety to employees, I get active resistance. "No way," say some workers. The thought of a co-worker watching them work and tallying their observations of safe and at-risk behaviors for a subsequent feedback session sounds intrusive and aversive to these individuals. Similarly, several managers and safety directors have expressed agreement with the basic observation and feedback process of behavior-based safety, but then they say, "Our plant is not ready for such a process."

What causes such active resistance to an interpersonal observation and feedback process? What is missing in a work culture that is not "ready" for this basic component of behavior-based safety? Actually, I have heard both managers and line workers voice the same answer to these questions many times. One word says it all—trust. Lack of interpersonal trust causes resistance to an observation and feedback process, and interpersonal trust is what's missing in a culture deemed unready for behavior-based safety. This paper explores the concept of interpersonal trust, introduces an assessment tool for measuring it, and reviews some basic strategies for building trust in a work culture.

What is Interpersonal Trust?

The first definition of "trust" given in my *American Heritage Dictionary* (1991) is "confidence in the integrity, ability, character, and truth of a person or thing" (p. 1300). "Interpersonal" merely limits this definition to "person" or "situations between people." This definition refers to behavior (as in "ability") as well as internal or person-based dimensions (as in "integrity" and "character"). As such, this definition is consistent with a measurement tool my colleagues and I use to assess interpersonal trust in a work culture. This survey device (see Figure 6.3.1) is based on research by John Cook and Toby Wall (1980).

The survey distinguishes between intention and ability. In other words, you could be confident a person means well, but you might doubt his or her ability to complete the intended task. In this case, you trust the individual's intentions but you are not so sure the stated outcomes will occur. You lack confidence in the capability of the person to make good on his or her promises. This perception is common in situations in which well-intentioned managers or safety leaders verbalize missions, goals, or policies that are viewed as idealistic or unrealistic. When punishment or reward contingencies are not carried out consistently and fairly, for example, interpersonal trust can be limited to intentions and not actions.

It is also possible for individuals to have faith in the ability of others but to mistrust the intentions of others. This happens, for example, when people are not informed of events or decisions leading up to a policy change that affects their work life. When a policy change is sprung on workers without warning or rationale, they might become suspicious of management's intentions. They might believe their leaders have the intellect and skills to make things happen, but they might be concerned about what particular things will happen. Do they really have the workers' welfare in mind when they deliberate about changes in equipment design or production quotas?

Could there be a disconnection between intention and capability when management establishes a safety incentive program? Consider, for example, the program that offers everyone a prize if no one is hurt over a certain period of time. This translates into a contingency that puts pressure on individuals to cover up an injury. After all, if anyone reports an injury, then everyone loses their reward. It is logical for workers to perceive this kind of program as a scheme to keep the numbers down without really caring about their personal welfare.

Interpersonal Trust Among Co-workers

All of the examples I have used so far to distinguish between trust in "intention" versus "ability" use management as the target. In other words, the focus has been on whether a worker has confidence in the intention and/or capability of a manager or supervisor. Interpersonal trust in a work culture, however, also refers to the extent that people ascribe good intentions and abilities to their peers. In other words, a line worker might have confidence in the ability of a co-worker to perform a job safely and competently but might be wary of telling him or her certain things because of mistrust in that person's intentions. Such mistrust is implied in the statement, "My partner might use the information against me for personal gain—to get the promotion before me." Similarly, you might trust the intentions of a co-worker ("He would never take advantage of me") but lack confidence in his capability on a particular job assignment ("I'm sure he will do his best, but I'm afraid his lack of experience means his best will not be good enough").

The success of a behavior-based observation and feedback process certainly requires a high degree of interpersonal trust among co-workers. To accept and use behavioral feedback, a person must believe the feedback is accurate. And if this feedback comes from another person, as in the case of a behavior-based coaching process, the worker being observed must have confidence in the coach's ability to obtain the information used for behavioral feedback. This boils down to having confidence in the tool used to record behavioral information *and* in the observer's ability to use this tool correctly.

But it is not enough to believe in the validity of an observation and feedback process and in the ability of a co-worker to carry it out. Trusting the intentions of an observer is critical. The person whose behavior is being observed must believe the information will be used *only* for personal protection against injury and never as grounds for punishment. That is why a one-to-one behavioral feedback process should begin with the coach asking for a co-worker's permission to make the observations.

Interpersonal Trust Scale

This discussion of interpersonal trust identified two dimensions of trust: faith in the *intentions* of others and confidence in the *ability* of others. From the viewpoint of a line worker evaluating the overall interpersonal trust in a work culture, these dimensions can refer either to other co-workers or to management. This fourfold classification system was used by Cook and Wall (1980) to derive a 12-item questionnaire to assess "interpersonal trust at work."

This interpersonal trust scale is given in Figure 6.3.1. This scale can be administered to a discussion group or work team to stimulate interesting and instructive group discussion. It is an excellent way to increase people's understanding of interpersonal trust.

Ask the participants to score their own surveys as follows. Except for two items (2 and 12), the higher the scale value, the greater the perceived interpersonal trust. Since items 2 and 12 are negatively phrased, these need to be reverse scored. In other words, for these two items, the number selected is first subtracted from 8, so a "1" becomes a "7," a "2" becomes a "6," and so on.

Totaling the 12 item scores for a particular survey (with items 2 and 12 reverse scored) yields an estimate of an individual's perception of overall interpersonal trust in his or her work culture. You can obtain an overall trust index by calculating the mean survey score from many respondents. You can also estimate the different dimensions of trust introduced here. Specifically, items 3, 5, and 8 assess faith in intentions of peers (or co-workers) and items 1, 7, and 12 address faith in the intentions of management. Confidence in the ability or actions of peers is measured by items 9, 10, and 12, whereas items 2, 4, and 6 refer to confidence in the capability of management.

Figure 6.3.1 A Scale to Assess Interpersonal Trust

The statements below express opinions that people might hold about the confidence and trust that can be placed in others at work, both fellow workers and management. Circle the scale numbers next to each statement to indicate how much you agree with it, with "1" meaning strongly disagree, "3" neutral, and "5" meaning strongly agree.

1. Management in my company is sincere in its attempts to meet the workers' point of view.	1	2	3	4	5
2. Our company has a poor future unless it can attract better managers.	1	2	3	4	5
3. If I got into difficulties at work I know my co-workers would try and help me out.	1	2	3	4	5
4. Management can be trusted to make sensible decisions for the company's future.	1	2	3	4	5
5. I can trust the people I work with to lend me a hand if I need it.	1	2	3	4	5
6. Management at work seems to do an effective job.	1	2	3	4	5
7. I feel quite confident that the company will always treat me fairly.	1	2	3	4	5
8. Most of my co-workers can be relied upon to do as they say they will.	1	2	3	4	5
9. I have full confidence in the skills of my co-workers.	1	2	3	4	5
10. Most of my fellow workers can do their work well even when supervisors are not around.	1	2	3	4	5
11. I can rely on other workers not to make my job more difficult by careless work.	1	2	3	4	5
12. Our management would be quite prepared to gain advantage by deceiving the workers.	1	2	3	4	5

Please note there is nothing special or magical about the wording of these items. You should feel free to reword a particular item if different language fits better with your culture. And you could decide to add or substitute new items. Note also that the scale was developed to measure interpersonal trust from the viewpoint of an hourly worker. With only slight adjustments, the scale can also estimate interpersonal trust from a manager's perspective.

The fourfold classification system for interpersonal trust is an important contribution of the Cook and Wall survey tool provided here. In industrial settings, it is important to differentiate perceptions of interpersonal trust as targeting management or co-workers and with regard to other people's intentions versus their actions. It would be nice if everyone trusted each other's intentions with respect to health and safety. Then, when at-risk behavior, a "near-miss," property damage, or an injury occurs, the focus would be on an individual's capability to act safely under particular circumstances. This would lead to fact finding (not fault finding) regarding ways to improve behavior (rather than to correct a "bad attitude"). Then an observation and feedback process would be trusted as a method for obtaining information relevant to correcting environmental and system factors that facilitate at-risk behavior or human error.

Building a Trusting Culture

Now that we have an operational definition of "interpersonal trust," the more important issue can be addressed. How can we increase interpersonal trust throughout a work culture? I searched the research literature for answers to this question and found none; therefore, I attempted a common-sense approach. I called together a group of my research students and colleagues and asked the question: How can we increase interpersonal trust among the members of our various research teams? First I explained the concept of interpersonal trust, as defined here, and then facilitated a "brainstorming" discussion of the trust-building question among these nine individuals. I tried to characterize answers according to the intention versus ability dichotomy. I wrote each suggestion on the blackboard and then attempted to assign an "intention" or "ability" label to each.

After arriving at a rather large list of comments related to the building of interpersonal trust, we refined our list with a consensus process. I asked for suggestions regarding the elimination or combination of list items. During this process we had to remind ourselves we were not looking for items that made us believe another person could be trusted in intention or ability. Rather, our assignment was to suggest ways for increasing perceptions of interpersonal trust.

After almost two hours, we arrived at an interesting and seemingly useful list of proposals for increasing interpersonal trust, as well as a beneficial approach for teaching our recommendations to others. Before sharing our insights here, I need to make a very important point. The process of brainstorming was extremely valuable. We not only arrived at a list of trust-building principles and procedures, but we also increased feelings of group cohesion and ownership around this critically important concept. We put the concept of interpersonal trust and the need to increase it on the table. We faced the reality that there are many things each of us can do to build interpersonal trust in our work culture. A summary of our brainstorming session—as summarized in the "Seven Cs of Trust-building below—is now posted in our meeting room, and we review this list at the start of every semester when new students join our research group.

I urge you to have a brainstorming session similar to the one we experienced. Your list of proposals will likely be similar to ours, but it will be owned by your group. This increases the likelihood people will heed the recommendations for increasing interpersonal trust. You might find it useful to use the following summary of our session as a "kick off" for your meeting. You could ask for reactions to the key words that emanated from our discussion, and then make your own additions, substitutions, or refinements. Owning a set of recommendations for building interpersonal trust is a good way to encourage trustworthy behavior.

The Seven "Cs" of Trust-Building

Near the end of our two-hour brainstorming session on interpersonal trust, I asked the group if we could encapsulate our various suggestions for increasing trust with certain words. It seemed, in fact, many words beginning with the letter "C" reflected specific recommendations for building interpersonal trust in a work culture. We reached a consensus that seven C-words capture the essence of our group exercise. The phrases associated with these words summarize the key definitions given in my *American Heritage* (1991) and *New-Merriam-Webster* (1989) dictionaries.

1. Communication: *exchange of information or opinion by speech, writing, or signals*

How we interact with others is obviously a key determinant of interpersonal trust. What people say and how they say it influences our trust in both their capability and their intentions. An individual's expertise is displayed by his or her words, spoken or written, and by the confidence and credibility linked to the words. I am sure you have heard many times that the way something is said, including intonation, pace, facial expressions, hand gestures, and overall posture, has greater impact than what was actually said. And, you have certainly experienced personal feelings of trust toward another person change as the result of *how* that individual communicated information. It is possible the high-tech electronic approaches to communication decrease face-to-face interaction and hence the development of interpersonal trust.

2. Caring: *showing concern or interest about what happens*

When you take the time to listen to another perspective, you send a most important message that you care. And when people believe you sincerely care about them, they will care about what you tell them. They trust you will look out for them when applying your knowledge, skills, and abilities. They trust your intention because they believe you care.

You also communicate caring and build interpersonal trust when you ask questions. I am not talking about the probing questions we sometimes pose instead of actively listening. I am referring to the use of a question to initiate conversation. And, I am not talking about the typical general questions we often ask people we have not seen for awhile. "How are you doing?" we ask, and we get the standard reply, "I'm doing fine, how about you?" No, I am referring to inquiry about a particular task or set of circumstances. Questions targeting a specific aspect of a person's job send the signal you care about him or her. This communication is more than a general greeting. It is a statement of genuine interest in what people are doing and how they feel. It is especially powerful when your words reflect concern about health and safety.

In order to show caring with specific behavior-based questions, you need to take the time to learn what others are doing. This comes from active listening and behavior-based observation. I am sure you have heard the phrase "Walk the talk." Well, here I am talking about "listening to the talk and watching the walk." This shows you care and gives you an opportunity to "talk the walk" so people will trust your intentions.

3. Candor: *straightforwardness and frankness of expression; freedom from prejudice*

We trust people who are frank and open with us. They get right to the point, whether asking for a favor or giving us behavior-based feedback. When these individuals do not know an answer to our questions, they do not ignore us or hem and haw about possibilities. They tell us outright when they do not know something, and they tell us they will get back to us later. And when they get back to us soon with an answer, our trust in both their intention and ability increases.

The second dictionary definition of "candor"—freedom from prejudice—reflects another important aspect of trust-building. When someone gives an opinion about another person because of race, religion, gender, age, sexual orientation, or birthplace, you should doubt this person's ability to make people-related decisions. You should wonder whether this person's intention to perform on behalf of another person will be biased or tainted by a tendency to prejudge others on the basis of overly simple and usually inaccurate stereotypes. You are rightfully suspicious about this person's ability to evaluate others and in his or her intention to treat people fairly. And even when the prejudice is not directed toward you, your trust in this person decreases. Right?

4. Consistency: *agreement among successive acts, ideas, or events*

We usually trust the intentions of people who confess openly their inability to answer our question. Our trust in them might increase when they tell us they will get back to us right away with a response. But what happens when we must wait a long time for a reply? Or suppose you never hear from them again. Now what happens to interpersonal trust? Consistency is a word everyone in our discussion group mentioned as a determinant of interpersonal trust.

Perhaps the *quickest* way to destroy interpersonal trust in intention and ability is not to follow through consistently with our agreements. This is also the *easiest* way to stifle trust. How often do we make promises we do not keep? Most promises are behavior-consequence contingencies. We specify that a certain consequence will follow a certain behavior.

Whether the consequence is positive or negative, trust decreases when the behavior is not rewarded or punished as promised.

One of the problems with punishment contingencies is that they are difficult to implement fairly and consistently. It is easy to state a policy that anyone not using appropriate personal protective equipment will be "written up," but it is quite difficult or impossible to carry out this contingency consistently. Consider the "safe employee of the month" program that selects winners according to nonobjective criteria or does not consider everyone consistently as a potential award recipient.

5. Commitment: *bound emotionally or intellectually to a course of action*

People who are dependable and reliable are not only showing consistency, but they are also demonstrating commitment. When you follow through on a promise or pledge to do something, you prove to others they can count on you. You can be trusted to do what you say you will do. Making a commitment and honoring it builds trust in both intention and ability.

Telling a personal anecdote to illustrate a point is often a good way to demonstrate commitment for something. When this commitment is consistent with the theme of your speech or written presentation, your credibility increases. The audience has reason to trust your intention to give accurate and useful information.

Whenever I give presentations on the psychology of safety, I try to work in one or more personal experiences. I have done many things over the years in an attempt to promote safety, and when a story fits the theme of my talk, the audience hears firsthand evidence of my commitment. When I discuss the value of using safety belts, for example, I can relate any of a number of interventions I have tried over the years to encourage people to buckle up—from setting up safety-belt incentive programs at numerous industrial and community sites to teaching my young daughters to hold up a "Please Buckle Up—I Care" flash card whenever we stopped at an intersection.

These stories show I am committed to promoting safety-related behavior. They show I am well-intentioned, and I try to share the best safety-related information. Taken alone however, my personal stories of commitment say nothing about the accuracy of the information. Only reference to research should do that. But, this story-telling strategy for building trust in intention also builds some trust in ability.

6. Consensus: *agreement in opinion, testimony, or belief*

Demonstrating personal commitment to a mission, purpose, or goal helps to build group consensus. When a group of people reach consensus about something, all group members agree on a decision or course of action and are willing to support it. Leaders or group facilitators who develop consensus among people build interpersonal trust. Consensus-building is the opposite of top-down decision making and is not the same as negotiating, calling for a vote and letting the majority win, or working out a compromise between two differing sets of opinions

Whenever the results of a group decision-making process comes across as "win-lose," some mistrust is going to develop. A majority of the group might be pleased, but others will be discontented and might actively or passively resist involvement. And even the "winners" could feel decreased interpersonal trust. Without solid backup support for the decision, the outcome will be less than desired. Without everyone's buy-in, commitment, and involvement, we cannot trust the process to come off as expected.

So how can group consensus be developed? How can the outcome of a heated debate on ways to solve a problem be perceived as a win-win solution everyone supports, instead of a win-lose compromise or negotiation headed to depreciate interpersonal trust? Consensus-building takes time and energy. It requires candid, consistent, and caring communication among all members of a discussion or decision-making group. In other words, when people demonstrate the C-words discussed here for building trust in interpersonal dialogue, they also develop consensus and more interpersonal trust regarding a particular decision or action plan.

Fran Rees (1997) describes six basic steps to reaching a consensus decision:

- *Set the decision goal*. What is the aim or purpose of the consensus-building exercise? What will be the end result, one way or another, of the group's decision process?

- ⋇ *Spell out the criteria needed to make the group decision worthwhile.* What qualities or characteristics of the decision are needed to reach a particular goal? Which criteria are consistent with the team's ultimate purpose or mission? What are the budget constraints? What principles or quality standards are relevant? What will a decision provide for the team? What criteria or restrictions are essential, and what criteria are desirable but not absolutely necessary?
- ⋇ *Gather information.* What information is useful for making the decision? Where is this information, and who can provide it? How should the information be summarized for optimal understanding and consideration?
- ⋇ *Brainstorm possible options.* What are the variety of possible solutions? Does everyone understand each option and its ramifications? How does everyone feel about the possibilities? Has everyone had a chance to voice a personal opinion?
- ⋇ *Evaluate the brainstormed options against the group's criteria.* Which solutions appear to meet the "must have" criteria? Which options seem to meet the "nice but not necessary" criteria? To what degree will each option meet both the "necessary" and "desirable" criteria? Can certain solutions be combined to meet more criteria?
- ⋇ *Make the final decision as a team.* Which option or combination best meets all of the "necessary" criteria and most of the "desirable" criteria? Who has reservations and why? How can we resolve individual skepticism? Can everyone support the most popular option? What can be altered in the most popular action plan to attract unanimous support and ownership?

I am sure you see that building consensus around a group process or action plan is not easy. It requires plenty of interpersonal communication, including straightforward opinion sharing, intense discussion, emotional debate, active listening, careful evaluation, methodical organization, and systematic prioritizing. On important matters, the outcome is well worth the investment. When you develop a solution or process every potential participant can get behind and champion, you have cultivated the degree of interpersonal trust needed for total involvement. Involvement in turn builds personal commitment, more interpersonal trust, and then more involvement.

7. Character: *the combined moral or ethical structure of a person or group; integrity; fortitude*

This final C-word means different things to different people. Generally, a person with "character" is considered honest, ethical, and principled. People with character are credible or worthy of another person's trust because they display confidence and competence in following a consistent set of morally-sound beliefs. They are believable and trusted because they know who they are, they know where they want to go, and they know how to get there.

All of the strategies discussed here for cultivating a trusting culture are practiced by a person with character. Therefore, this C-word epitomizes interpersonal trust from both the intention and capability perspective. I would like to add a few additional trust-building methods that also fit this category, although they naturally overlap with other C-words discussed here.

First, individuals with character are willing to admit *vulnerability*. They realize they are not perfect and need behavioral feedback from others. They know their strengths and weaknesses and find exemplars to model. By actively listening to others and observing their behaviors, individuals with character learn how to improve their own performance. If they are building a high-performance team, they can readily find people with knowledge, skills, and abilities to complement their own competencies. They know how to make diversity work for them, their group, and the entire organization.

Having the courage to admit your weaknesses means you are willing to apologize when you have made a mistake and to ask for forgiveness. There is probably no better way to build trust between individuals than to own up to an error that might have affected another person. Of course you should also indicate what you will do better next time, or you should ask for specific advice on how to improve. This kind of vulnerability enables you to heed the powerful enrichment principle I learned from Frank Bird and George Germain (1987, p. 111).

> *Good, better, best. Never let us rest, until good is better, and better best.*

What is your trust level for a group leader who not only admits failure but continually seeks ways to improve? This is the kind of person you want on your team. Right! You can level with this person about your own incompetencies or insecurities without fear of ridicule or reprisal. You trust that this person will appreciate your desire to improve and will offer the guidance you need to do better. You also trust this individual to maintain the confidentiality of any disclosure of personal failure or vulnerability.

While admitting personal vulnerability is a powerful way to build interpersonal trust, the surest way to reduce interpersonal trust is to tell one person about the weakness of another. In this situation it is natural to think, "If he talks that way about her, I wonder what he says about me behind my back." It is obvious how criticizing or demeaning others in their absence can lead to interpersonal suspiciousness and mistrust and eventually to a work culture of independent people doing their own thing, fearful of making an error and unreceptive to any kind of performance-based feedback. Key aspects of behavior-based safety—team building, interpersonal observation, and coaching—are extremely difficult or impossible to implement in such a culture. Under these circumstances it is necessary to first break down barriers to interpersonal trust before implementing a behavior-based observation and feedback process.

Start to build interpersonal trust by making a personal commitment and implementing a team policy of no "back-stabbing." People with character, as defined here, always talk about other people as if they can hear what is said. In other words, to replace interpersonal mistrust with trust, never criticize other individuals behind their backs.

Conclusion

The fourfold classification system for interpersonal trust is an important contribution of the Cook and Wall survey tool. In industrial settings, it is instructive to differentiate perceptions of interpersonal trust with regard to targeting management or co-workers and with regard to another person's intentions versus his or her actions.

Research literature offers limited guidance on how to build interpersonal trust throughout a work culture. Therefore, it is necessary to allow common sense or personal experience to be your guide. The most frequent recommendation I have heard from consultants and practitioners regarding the facilitation of interpersonal trust is to be trustworthy. "If you want others to trust you," they say, "you need to trust others and behave toward others in ways that warrant their trust." This is certainly sound advice, but a critical question remains: What kinds of behaviors are trustworthy?

The seven C-words reviewed here are easy to remember, and although their meanings overlap to some extent, each offers distinct directives for trust-building behavior. *Communicating* these guidelines to others in a *candid* and *caring* way opens up the kind of dialogue that starts people on a journey of interpersonal trust-building. Then people need to give each other *consistent* and *candid* feedback regarding those behaviors that reflect these trust-building principles. With *character* and *commitment*, they need to recognize others for doing it right and offer corrective feedback when there is room for improvement. And of course it is critical for the recipient of such behavior-based feedback to accept it with *caring* appreciation and a *commitment* to improve. Then the feedback recipient needs to show the *character* to thank the observer for the feedback, even when the *communication* is not all positive or is not delivered well. The recipient might then offer feedback on how to make the behavior-based feedback more useful. Dialogue like this is necessary to build *consensus* and sustain a journey of continuous trust-building. Such a journey is essential to reap the many benefits of behavior-based safety.

References

American Heritage Dictionary. 1991. 2nd college ed. New York: Houghton Mifflin Company.

Bird, F. E., Jr., and G. L. Germain. 1987. *Commitment*. Loganville, GA: International Loss Control Institute, Inc.

Cook, J., and T. Wall. 1980. New work attitude measures of trust, organizational commitment and personal need non-fulfillment. *Journal of Occupational Psychology*, 53:39–52.

Geller, E. S. 2001. *The psychology of safety Handbook*. Boca Raton, FL: CRC Press.

Krause, T. R., J. H. Hidley, and S. J. Hodson, 1996. *The behavior-based safety process.* 2nd ed. New York: Van Nostrand Reinhold.

McSween, T. E. 1995. *The values-based safety process: Improving your safety culture with a behavioral approach*. New York: Van Nostrand Reinhold.

New Merriam-Webster Dictionary. 1989. Springfield, MA: Merriam-Webster Inc., Publishers.

Rees, F. 1997. *Teamwork from start to finish*. San Francisco, CA: Jossey-Bass, Inc.

Notes

[1] Originally appeared in *Professional Safety* 45:5 (2000):38–41.

[2] Original version appeared as "How to sustain involvement in occupational safety: From research-based theory to real-world practice." In *Proceedings of the 39th Annual Professional Development Conference and Exposition*. (Des Plaines, IL: American Society of Safety Engineers, 2000).

[3] Originally appeared in *Professional Safety* 44:4 (1999):16–19.

7

Maximizing Behavior-Based Safety Efforts

Behavior-based safety (BBS) is founded on empirical research and goes beyond common sense. This has been a theme throughout this book. This final chapter includes important and useful papers that support this and illustrate that understanding the human dynamics of occupational safety is not easy.

As BBS becomes more popular, there will be more and more abuses of the principles and procedures described in this book. This occurs because some people believe BBS is simple. They do not understand the various complexities of a work culture that need to be addressed in order to make BBS work optimally. Failures of weak implementations of BBS will lead to more critics who complain because they do not know the facts.

Two papers in this chapter address myths that slow the progress of understanding the true dynamics of BBS. The first paper addresses the myths spread by pop psychologists or motivational speakers who don't understand research-based psychology while the second paper explains fallacies about BBS disseminated by safety consultants who don't understand BBS as explained in this text. The third paper covers barriers to overcome when implementing BBS.

The next three papers in this chapter provide research-based guidelines for increasing the long-term impact of BBS, from the simple application of signs to activate safety awareness to obtaining the right amount of external accountability and internal responsibility among employees in order to achieve sustained participation in a BBS process. The final paper reports the extensive evaluation of 20 successful applications of BBS in an attempt to discover the critical success factors for BBS.

Watch Out for Pop Psychology[1]

E. Scott Geller

Beginning in 1992, my monthly contributions to *Industrial Safety & Hygiene News* have addressed the psychology of injury prevention. I have attempted to teach principles and procedures relevant to understanding or improving the human dynamics of occupational safety and health. The information in each article was based on scientific investigation rather than speculation or hearsay. In fact, I have frequently made the point that the psychology of safety is not just common sense but should be based on results of rigorous behavioral research.

Many presentations of human factors related to safety, however, base assertions only on common sense. They make statements that sound good but are actually incorrect or unfounded. Unfortunately, it is not always easy to determine which psychology-related statements are valid. In fact, some invalid statements are given so often by motivational speakers or pop psychologists that they are accepted as fact and subsequently misdirect action plans. Let's consider a few of the more prevalent of these fallacies, which I am sure you have heard before, perhaps at a safety conference.

Fallacy 1: Reprimand Privately but Recognize Publicly

This frequent statement from motivational speakers sounds good, but even common sense should tell you it is not true. Whether correcting behavior or giving recognition, it is usually better to communicate one-on-one in private. I urge you never to recognize a person in front of a group without that person's permission. Some people are embarrassed by public commendation, while others fear verbal harassment from peers. They imagine someone saying, "Why did you deserve that 'safe-employee-of-the-month' award? I have done as much around here for safety as you. Have you been kissing-up to the boss again?"

Fallacy 2: We Learn More from Our Mistakes Than from Our Successes

This is another pop psychology statement we hear so often we might actually believe it. But think about it. What do you learn when you make a mistake? You learn what *not* to do. That is something worth learning, but consider how much more you learn when you do something correctly and receive feedback that you are correct. You learn what you need to continue in order to be successful. We need to take this to heart and use recognition more often than reprimands to teach and motivate others.

Fallacy 3: Seventy-seven Percent of our Mental Thoughts Are Negative

I have heard more than one pop psychologist make this statement to justify the need to give more positive feedback than negative. I like the conclusion. We do need more recognition for correct behavior than correction for incorrect behavior, primarily because we learn more from correct than incorrect performance. But to claim a certain proportion of negative thoughts pervade the minds of human beings is absurd. Ask yourself the question "How could they know?" and you will see the ridiculousness in this and some other pop psychology statements.

Fallacy 4: Do Something 21 Times and It Becomes a Habit

I am sure you have heard this foolish assertion before. It is even the title of a self-help book. For years I have wondered where the "21" came from. Then a local farmer reminded me it takes 21 days for an egg to hatch into a chicken.

Sure enough, behavior needs to be repeated many times to develop fluency and then become a habit. That is why it is important to support the safe work practices of our friends and teammates. But to presume there is a set number of repetitions needed for habit formation is downright silly and frankly insults the intelligence of most people.

Many people have developed the habit of safety-belt use, for example, but the number of times belt use occurred before it became automatic varied dramatically across individuals and depended partly on the strength of the old bad habit. Specifically, how ingrained was a person's routine of entering and starting a vehicle without buckling up? And how inconvenient is the simple buckle-up behavior for a certain individual in a particular vehicle?

Some behaviors are so complex or inconvenient they never become habitual. Consider, for example, the chain of behaviors needed to lock out a power source, complete a behavioral audit, or follow a stretch and exercise routine. These important safety-related behaviors will likely never become automatic. Their occurrence will probably always require some deliberate motivating influence, whether external or self-imposed.

Fallacy 5: We Can Only Motivate Ourselves, Not Others

It is a good thing this frequent pop psychology statement is untrue, or we could not motivate others to choose the safe way of doing something when the at-risk alternative is more comfortable, efficient, convenient, or perhaps habitual. External motivational intervention has an essential role in increasing safe behavior and decreasing at-risk behavior. Here I just want to discredit the myth that only self-motivation works, as spread by a number of naïve speakers and writers.

Fallacy 6: Incentives and Rewards Are Detrimental to Self-motivation

This unfounded statement is related to the previous pop-psychology myth and is frequently used by safety professionals to criticize traditional safety incentive programs. A popular book titled *Punished by Rewards* has been read by too many safety professionals (one is actually too many). Now this myth is getting dispersed at safety conferences and in safety magazines.

Yes, you can insult another person's intelligence or self-motivation by using the wrong type of intervention to improve a particular behavior.

This can have a temporary detrimental effect on individual performance. But research indicates this negative impact is relatively infrequent, and if it does occur, it is short-lived. Incentives and rewards are far more likely to benefit desired performance and even self-motivation for long-term behavior change if they are used correctly. The key is to use a behavior-based approach to incentive/reward programs.

Fallacy 7: For Every 300 Unsafe Acts There Are 29 Minor Injuries and One Major Injury

I could not end this brief list of unfounded claims without adding the most popular myth in the field of industrial safety. H. W. Heinrich first proposed this ratio between at-risk behavior and injury in the 1930s. It has been repeated so often that some safety pros refer to it as "Heinrich's Law." It started as a mere estimate, and after years of use in safety speeches and publications, without any empirical verification, its status was elevated to "basic principle" or "natural law."

This statement does show the connection between behavior and injury and implies that a proactive approach to injury prevention requires attention to behaviors and near-hits. However, the number of at-risk behaviors per injury is much larger than 300, as verified empirically by Frank Bird in 1966, who also found property damage to be a reliable predictor or leading indicator of personal injury.

My point here, however, is not about a proactive versus reactive approach to safety but rather how the long-term repetition of an unfounded proclamation that sounds good can become a common myth with presumed validity. This happens all too often when dealing with a topic like psychology that is regularly relevant to our everyday activities.

Conclusion

I reviewed seven myths, the first six of which are spread by some motivational speakers and safety consultants as if they were fact rather than fiction. It is interesting that many of us do not need research evidence to accept these when they are delivered well. They sound good, so they must be true. Never mind the fact that some are not even testable.

Most disappointing to me is that the key factor influencing belief in a pop psychology statement like these is large-scale marketing rather than the credibility of the source. "Say you're the greatest and you'll be the greatest," they tell us. Well, that is only if their audiences continue to accept their incredible statements at face value without asking them what objective scientific research supports their claims.

The Ten Myths of Behavior-Based Safety[2]

E. Scott Geller

I have read and heard numerous incorrect statements associated with behavior-based safety. The more popular the behavior-based safety approach becomes, the more it is misrepresented. Unfortunately, this can lead to behavior-based safety being misunderstood and misused, which hinders its overall potential.

Let's set the record straight by dispelling these ten common myths of behavior-based safety:

It's Only Common Sense

You know this is wrong. The principles and procedures of behavior-based safety have been developed and verified through many years of rigorous research. If a particular behavior-based safety project does not work, the problem is not with the principles. It is how the principles are translated for a particular work culture.

It's Just a Fad

Misuse of behavior-based safety can lead to failure or less than desired outcomes. If this happens, support for behavior-based safety will wane. This is why it is so important to clear up the misperceptions that cause misuses of behavior-based safety. If we do not communicate what behavior-based safety is really all about, the odds are greater that it will be considered a fad.

It's a Magic Bullet

In fact, there is 'magic' involved. Behavior-based safety stimulates and facilitates interdependent teamwork, which leads to innovation and creative synergy. To watch this transformation at work is a magical process. But this magic does not come easily or quickly. It happens with proper top-down support and bottom-up involvement. Expecting too much too soon from behavior-based safety can result in disappointment and a label of "failure."

Employees Get the Blame

To talk about "behavior" sounds threatening to many people, as in "Let's talk about your behavior last night." But the names of the individuals observed in a behavior-based coaching process are never recorded. Behaviors of work groups are examined to identify environmental or system factors that influence behavior. Then we change those factors to improve behaviors.

It's Only Observation and Feedback

Behavior-based safety coaching depends on interpersonal observation and feedback in order to succeed. But the principles of behavior-based safety are applicable to many other aspects of safety performance, including incident analysis, process evaluation, corrective action, education and training, and the design of incentive/reward programs.

Management Is Off the Hook

If you implement behavior-based safety with this idea in mind you are destined for failure. Even an observation and feedback process among work teams will fail without proper management support. And the many other applications of behavior-based safety require managers to be actively involved. What management measures gets done, and what management recognizes and rewards gets done well.

Environmental Fixes Are Not Needed

At first glance, behavior-based safety can appear to substitute a behavioral—or human factors—approach for environmental or engineering solutions. But anyone who has implemented this approach effectively knows this could not be further from the truth. When at-risk behavior is identified, the focus turns to finding environmental fixes that can reduce or prevent it.

It's "Touchy-feely" Psychology

When you think of the human dynamics of safety, concepts like attitude, intelligence, emotion, motivation, and self-esteem come to mind. We see and hear these words used in the media to refer to psychology. Such 'touchy-feely' human factors are only studied objectively and scientifically when they are defined according to observable behavior. So behavior-based safety represents the objective and reliable aspect of psychology. It is as close to engineering as any human factors approach can be.

Attitude Change Must Come First

Some psychologists who work with the touchy-feely aspects of human dynamics justify their efforts by claiming the focus of their study—whether it is attitude, self-esteem, or intrinsic motivation—must change before behavior can change. Such a notion is contrary to both research and common sense.

For example, when would you feel the emotion of being scared? Before or after running from a large wild animal? Research and common sense tell us the behavior precedes the internal touchy-feely state. First you see the animal, next you run, and then you experience the fear.

Likewise, when we contribute voluntarily to a successful process, we adjust our attitude to be consistent with our effort. We act ourselves into the right attitude.

There's No Bottom-line Payoff

If most injuries are caused in part by at-risk behavior, then reducing at-risk behavior must eventually lead to injury prevention and financial benefits. "But how long will it take?" asks the plant manager, "And what's the return on investment?" The reply should be: That depends on many factors, including your current injury rate, your particular work system, your readiness to implement a behavior-based safety process, and a bunch of touchy-feely characteristics of your work culture like degree of trust, optimism, belonging, interdependency, and systems thinking.

Money saved by reducing injuries and workers' compensation claims is not the only bottom-line issue here. What about workers' morale, self-esteem, and sense of accomplishment? What about workers' perceived empowerment for safety or the belief that they contribute to preventing their friends and co-workers from getting hurt?

These are bottom-line feeling states that lead to continuous and long-term improvement in safety and in both the quality and quantity of production. Yes, these are touchy-feely aspects of human experience, but they do exist inside all of us and they do influence our everyday behavior.

Organizational Change

How do we improve these feeling states most effectively in an organization? You know that one answer to this important question is behavior-based safety. When people are put in control of a process that visibly contributes to preventing themselves and others from getting hurt, they feel responsible. They go beyond the call of duty to make the process work. Working together this way boosts morale, trust, belonging, and optimism. Obviously, the organization benefits as do those individuals experiencing these feelings.

Culture change led by individual and group behavior does not come quickly. It takes time and a continuous commitment from everyone. Management needs to give the kind of consistent support that encourages employee involvement. Understanding the fallacies in the ten common myths reviewed here is a start.

Barriers to Breakthrough Performance: Seven Reasons Why Programs Can Fail to Produce the Improvements You Want[3]

E. Scott Geller

Over the years, my colleagues at Safety Performance Solutions and I have learned, often the hard way, what it takes to translate the principles of behavior-based safety into effective action plans and long-term improvement. This article points out seven common barriers that can limit your chances of success. In fact, these pitfalls are universal in nature; you should watch out for them when implementing almost any type of program in the workplace.

Failing to Teach Principles

Let's consider how safety programs are typically rolled out. A corporate official, often the safety director, learns about a new program at a conference or through advertising. The appropriate materials—workbooks, videotapes, and a facilitator's guide, for instance—are ordered. An outside consultant or trainer might be hired to teach procedures to certain personnel. These employees then train others while on the job. And a new safety program is born. But to many employees this is just another temporary fix, designed to reduce recordable injuries and make management look good. They do not believe it will work, so they wait for the next "flavor-of-the-month" to come along.

This cynical attitude occurs when people are not taught the principles, rationale, research, and corporate mission behind a program. They are just shown how to do it, perhaps by a co-worker who has just learned himself. I have heard, for example, that some behavior-based safety consultants advocate teaching observation and feedback procedures to a select few employees and expect that the steps will eventually spread throughout the organization.

But in a Total Safety Culture everyone participates because everyone understands the principles behind observation and feedback.

You see, when people grasp the theory underlying a method, they develop their own belief system to rationalize why they should comply. They can also use the flexibility of general principles and guidelines to achieve goals in the manner best suited for their jobs. Finally, when employees make these personal contributions, it helps overcome the next barrier you can encounter.

Lack of Perceived Ownership

Companies often buy copyrighted safety programs. Of course they cannot stamp their own name on them, and therein lies a problem. Employees can and do reject these commercial canned approaches because they feel no personal ownership. When employees learn and come to believe in the principles underlying a program, they willingly customize procedures for their work culture. They give the program its own label and work to keep it relevant and evergreen. Perceptions of ownership and commitment help avoid the third common pitfall in safety programming.

Insufficient Worker Involvement

How can we get more people actively involved in our safety program? I hear this question time and again at seminars. It is obvious to most of us that safety is not the sole job of the plant safety director but a shared responsibility among all employees. After all, line workers and operators are the true safety experts. They know where daily hazards lie and how to avoid them. They also know who takes risks. With proper training, they can increase safe behaviors and decrease risky behaviors of co-workers.

We need to be constantly thinking about how to maximize employee involvement along every step of the way in devising and implementing safety programs.

Each of the seven points in this article relates to engaging employees in the process. For example, a program attracts more participation when it is perceived as founded on the right principles, customized and owned by the work force, and fueled by a proactive need to achieve rather than a reactive need to avoid failure.

Regarding this last point, employees should be encouraged to participate with positive consequences—personal recognition, group celebrations, and trinket rewards—and never forced with threats of punishment, especially in the form of "discipline."

Invisible Top-down Support

Whenever a company launches a new safety program, some amount of management support is implied. If nothing else, management is bankrolling the expense of the endeavor. But it is critical that managers and supervisors get personally involved. They need to talk about their own understanding and belief in the principles. And they must recognize individuals and work teams for accomplishing program objectives.

When I am at a company to "kick off" a behavior-based safety effort, it is common for the plant CEO to introduce me. It is also common for this VIP to turn around and head back to his office after I take the stage. Even if his introduction expresses support for behavior-based principles and processes, actions speak louder than words. To some employees, his quick exit will signal a lack of personal commitment. I have noticed program implementation is usually more effective and the benefits are more long-lasting when the CEO and key supervisory staff sit through my kick-off presentations.

Too Few Champions

Managers and supervisors are making a statement to employees when they willingly sit through a series of three consecutive two-hour presentations to launch a behavior-based program. This will not be the only display of their commitment and leadership. They are champions of the process. Leaders like this are needed throughout an organization—especially at the operator level—for a process to be sustained over the long haul. When a few key individuals believe deeply in the principles and procedures being implemented, they "walk the talk" and make sure the program continues.

I have seen no better way to develop champions of a process than to:

- teach potential leaders principles and procedures,
- teach them how to relay the information to others, and
- provide them opportunities to teach colleagues and co-workers.

When people teach, they develop optimal understanding, commitment, and ownership for the principles and procedures being discussed. Teaching develops internal mental scripts that support the material being taught, and this translates to behaviors that set the right example.

Confusing Goals With Purpose

Champions of behavior-based safety understand the difference between a purpose and a goal. Purpose reflects the overall mission of a safety program; a goal defines a specific outcome targeted with a particular activity or process.

For example, the purpose of a safety program might be to achieve a Total Safety Culture and experience no injuries. Goals are used to keep the mission on track. Achieving a goal should be celebrated because it reflects progress toward the ultimate aim. Indeed, overall success is demonstrated by the number of goals that are reached.

Confusing a program's goals and purpose can drain the life out of a safety program. A champion of behavior-based safety recently told me that he feared for his job because his boss did not understand this distinction. This safety leader felt more pressure to avoid outcome numbers—recordable injuries—than to achieve higher program participation and impact. His motivation, confidence, and optimism were being sapped because he feared being evaluated on the basis of numbers beyond his control. This problem occurs whenever managers disregard the critical distinction between a goal and a purpose, expect quick-fix solutions to large-scale problems, and fall victim to the next pitfall.

Poor Measures of Success

Confusion between a purpose and a goal is especially prominent in safety because numbers—total recordable and lost-time injuries—are the standard measure of success. In fact, some companies use these rates as the sole determinant of success. Safety awards, promotions, and pay increases can hinge on these outcomes.

Judging safety performance solely on injury rates has several drawbacks, including the fact that numbers can be readily manipulated to "look good." Employees often hide an OSHA recordable, perhaps to win a safety reward based on remaining injury-free. It is even possible to report that a lost-time injury happened off the job. Supervisors and safety directors might actually encourage such cheating to influence the numbers on which their bonus depends.

Once employees discover attempts to hide injuries, they will no longer trust the system or believe that they can control the organization's injury rate. Obviously, this does little to support a safety program.

Behavior-based safety provides opportunities to systematically track a variety of success indicators. In fact, the ability to measure program impact objectively and continuously is a special strength of behavior-based safety. The progress of a behavior-based program can be assessed by recording and tracking the number of behavioral observations, the percentage of employees volunteering to be observed, the number of coaching sessions conducted per week, and the percentage of safe behaviors per critical behavior category or per work area.

This is only a partial list of process measures. (See Chapter 5 in this book for more specifics on measuring safety progress.) Achievable goals can be set for these indices, and when employees see that they are making progress toward reaching the goals, it boosts their belief that they can make a difference. This increases feelings of program ownership and commitment to keeping the process going.

I have not given you a complete list of roadblocks to effectively implementing and sustaining a safety program, but these seven pitfalls are key factors that determine if teaching the principles and procedures of behavior-based safety will lead to long-term and large-scale success. The next paper discusses additional principles relevant to maintaining a behavior-based safety process.

Increasing the Impact of Signs and Slogans[5]

E. Scott Geller

W. Edwards Deming told us to eliminate signs, slogans, exhortations, and objectives from the workplace. What did he mean? Did he mean we ought to stop giving people directions to follow and goals to shoot for? If he did, he was denying the basic Activator-Behavior-Consequence (ABC) framework for behavior-based safety. Substantial research has verified that behavior is influenced markedly by activators (or stimuli like signs preceding behavior) and by consequences (pleasant or unpleasant events following behavior).

I do not think Dr. Deming was disregarding behavioral science principles with his point about eliminating signs, slogans, and exhortations. Rather, he was criticizing the standard top-down development and display of performance activators. If we keep attempting to activate behavior with standard practices, we might as well eliminate them all together. As they are currently used, signs, slogans, and exhortations for safety might only raise expectations without giving relevant and meaningful direction. As such, they could do more harm than good. The ABC principle of direction from activators is not wrong, however. The way the principle gets translated into procedures or operations needs elimination or improvement.

In this article, I offer six guidelines for increasing the impact of activator techniques. These guidelines might imply that some of your current signs, slogans, goals, and exhortations ought to be eliminated, as Deming suggested. But, if you follow these guidelines when developing new activators you will not only increase safe behavior and decrease at-risk behavior, you will also help to develop the kind of attitudes needed to sustain behavior change.

Guideline 1: Specify Behavior

Behavioral research demonstrates that signs with general messages and no specification of a desired behavior to perform (or an undesirable behavior to avoid) have very little impact on actual behavior. But signs that refer to a specific behavior can be beneficial. Therefore, don't expect signs or slogans with general exhortations like "Think Safe," "Eliminate Accidents," and "Meet Our Goal of Zero Injuries" to have much impact. On the other hand, activators that give specific direction can influence behavior and thus prevent injuries. Specific messages might tell people where certain behaviors are needed or appropriate (such as "Hard Hat Zone," "Designated Smoking Area," and "Ear Protection Required") or give direction on how to perform safely (such as "Lift With Legs," "Walk Within the Yellow Lines," and "Buckle Up for Safety").

Although activators should specify a response to perform or avoid, they should not be overly complex. Signs with a lot of words are easy to overlook—with time complex signs just blend into the woodwork. But, of course, even simple and specific messages will also lose their impact with time. Following the next guideline explains why this is the case.

Guideline 2: Maintain Salience with Novelty

It is perfectly natural for activators like sign messages to lose their impact over time. This process is called habituation, and it is considered to be the simplest form of learning. Through habituation we learn *not* to respond to an event that occurs repeatedly. If there is no obvious consequence (good or bad) from responding to a stimulus, the organism stops reacting to it. It is a waste of time and energy to continue responding to an activator that seems to be insignificant.

What is the relevance of habituation for safety? It is human nature to habituate to everyday activators in our environment that are not supported by consequences. And this is the case with many safety activators. So we should expect the same sign or slogan to lose its influence over time unless consequences are in place to support the message.

A sign requesting the use of personal protective equipment might eventually be ignored, for example, if consequences are not in place to support the activator. And, I do not mean only the availability of penalties if the message is not followed. Given the natural negative consequences for most safe behavior (such as inconvenience, discomfort, or inefficiency), it is essential to add some positive consequences (like positive feedback or recognition) to support compliance with safe-behavior activators.

Guideline 3: Vary the Message

What does habituation tell us about the design of safety activators? Essentially, we need to vary the message. When an activator changes it can become more salient and noticeable. Over the years I have noticed a variety of techniques for changing the message on safety signs. Some signs have removable slats on which to place different messages. I am sure most of you have seen computer-generated signs with an infinite variety of safety messages. Some plants even have video screens in break areas, lunchrooms, visitor lounges, and hallways that display many kinds of safety messages, conveniently controlled by user-friendly computer software.

Who determines the content of these messages? I know who—the target audience for these signs. The same people expected to follow the specific behavioral advice should have as much input as possible in defining message content. Many organizations can get suggestions for safety messages just by asking. But if employees are not accustomed to giving safety suggestions, they might need a positive consequence to motivate their input.

Guideline 4: Involve the Target Audience

This guideline should be obvious by now. It is relevant for developing and implementing any behavior-change intervention. When people contribute to a safety effort, their ownership and commitment to both safety and the improvement process increase. Of course, this guideline works both ways. When individuals feel a greater sense of ownership and commitment, their involvement in safety achievement is more likely to continue. Thus, involvement feeds ownership and commitment, and vice versa.

When public trash receptacles include the logos of nearby businesses, the merchants whose logos are displayed typically take care of the receptacle and keep the surrounding area clean. This guideline is also supported by the success of "Adopt-a-Highway" programs that have groups keep a certain roadway clear of litter and perhaps beautified with plants, shrubs, or flowers. Group ownership of public space typically leads to actively caring for its appearance.

Guideline 5: Activate Close to Response Opportunity

It is intuitive that the shorter the delay between a direction and the opportunity to follow the direction, the greater the probability of compliance. Thus, safety messages placed in work areas where the activated behavior should be performed have greater impact than similar messages in memos, newsletters, and safety talks. Likewise, "point-of-purchase advertising"—or placing ads at locations where the target products can be purchased—is an optimal form of product marketing.

This simple guideline explains why researchers find greater increases in vehicle safety-belt use when "Buckle Up" messages are located at parking-lot entrances/exits and intersections than on television. Similarly, road signs that give drivers feedback on the percentage of vehicles exceeding posted speed limits are more effective at reducing speeding, whereas public service announcements on radio and television about vehicle safety have minimal behavioral impact. (See Geller 2001 for a comprehensive review of the research in this area.) Of course, one of the most effective activators of speed reduction is the beeping sound of a radar detector. This activator is not only salient, response specific, and proximal to response opportunity, it also has the next guideline going for it.

Guideline 6: Implicate Consequences

The salient beep of a radar detector effectively motivates reductions in vehicle speed because it enables drivers to avoid a negative consequence—an encounter with a police officer. Field research has shown that activators that do not implicate consequences can influence some behavior when they are salient and implemented in close proximity to an opportunity to perform the specified target behavior. It is important to realize, though, that the target behaviors in these studies were all relatively convenient to perform: depositing handbills in a particular receptacle, choosing certain products, using available safety glasses and safety belts. There is plenty of evidence that activators alone will not succeed when the target behavior requires more than a little effort or inconvenience. (See Geller, Winett, and Everett 1982 for a review of the research.)

Activators that signal the availability of a consequence are either incentives or disincentives. An incentive announces to an individual or group, in written or oral form, the availability of a reward. This pleasant consequence follows the occurrence of a certain behavior or an outcome of one or more behaviors. In contrast, a disincentive is an activator announcing or signaling the possibility of receiving a penalty. This unpleasant consequence is contingent on the occurrence of a particular undesirable behavior.

Research has shown quite convincingly that the impact of a legal mandate varies directly with the amount of media promotion or disincentive. Similarly, the success of an incentive program depends on making the target population aware of the possible rewards. In other words, marketing the availability of positive or negative consequences with activators (incentives or disincentives) is critical for the motivating success of a consequence intervention.

Some people add their own internal consequence to an activator. They might follow a safety rule to set an example for others or to avoid an embarrassing encounter with another person. Thus, whether talking about consequences from our external world or from our own internal dialogue, when activators implicate them, they are more powerful. Goal setting will motivate as well as direct behavior to the extent that the consequences achieved by reaching the goal are realized. (See Geller 1996 and 2001 for more on the role of using consequences to make activators more effective.)

Conclusion

We are constantly bombarded with activators. At home we get telephone solicitations, junk mail, television commercials, and verbal requests from family members. At work it is phone mail, e-mail, memos, policy pronouncements, and verbal directions from supervisors and co-workers. On the road there is no escape from billboards, traffic signals, vehicle displays, radio ads, and verbal communication from people inside and outside our vehicles. Only a portion of the activators we perceive actually influences our behavior. Understanding the six principles discussed in this article can help you predict which ones will influence behavior change.

Obviously, we do not need more activators in our lives. We certainly do need *more effective* activators to promote safety and health. It would be far better to make a few safety activators more powerful than to add more activators to a system already overloaded with information. We need to plan our safety activators carefully so the right safety directives receive the attention and ultimate action they deserve.

References

Geller, E. S. 1996. The truth about safety incentives. *Professional Safety*, 42(1).

———. 2001. *The psychology of safety handbook*. Boca Raton, FL: CRC Press.

Geller, E. S., R. A. Winett, and P. B. Everett. 1982. *Preserving the environment: New strategies for behavior change*. Elmsford, NY: Pergamon Press.

Sustaining Participation in a Safety Improvement Process: Ten Relevant Principles from Behavioral Science[5]

E. Scott Geller

Behavior-based safety (BBS) has contributed significantly to the safety profession. Defined in several other sources (Geller 1998c; Krause, Hidley, and Hodson 1996; McSween 1995; Sulzer-Azaroff 1998), applications of the basic principles and procedures of BBS have dramatically benefited safety-related behavior when they are implemented correctly (see reviews in Geller 2001a; Petersen 1989). But the offerings of BBS extend beyond behavior change and the reduction of injuries. As reviewed elsewhere (Geller 2001b), BBS has helped safety professionals realize that understanding and improving the human dynamics of safety require more than common sense. Behavioral safety has shown the value of psychological research and provided tools for obtaining objective data for analyzing barriers to safe behavior and deciding what to do about them.

Behavior safety has also stimulated controversy; some critics argue that a behavioral focus puts excessive responsibility on the worker and gives managers an excuse to shirk their safety responsibilities (Frederick and Howe 2001; Howe 1998; Hoyle 1998; Manuel 1998), and others claim that BBS is too limiting and should be abandoned for a more holistic or culture-focused approach (Simon 2001; Topf 1998, 2001). Other safety professionals have recognized utility in BBS but have appealed for breakthroughs including more efficiency, flexibility, and effectiveness with regard to producing long-term change (Kamp 2001; Sarkus 2001).

This is all good news for the safety profession. Behavioral safety has provided a platform for constructive debate, and the conflicting opinions have challenged safety professionals to learn more about the psychology of injury prevention. Resistance to accepting a principle or procedure on hearsay alone is commendable and inspires objective research and dissemination of practical findings. As a result we learn more about the human dynamics of safety and feed our subjective and prejudiced common sense with unbiased research-based wisdom.

This article is presented in the spirit of expanding common behavior-based knowledge about the human dynamics of safety. More specifically, this article presents ten research-based principles relevant to meeting the challenge of producing more "durability of positive change" in safety-related attitudes and behaviors (Sarkus 2001, p. 24). This requires a discussion of concepts not typically linked to BBS. The presentation is not inconsistent with BBS but rather provides psychological principles that should be considered whenever long-term change in behavior and attitude is desired. When operationalized appropriately, the basic principles given here can benefit any safety process requiring lasting participation.

1. Self-Perception Is Defined by Behavior

We begin with a basic premise in BBS: the need to get people to change what they do. This is the principle behind the common BBS slogan that it is possible to "act a person into safe thinking" (Geller 2001a, p. 18). I will show later however, that the method used to activate or motivate behavior will determine how much thinking or attitude change will follow the behavior change.

Our behaviors define who we think we are. The notion that we define who we are from our behaviors is founded in the teaching and research of B. F. Skinner and the follow-up scholarship of Daryl J. Bem (1972). Dr. Bem developed a comprehensive theory of self-perception on the basic premise that "[i]ndividuals come to 'know' their own attitudes, emotions, and other internal states partially by inferring them from observations of their own overt behavior and/or the circumstances in which this behavior occurs" (p. 2).

When we want to learn what another person is thinking or feeling, we look to see how that person acts in a particular environmental context. Similarly, according to Bem's self-perception theory, when we want to know how we feel, we look at our own behavior and the circumstances surrounding it. For example, a person eating an excessive amount of food may then state, "I must have been hungrier than I thought." Or, the individual who performs a certain task below par may conclude, "I'm not as good at this as I thought I was." Then there's the worker who goes out of her way to help another person and thinks, "I must care more than I thought I did." These are obvious cases of behavior influencing thought processes.

Supportive Research

Classic research by Stanley Schachter and Jerome Singer (1962) supports the claim that self-perception is determined in part by overt behavior. Experimental subjects were given injections of epinephrine or adrenaline, which made them feel physically aroused. Then these subjects waited individually with one other person who presumably had received the same injection. Moments later the experimenter returned and asked the two individuals to complete a questionnaire about their feelings.

Each subject experienced physiological arousal, but some reported extreme anger while others said they were very happy. What made the difference? The subject's own behavior and that of the other person in the waiting room—who was actually a research assistant—determined whether the participant felt extreme anger or joy.

The research assistant, posing as another subject, acted in one of two ways during the waiting period. In the Euphoric condition, he threw paper airplanes, shot crumpled balls of paper into a wastebasket, and twirled in a hula hoop. He encouraged the subject to join in the fun, and most did.

In the Anger condition, both individuals were asked to complete a questionnaire while waiting for the experimenter to return. The questions were intimate and quite inappropriate, especially the question "With how many men has your mother had extramarital affairs – four and under, five to nine, or ten and over?" When the research assistant read this question he ripped up his survey form in a fit of rage.

Perhaps the finding that behavior influenced the emotional state in this study seems obvious or just common sense. But realize what the results mean. The participants' emotional reactions were determined by the behavior they observed in themselves and another person. After being physically aroused, they felt joy or anger depending on external circumstances and accompanying behavior.

Actually, this is only one example of numerous experiments that have demonstrated the crucial role of external events and behaviors on perception of personal emotions, attitudes, and moods. For example, in one series of studies, participants' emotions were manipulated by giving them false auditory feedback about their heartbeat (Valins 1966). When they presumably heard their heart beating faster, they felt more sexually aroused or fearful, depending on other external events. Thus, it is possible to change self-perception by altering external conditions. In other words, we can act ourselves (and others) into thinking or feeling a certain way.

Acting People Into Certain Thinking

Suppose you want to motivate someone to do something. How would you do it? Well, even if you believe people can only truly motivate themselves from within, you can at least establish an external condition or environmental context that facilitates intrapersonal motivation. What does this mean? It means you can do things to increase the probability people will do what you want. Attempts to make this happen are considered interventions, and they can vary from developing an external accountability system or incentive program to initiating opportunities for personal choice, ownership, or constructive interpersonal conversation.

The most efficient way to motivate a certain action in others is to create an environmental context or behavior-consequence contingency that facilitates the occurrence of desired performance. In other words, establish the conditions for people to act themselves into new ways of thinking. Their new behavior can influence a new way of perceiving themselves, which can lead to a new personal label and then to more behavior consistent with that label.

Personal change can be viewed as a continuous spiral of behavior causing thinking, thinking inducing more behavior, and then this additional behavior influencing more thinking consistent with the behavior, and so on. It does not matter which came first—behavior or thinking. What matters is that we can affect beneficial change in others by focusing on their behavior. Then, when people see themselves doing what you want, they might change how they view themselves. They might act themselves into thinking differently and thus motivate themselves to sustain the new behavior.

I emphasized "might" in the prior two sentences because some interventions to change behavior do not facilitate an attendant change in thinking; not all behavior change leads to relevant and supportive change in thinking and self-perception. Let's consider some research-based principles for increasing the likelihood that an intervention to alter behavior will also influence thinking consistent with that behavior change.

2. Direct Persuasion Has Limited Impact

Advertisers use direct persuasion. They show us people enjoying positive consequences or avoiding negative consequences by using their products. As such, they apply the ABC contingency of BBS to sell their wares or services. The activator (or "A" of the ABC contingency) announces the availability of a reinforcing consequence (the "C" of the ABC contingency) if the purchasing behavior is performed (the "B" of the ABC contingency).

Advertisers also apply research-based principles from social psychology to make their messages more persuasive. Specifically, social scientists have shown advantages of using highly credible communicators and arousing the audience's emotions (Aronson 1999; Hovland and Weiss 1951). Sales pitches—often given by authority figures—attempt to get viewers emotionally involved with product-related issues.

Note, however, that these attempts at direct persuasion are not asking for behavior that is inconvenient or difficult. Normally, the purpose of an advertisement is to persuade a consumer to select a certain brand of merchandise. This boils down to merely choosing one commodity over another at the retail store. While shopping, consumers only need to move their hands a few inches to select one product over another. This is hardly a burdensome change in lifestyle.

Safety-related behavior is usually more inconvenient and requires more effort than switching brands at a supermarket. It often requires significant adjustment in a highly practiced and regular routine at work, at home, or on the road. Adopting a safe way of doing something might first require the elimination of an efficient and convenient at-risk habit. Furthermore, participation in a safety-promotion effort usually requires the regular performance of several inconvenient safety-related behaviors.

Here's my point: long-term participation in a safety-related work process is far more cumbersome and lifestyle-changing than the consumer behavior targeted by advertisers. As a result, direct persuasion is frequently not the best approach to increase safety-related behavior or to promote long-term participation in a safety process.

Direct attempts to persuade people to make inconvenient changes in their lifestyle usually yield disappointing outcomes. For example, communication strategies have generally been unsuccessful when designed to persuade smokers to quit smoking (Elder, Geller, Hovell, and Mayer 1994), drivers to stop speeding (Geller 1998a), homeowners to conserve water (Geller, Erikson, and Buttram 1983) or insulate their water heaters (Geller 1981), bigoted individuals to cease prejudicial behavior, or sexually active people to use condoms (Aronson 1999). Similarly, the "Just Say No to Drugs" campaigns have not influenced much behavior change.

The problem with direct persuasion is that it comes across as someone else's idea. It could even give the impression that the behavior is actually for someone other than the performer. This can cause a disconnection between the behavior and self-perception. There is no self-persuasion—a state or mindset needed for lasting change.

3. An Indirect Approach Is More Likely to Influence Self-Persuasion

Self-persuasion is more likely to occur when the motivational strategy is less obvious. Have you ever received flattery or a favor from someone and thought, "That person is only trying to get something from me?" In this case, your self-perception will not change, because you're suspicious of the other person's intentions. You might think, "She doesn't really mean that, she's only trying to win my favor." My students use the term "kissing up" for this kind of behavior.

Compliments regarding a person's performance are often more powerful when they are more indirect than direct (Allen 1990; Geller 1997). Your personal experience probably verifies this. Consider that you overhear a person tell someone else about your superb achievement on a particular assignment. Or, suppose a friend gives you secondhand recognition by sharing what another person said about your special talents. Both of these situations reflect indirect commendation and would likely have more influence on your self-perception than a direct interpersonal statement of praise. Why? Because, the direct approach is tainted by the possibility the flattery is given for an ulterior motive.

Indirect persuasion deviates significantly from the standard direct and top-down method of attempting to obtain compliance with safety regulations. Both approaches might be equally effective at motivating behavior change, but the indirect approach will be far more successful at enhancing the kind of internal dialogue needed to maintain behavior in the absence of an external motivator or accountability system.

Defining intervention conditions that can make this happen is not easy, but start by asking yourself these questions:

- ※ "Does the situation promote individual choice, ownership, and personal accountability?"
- ※ "Does the context in which safety participation is desired contribute to connecting or disconnecting the link between what people do and what they think of themselves?" and,
- ※ "Are the safety-related activities only behaviors or do they stimulate supportive cognitive activity or self-persuasion?"

4. Self-Persuasion Is Key to Long-Term Behavior Change

I define self-persuasion as an internal dialogue or mental script that supports our ongoing behavior and motivates continued participation in the absence of external contingencies. Such behavior is self-directed (as opposed to other-directed) and self-motivated. Self-directed behavior is more likely than other-directed behavior to influence self-perception, and in turn help to sustain behavior change.

Most safety-related behavior starts as other-directed, meaning the safe way of doing something is told or showed through instructional manuals or training sessions (Geller 1998b, 1999). We continue to practice the more inconvenient but safe way of doing something, either because someone is holding us accountable (the behavior is still other-directed) or because we are holding ourselves accountable (the behavior has transitioned to self-directed).

Obviously, when it comes to injury prevention, self-directed behavior is more desirable than other-directed behavior, because in this state individuals choose the safe way when they perform alone. Consequently, it is important to define situations and contingencies that promote self-persuasion and a transition from other-directed to self-directed behavior.

Let's consider the impact of certain intervention approaches at facilitating or inhibiting self-persuasion and self-directed behavior. Behavioral research suggests that some standard ways of promoting safe work practices leave much to be desired; that is, they hinder self-persuasion, self-directed behavior, and thus enduring change.

5. Large Incentives Can Hinder Self-Persuasion and Lasting Change

Consider that someone offers you a large sum of money—an incentive—to do something for safety. While the likelihood you will perform the desired activity will increase, the incentive is apt to stifle self-persuasion and a self-directed state. You will be less likely to persuade yourself that the behavior is a reflection of your personal values than if you did the safety-related behavior for little or no external incentive.

The scenario outlined above has been evaluated in numerous experiments, and the results have demonstrated the superior influence of small over large incentives. The classic study in this domain was conducted by Leon Festinger and Merrill Carlsmith in 1959. They paid college students either $20 or $1 to tell another student a boring task they just performed was actually fun. Afterwards, they were asked to offer their personal opinion of the task.

Which group was more likely to develop a self-perception consistent with their verbal behavior? In other words, which incentive condition influenced more self-persuasion that the task was not as boring as it seemed? Yes, the lower incentive contingency facilitated more self-persuasion, presumably because these subjects had less external motive to call a dull task fun. As a result, they provided themselves internal motivation or justification for their verbal behavior. With only minimal incentive to tell a lie, they convinced themselves the task was really not that boring. In contrast, the $20 group had an excuse for lying and thus did not need to change their perceptions of the task.

The same kind of self-persuasion occurs when we put a lot of effort into a special assignment without extra compensation. Without an external reinforcer for our behavior, we move inside our heads for justification. We persuade ourselves the effort is especially worthwhile and deserves our "blood, sweat, and tears." Analogously, the more we need to go through to join a group (as in the infamous fraternity "Hell Week"), the more self-persuasion will occur to convince ourselves it was worth it. Research by Eliot Aronson and Judson Mills (1959) supports this conclusion, finding that students who went through a severe initiation to become a member of a "special" discussion group rated the group's silly and boring discussion as significantly more interesting than did students who gained admission to the same group with only a mild initiation.

What is the message here for using incentives to motivate safe work practices? Most obviously, safety incentives must not be presented nor perceived as a "payoff" for behavior. When the only justification people give for their behavior are external consequences, they are not likely to develop an internal rationale for their actions. Thus, behavior-based rewards for safety should be given as only "tokens of appreciation" for the many things people do to keep themselves and others injury-free. Preventing injury is the big payoff, and that warrants plenty of internal justification or self-persuasion.

6. Mild Threats Influence More Self-Persuasion Than Severe Threats

Now let's consider the use of disincentives or threats to motivate behavior. Should a threat be severe or mild? You know the answer to this question. If you want self-persuasion to occur consistent with the desired behavior, you should use the smallest disincentive possible to initiate the behavior you want. Then, through self-persuasion, the behavior has a chance of continuing when the intervention is no longer available.

The superiority of a small over a large disincentive to prevent undesirable behavior was demonstrated in a series of experiments referred to as "the forbidden toy studies" (Aronson and Carlsmith 1963; Freedman 1965). Children were asked not to play with an attractive toy and then were given either a mild or severe threat of punishment for disobeying (Aronson and Carlsmith 1963; Freedman 1965).

In the Mild Threat condition the experimenter said something like, "It is wrong to play with that toy." An additional statement was added in the Severe Threat condition, like, "If you play with that toy, I shall be very angry and will have to do something about it." Then, the experimenter left the room and stepped behind a one-way mirror to record whether the subject played with the forbidden toy or with a number of other less attractive toys that were available.

Regardless of the disincentive condition, very few children played with the forbidden toy. That is a critical point. The mild threat was sufficient to prevent the undesirable behavior. Then, the experimenter tested which condition produced the most self-persuasion by assessing the children's preference for the toys or providing them an opportunity to play with the forbidden toy later without the disincentive.

In a classic study by Jonathan Freedman (1965), for example, another experimenter returned to the school where 44 boys had participated in a Mild or Severe Threat condition six weeks earlier. The experimenter took the boys out of class individually, and with no reference to the prior study, instructed each boy to take a drawing test. While scoring the test, the experimenter told the boy he could play with any toy in the room.

The same five toys from the previous study were available, including the forbidden toy.

Of the boys from the Severe Threat condition, 17 (77 percent) played with the forbidden toy, compared to only 7 (32 percent) from the Mild Threat condition. Presumably, more children given the mild disincentive adopted a self-perception consistent with their avoidance behavior during the earlier session. Through self-persuasion, they developed a rationale for avoiding the previously forbidden toy in the absence of an external punishment contingency.

In an instructive follow-up experiment in 1971, Marc Lepper (as cited by Bem 1972) tempted young boys (with an attractive prize) to falsify their scores on a test he gave them. Three weeks earlier in another setting, these same subjects had resisted playing with the forbidden toy following a mild or severe threat. Those boys who had earlier received the mild threat were significantly less likely to cheat than those who had received the severe threat. Apparently, the boys who earlier complied with only a mild threat were more likely to develop a self-perception like "I'm a good boy who resists temptation." Such internal dialogue or self-persuasion influenced resistance to temptation to cheat three weeks later.

Is there a message here for workplace programs designed to motivate compliance with safety regulations? Language like "compliance" and "regulation" clearly puts the control outside of people. Now add the threat of a fine or losing one's job. These kinds of contingencies will certainly motivate people to follow rules that are enforced, but don't expect such disincentives to encourage the kind of self-persuasion needed when people work alone with no one to hold them accountable except themselves.

7. The More Obvious the External Control the Less the Self-Persuasion

To decide how a particular situation might influence your own or another person's self-persuasion, try this. Imagine you are watching the individual (either yourself or another person) performing a particular behavior under a particular accountability system or set of circumstances. Then ask this question: Are there sufficient external consequences to justify the amount of effort demonstrated? If yes, then the performer is not likely to develop an internal justification for the behavior. If your answer is no, then you could assume some internal dialogue or self-persuasion has occurred or is occurring.

Bottom line: It is necessary to promote self-persuasion and self-directed behavior whenever resources are insufficient to keep incentives or disincentives in place to sustain desired and effortful behavior over the long term. This means the ABC contingency (activator-behavior-consequence) of BBS must be strong enough to get the behavior started but not sufficient to provide complete justification for the effort. This allows for self-persuasion and maintenance of participation when an external accountability system is no longer available.

8. Self-Efficacy Is Key to Empowerment and Long-Term Participation

There is empirical and theoretical support for the assertion that believing you can do something is the first step toward doing it. The academic term for this belief is "self-efficacy," and it has been the topic of many research articles and theoretical proposals. Most notable is Albert Bandura's 1997 book, entitled simply "Self-Efficacy." Here, in 604 pages of fine print, the author makes a strong case for self-efficacy being the most central and critical concept in applied psychology. Let's consider the main features of this concept.

Can I Do It?

Self-efficacy reflects a "can do" attitude. It refers to a person's perception that he or she can organize the relevant resources and execute the procedures necessary to reach a certain goal. Numerous studies have shown that people who score relatively high on a measure of self-efficacy gain more from psychotherapy and perform better at a wide range of tasks. They show greater ability and motivation to solve complex problems at work, and they have better health and safety habits.

Self-Efficacy vs. Self-Esteem

Self-efficacy is not the same as self-esteem, but self-efficacy contributes to self-esteem, and vice versa. Self-esteem reflects a general sense of self-worth as in "I am valuable," whereas self-efficacy refers to feeling successful or competent at a particular task. While self-esteem remains rather constant across situations, self-efficacy is task-focused and therefore can vary markedly from one circumstance to another.

Self-Efficacy and Training

The aim of good training is to give participants certain skills or competencies. For example, effective training in BBS safety coaching (Geller 1995) involves teaching specific procedural steps for observing and analyzing another person's behaviors, delivering supportive and corrective feedback, developing an improvement plan, and obtaining commitment for change and follow-up. But effective safety coaches need more than these skills. They need self-efficacy.

Participants in a training program show what they know on written exams, and they demonstrate their behavioral expertise through role-play. Skillful trainers improve skills through behavior-based feedback. In other words, through practice and feedback, participants develop competence.

Mastery at something usually builds self-efficacy, and vice versa. But having one does not mean the person has the other. People may have the skills to excel at a certain task, but lack the personal confidence to execute the required behaviors. And, it is possible for people to have more self-efficacy than warranted by their ability to perform. That is what we mean when we label a person "over-confident." But, even when people have both self-efficacy and adequate skills to implement an intervention, they will not do it unless they also have *response*-efficacy.

9. Response-Efficacy Is Key to Empowerment and Long-Term Participation

Response-efficacy refers to one's belief that a certain technique or strategy will actually produce a desired outcome. Thus, it is not enough to know what to do and have the confidence to do it. The participant must believe it is worth doing. People can have the skills and self-efficacy to perform safety coaching, for example, but they will not actually coach on a regular basis unless they believe the coaching process will improve safety.

Response-Efficacy and Training

This concept of response-efficacy has critical implications for safety training. Specifically, it is not enough to teach participants the procedural steps for conducting a certain safety process. Trainees need to be convinced the technique has the potential to prevent personal injury. How can we accomplish this?

Response-efficacy is most commonly taught with group numbers or outcome statistics. Data are presented from research that show significant improvement as a result of a particular intervention strategy. This approach is effective for research presentations, but it might not convince the average employee.

Case Studies Over Statistics

I do not mean to imply that employees can't understand statistics. My point is, they will not necessarily relate to such numbers. It is better to get more personal and relevant when trying to "sell" the value of an intervention process to a work force. Research on risk perception, for example, has shown that people get more concerned or outraged about an issue when individual case studies are used in lieu of group statistics (Sandman 1991; Slovic 1991). That is the rationale behind politicians pointing to specific individuals in their audiences when attempting to gain support for a particular issue or plan of action.

Personal testimonies provide a powerful image. Listeners can relate to an individual's story and put themselves in the same situation. Two kinds of testimonies can increase response-efficacy: 1) a personal account of an injury that could have been prevented by the safety technique, and 2) an anecdote about someone who avoided injury by using the particular strategy or safety process.

10. Motivation to Act Comes from Outcome Expectancy

The third type of belief entertained in Dr. Bandura's text on self-efficacy is "outcome expectancy." This is about the consequence one expects to receive when practicing the intervention process. One of the most important legacies of B. F. Skinner (1953, 1971, 1974) is "selection by consequences," which means behavior is motivated by events or conditions that follow it. In other words, we motivate ourselves to do or not do something by anticipating what positive consequences we expect to gain from our participation and/or what negative consequences we expect to avoid.

We might believe we can do something and believe what we do will have a certain effect, but we will not perform unless we also believe the effect is worth working for. Will our labor reap sufficient benefits to outweigh the costs, including our effort? In safety, for example, a group might believe their safety record is good enough, given that they see very few co-workers getting seriously injured. The potential gain from an inconvenient safety process might seem too small to justify the amount of extra effort required for implementation.

Here again, it is more useful to "sell" with a case study than a statistic like "total recordable injury rate." It is also important to point out the difference between an individualistic mindset and a systems perspective. From an individual outlook, the probability of an injury is minuscule. As a result, the outcome from participation in a safety effort can seem insignificant. In contrast, systems thinkers take a wider view. They recognize that someone will benefit from large-scale participation in a safety process.

You could show, for example, the details of a single injury that occurred in your plant and explain how an intervention like the one being taught could have prevented that incident. In this way, you are inspiring both response-efficacy and outcome expectancy—the belief that the intervention process can work and produce an outcome everyone should want and be willing to work for.

Conclusion

The pop psychologists and motivational speakers are right. Beliefs *are* important in determining success. However, these advocates usually only tell us about one kind of belief—self-affirmation or self-confidence in one's ability to meet a challenge. Clinical psychologists call this "self-efficacy."

Self-efficacy is a key to success, but it is not the only key. A belief that a particular action plan will be effective is also critical, as well as a belief that the ultimate outcome from the intervention is worth the effort. Psychologists call the former belief "response-efficacy" and the latter "outcome-expectancy."

Safety consultants and trainers need to understand and appreciate these three types of personal beliefs and attempt to influence them when teaching the steps of an intervention process. It is not enough to convince trainers they can perform a certain technique. Trainers must also believe the technique will work to produce a desirable outcome, and that the ultimate effects of the intervention will be worthwhile.

Systems thinking and an interdependent mindset can augment the perceived value of the potential consequences of an intervention. In other words, when people see the bigger picture and adopt a "collective community" perspective, they realize their participation in a safety process will eventually benefit someone in their workplace. This belief that it is meaningful to work for the potential benefit of others provides motivation to fuel efforts that develop self- and response- efficacy. This is the belief that it is preeminent to actively care for the safety and health of others.

Safety professionals also need to understand the concept of self-persuasion and its role in sustaining long-term behavior change. Situations most likely to facilitate self-persuasion are those which limit the salience of direct outside control. In other words, the more obvious the external control or accountability, the greater the disconnection between behavior and self-perception and the less self-persuasion and sustained participation when the intervention is removed.

References

Allen, J. 1990. *I saw what you did and I know who you are: Bloopers, blunders and success stories on giving and receiving recognition.* Tucker, GA: Performance Management Publications.

Aronson, E. 1999. The power of self-persuasion. *American Psychologist*, 54:875–884.

Aronson, E., and J. M. Carlsmith. 1963. Effect of severity of threat on the valuation of forbidden behavior. *Journal of Abnormal and Social Psychology*, 66:584–588.

Aronson, E., and J. Mills. 1959. The effect of severity of initiation on liking for a group. *Journal of Abnormal and Social Psychology*, 59:177–181.

Bandura, A. 1997. *Self-efficacy: The exercise of control.* New York: W. H. Freeman and Company.

Bem, D. J. 1972. Self-perception theory. In *Advances in experimental social psychology*, Vol. 6. L. Berkowitz, ed. New York: Academic Press. 1–60.

Elder, J. P., E. S. Geller, M. F. Hovell, and J. A. Mayer. 1994. *Motivating health behavior*. New York: Delmar Publishers.

Festinger, L., and J. M. Carlsmith. 1959. Cognitive consequences of forced compliance. *Journal of Abnormal and Social Psychology*, 58:203–210.

Frederick, J., and J. Howe. 2001. The employee's perspective on behavioral safety. Presentation at the *ASSE Behavioral Safety Symposium: The Next Step*, Orlando, FL.

Freedman, J. L. 1965. Long-term behavioral effects of cognitive dissonance. *Journal of Experimental and Social Psychology*, 1:145–155.

Geller, E. S. 1981. Evaluating energy conservation programs: Is verbal report enough? *Journal of Consumer Behavior*, 8:331–334.

———. 1995. Safety coaching: Key to achieving a Total Safety Culture. *Professional Safety*, 40(7):16–22.

———. 1997. Key processes for continuous safety improvement: Behavior-based recognition and celebration. *Professional Safety*, 42(10):40–44.

———. 1998a. *Applications of behavior analysis to prevent injuries from vehicle crashes.* 2nd ed. Cambridge, MA: Cambridge Center for Behavioral Studies.

———. 1998b. *Beyond safety accountability: How to increase personal responsibility.* Neenah, WI: J. J. Keller & Associates, Inc.

———. 1998c. *Understanding behavior-based safety: Step-by-step methods to improve your workplace.* 2nd ed. Neehan, WI: J. J. Keller & Associates, Inc.

———. 1999. Behavior-based safety: Confusion, controversy, and clarification. *Occupational Health & Safety*, 68(1):40, 42, 44, 46, 48–49.

———. 2001a. *The psychology of safety handbook.* Boca Raton, FL: CRC Press.

———. 2001b. What's so special about behavioral safety. *Proceedings of the ASSE Behavioral Safety Symposium: The Next Step* (11–25), Orlando, FL.

Geller, E. S., J. B. Erickson, and B. A. Buttram. 1983. Attempts to promote residential water conservation with educational, behavioral, and engineering strategies. *Population and Environment*, 6:96–112.

Hovland, C., and W. Weiss. 1951. The influence of source credibility on communication effectiveness. *Public Opinion Quarterly*, 15:635–650.

Howe, J. 1998. A union critique of behavioral safety. Presentation at the ASSE Behavioral Safety Symposium, Orlando, FL.

Hoyle, B. 1998. *Fixing the workplace, not the worker: A workers' guide to accident prevention.* Lakewood, CO: Oil, Chemical and Atomic Workers International Union.

Kamp, J. 2001. It's time to drag behavioral safety into the cognitive era. Proceedings of the *ASSE Behavioral Safety Symposium: The Next Step.* Orlando, FL. 45–52.

Krause, T. R., J. H. Hidley, and S. J. Hodson. 1996. *The behavior-based safety process: Managing improvement for an injury-free culture.* 2nd ed. New York: Van Nostrand Reinhold.

Manuel, F. 1998. Perspectives on behavioral safety: Observations of ASSE's behavior safety symposium. *Professional Safety,* 43(8): 32–37.

McSween, T. E. 1995. *The value-based safety process: Improving your safety culture with a behavioral approach.* New York: Van Nostrand Reinhold.

Petersen, D. 1989. *Safe behavior reinforcement.* New York: Aloray, Inc.

Sandman, P. M. 1991. *Risk = hazard + outrage: A formula for effective risk communication.* Videotaped presentation for the American Industrial Hygiene Association, Environmental Communication Research Program, Cook College, Rutgers University, New Brunswick, NJ.

Sarkus, D. J. 2001. Safety and psychology: Where do we go from here? *Professional Safety*, 46(1):18–25.

Schachter, S., and J. E. Singer. 1962. Cognitive, social, and physiological determinants of emotional state. *Psychological Review*, 69:379–399.

Simon, S. 2001. Implementing culture change – Three strategies. *Proceedings of the ASSE Behavioral Safety Symposium: The Next Step*. Orlando, FL. 135–140.

Skinner, B. F. 1953. *Science and human behavior.* New York: Macmillan.

———. 1971. *Beyond freedom and dignity.* New York: Alfred A. Knopf.

———. 1974. *About behaviorism.* New York: Alfred A. Knopf.

Slovic, P. 1991. Beyond numbers: A broader perspective on risk perception and risk communication. In *Deceptable evidence: Science and values in risk management.* D. G. Mayo and R. D. Hollander, eds. New York: Oxford University Press.

Sulzer-Azaroff, B. 1998. *Who killed my daddy? A behavioral safety fable.* Cambridge, MA: Cambridge Center for Behavioral Studies.

Topf, M. D. 1998. Behavioral safety: A multifactorial approach. *Professional Safety,* 43(8):34–35.

———. 2001. Behavioral? Holistic? Forget what you call it. Here's what works! *Proceedings of the ASSE Behavioral Safety Symposium: The Next Step*. Orlando, FL. 85–94.

Valins, S. 1966. Cognitive effects of false heart-rate feedback. *Journal of Personality and Social Psychology*, 4:400–408.

Critical Success Factors for Behavior-Based Safety: A Study of 20 Industry-Wide Applications[6]

Jason De Pasquale and E. Scott Geller

Abstract

One-on-one interviews and focus-group meetings were held at 20 organizations that had implemented a behavior-based safety (BBS) process in order to find reasons for program successes/failures. A total of 31 focus groups gave 629 answers to six questions. A content analysis of these responses uncovered critical information for understanding what employees are looking for in a BBS program. A perception survey administered to individual employees (n = 701) at these organizations measured a variety of variables identified in prior research to influence success in safety efforts. The survey data showed five variables to be significantly predictive of employee involvement in a BBS process:

- perceptions that BBS training was effective,
- trust in management abilities,
- accountability for BBS through performance appraisals,
- whether one had received education in BBS, and
- tenure with the organization.

Also, employees in organizations mandating employee participation in a BBS process (n = 8 companies) reported significantly higher levels of involvement, trust in management, trust in co-workers, and satisfaction with BBS training than did employees whose process was completely voluntary (n = 12 companies). In addition, employees in mandatory processes reported significantly greater frequency of giving and receiving positive behavior-based feedback.

Introduction

Occupational injuries due to at-risk work behavior remain a significant problem. On a daily basis, an estimated 16 workers are killed and 36,000 are injured (U.S. Department of Health and Human Services 1998). In addition to these sobering statistics are those which estimate manufacturing facilities annually incur an average of 101 lost workdays due to injuries per 100 employees (Leigh 1995; Miller 1997; National Institute Occupational Safety Health 1998; National Safety Council 1998; U.S. Bureau of Labor Statistics 1997). In sum, approximately 250,000 potential productive years of life are lost each year in the U.S. because of premature death due to work-related injuries (Baker et al. 1992).

Miller (1997) estimated that every year U.S. employers pay approximately $200 billion in direct costs associated with injuries that occur both on and off the job. Occupational injuries account for three-quarters of this total or nearly $155 billion annually. This amounts to over $1,400 per work-related injury. The majority of these costs are in the form of insurance premiums for workers and their families, as well as workers' compensation for days lost from work (Miller 1997). To make matters worse, these estimates are likely gross under-estimations because of unreliable surveillance techniques and the fact that many injuries are not reported (Baker et al. 1992; The National Committee for Injury Prevention and Control 1989).

In order to combat the ever-present threat of employee injury and associated losses, many organizations are implementing what is referred to as behavior-based safety (BBS). This approach to injury control has a number of advantages over more traditional approaches:

- it can be administered by individuals with minimal professional training;
- it can reach people in the setting where the problems occur (for example, school, workplace, or community at-large); and
- the leaders in these settings can be taught the behavior-change techniques most likely to work under specific circumstances (Baer, Wolf, and Risley 1968; Daniels 1989).

Research has also shown BBS to be cost effective primarily because behavior-change interventions are straightforward and relatively easy to administer, and their impact can be regularly monitored by indigenous personnel (Daniels 1989; Geller 1996; Geller, Lehman, and Kalsher 1989; Sulzer-Azaroff and De Santamaria 1980).

Research has demonstrated the cost-effectiveness of BBS and has been applied frequently and successfully in various industrial settings (Petersen 1989):

- participative education to increase safety-belt use (Kello, Geller, Rice, and Bryant 1988);
- incentives/rewards to increase safety-belt use (Geller 1984; Geller and Hahn 1984);
- goal-setting and behavioral feedback to increase safety-belt use, turn-signal use, and complete intersection stopping (Ludwig and Geller 1991, 1997) and to reduce driving speed (Van Houten and Nau 1983); and
- pledge-card commitment strategies to increase employees' use of safety belts (Geller, Rudd, Kalsher, Streff, and Lehman 1987) and other personal protective equipment (Streff, Kalsher, and Geller 1993).

Behavior-based safety starts by defining one or more critical behaviors to target. Then these behaviors are observed and recorded in particular work settings. When a relatively stable baseline measure of the frequency, duration, or rate of behavior is obtained, an intervention is implemented to change the behavior in beneficial directions. This intervention might involve removing environmental barriers, modifying a workstation, or adding antecedents or consequences to the situation to alter response probability. The frequency, duration, or rate of the target behavior is recorded during and after the intervention and compared to baseline measures of behavior to determine intervention impact (Daniels 1989; Geller 1996, 1998b).

Facilitating Employee Involvement in BBS

Although BBS methods are consistently effective at increasing the occurrence of safe behaviors, they can only work optimally if used throughout an organization. In other words, if employees do not "buy-in" to BBS principles, participate actively in observation and feedback sessions, and help to implement BBS intervention procedures, research describing the impact of this approach is academic (pun intended). Therefore, a primary objective of the current study was to identify organizational and interpersonal variables that inhibit or facilitate employee involvement in a BBS process. The research literature offered some guidance for our focus and for formulating hypotheses.

Providing adequate BBS training. Learning experiences can be a potent source of stimulation. Training programs are learning experiences designed to produce desired cognitive and/or behavior change among participants. An employee's first exposure to a BBS process often occurs during a training session. As such, employee perceptions regarding the quality and relevance of BBS training received may have great potential for determining the frequency and quality of involvement in a BBS process. However, our literature review found no empirical investigations of the relationship between perceptions of a training process and subsequent involvement in a safety program.

Mandatory versus voluntary participation. One obvious way to promote employee participation in a BBS process is to make such a process mandatory. In other words, some minimal level of participation in a BBS process could be a job requirement. A potential drawback of a mandatory approach, however, is that employees may feel a diminished sense of personal control which may cause them to engage in resistance behaviors (Geller 1998a). For example, monitoring systems which produced constant vigilance of production behaviors (Grant, Higgins, and Irving 1988), or management-set standards employees perceived as unreasonably high, led to the occurrence of false reporting to beat the system (Dose and Klimoski 1995). Research by Kelman (1958) suggests that organizations limiting personal control increase the occurrence of defensive strategies, including the motivation to do only what is required and no more.

Other research has found beneficial effects of elevated levels of personal control. For example, organizational structures which provide employees greater discretion were shown to correlate positively with employee satisfaction, perceived quality of work life, and actual job performance (Hackman and Lawler 1971; Sashkin 1984). In contrast, a lack of control over organizational outcomes can lead to a state of learned helplessness (Seligman 1975) and lowered effectiveness on the job. In this state, employees are said to possess "trained incapacity" (Dose and Klimoski 1995).

The role of interpersonal trust. Little research has examined the direct impact of interpersonal trust on performance outcomes (McAllister 1995). However, researchers have long recognized trust as a hallmark of favorable relationships (Hart 1988; Zand 1972) and organizational performance (Granovetter 1985). In fact, trust between individuals and between group members in an organization is considered a critical ingredient in the long-term stability of the organization and the well being of its members (Cook and Wall 1980). In other words, it is generally agreed that high levels of interpersonal trust are beneficial for most organizational functioning.

Yamagishi and Cook (1993) investigated the impact of interpersonal trust on individual participation in exchange relationships. They found strong support for the notion that interpersonal trust increases an individual's level of participation in a group situation. In fact, several other studies have shown a beneficial impact of interpersonal trust on the frequency and quality of group member interactions as well as participation in decision-making situations (Samuelson, Messick, Rutte, and Wilke1984).

The role of interpersonal trust may be particularly critical for successful BBS. An important BBS method involves employees in the observation of other employees' work behaviors. These observations are recorded systematically on a behavioral checklist and turned in for tracking and group feedback. If employees do not trust both the intentions and abilities of the behavioral observers, an observation and feedback process cannot work. In other words, if employees perceive the BBS process to be another method for management to monitor their behavior or for co-workers to "rat" on each other, participation in the process will likely be minimal. As concluded by Geller (1998c), "Lack of interpersonal trust causes resistance to an observation and feedback process, and interpersonal trust is what's missing in a culture deemed unready for behavior-based safety" (p. 14).

Hypotheses

Based on our review of the literature related to the present research, we derived the following hypotheses.

Hypothesis 1: The quantity and quality of BBS training will influence the amount of employee involvement in a BBS process.

Hypothesis 2: Mandated participation in a BBS process will decrease personal involvement and satisfaction with a BBS process.

Hypothesis 3: Trust in management abilities and intentions will facilitate employee involvement in a BBS process.

Hypothesis 4: Trust in co-workers abilities and intentions will facilitate employee involvement in a BBS process.

Method

Subjects and Setting

Participation was solicited through survey research conducted in a professional safety journal as well as through presentations and interactions at professional safety conferences and workshops. The participants were employees working at 20 different industrial sites. From these 20 organizations, 245 employees (221 males, 24 females) made up 31 different focus groups. About 80 percent of these participants were male, a male-to-female ratio that paralleled the workforce. All 20 of the sites visited were involved in an employee-driven BBS process that was in place for at least one year. Each BBS process included interpersonal observation and feedback with a checklist of specific safe and at-risk behaviors.

Measures

The culture survey administered in the current study used a combination of scales found in previous research, plus scales and items designed specifically for this investigation. Scale items were presented in a seven-point Likert format.

Interpersonal trust. Interpersonal trust was measured with the 12-item scale developed by Cook and Wall (1980). The interpersonal trust scale assesses employee levels of trust with both management and co-worker intentions and ability. (See the last paper in Chapter 6 for a copy of this measurement tool.) Cronbach Alpha levels for this interpersonal trust scale has been documented at over .80.

Impulsivity. Impulsivity is attributed to people who "act on the spur of the moment without being aware of any consequences or risk involved" (Clift et al. 1993, p. 404). Nine items from the Eysenck et al. (1985) L7 scale were used to assess impulsivity. This shortened scale has been shown to have acceptable inter-item estimates of .77 (Clift et al. 1993).

Involvement. The involvement scale was designed specifically for the current investigation. The three items that comprise the scale ask participants to rate their level of current and future involvement in the BBS process.

Perceptions of BBS training. The perceptions of BBS training items were also constructed specifically for this study. These three items ask participants to rate the BBS training they received with regard to relevance, importance, and quality.

Single item scales. A number of single-item measures asked respondents whether they had received BBS training, whether they had been trained to give behavior-based feedback, how often and how much they gave and received positive and negative feedback, and how satisfied they were with the BBS process.

Procedures

During scheduled site visits to all 20 participating organizations, two researchers collected data via two distinct methods: focus group discussions and perception surveys. One focus group involved only members of the safety steering committee, and the other involved a random selection of hourly employees. While this should have resulted in 40 focus-group sessions, nine of the sites visited did not have an intact safety committee.

Therefore, a total of 31 focus groups were formed (20 with hourly employees and 11 with members of the safety-committee). Focus groups ranged in size from four to 22 members.

The focus-group sessions lasted approximately 90 minutes, during which one of two researchers asked a series of six questions designed to solicit employees' opinions regarding the necessary ingredients for an effective employee-driven BBS process. All reactions to these questions were recorded simultaneously and independently on data collection sheets by both researchers.

After completing the focus groups, researchers left a perception survey with each organization. The contact person was instructed to get as many employees as possible to complete these surveys. When it was not possible to survey all employees, the contact person was asked to obtain a representative sample of every work area. When completed, the surveys were mailed to the research team for analysis.

Results

The dimensions measured by the survey and their scale alpha levels are displayed in Table 7.6.1.

Focus Groups. Two steps were used to analyze information gathered during focus groups. First, researchers looked at the employee responses for all questions as recorded by both focus-group facilitators. Only those responses that appeared on both of the data sheets were retained for analysis. Second, for the first five questions, each response was classified as a positive, negative, or neutral statement by two researchers. Neither researcher was aware of how the other was classifying any of the responses. Only if a statement was classified as positive, negative, or neutral by both researchers was it retained. A total of 12 responses were eliminated as a result of this process. For questions 6 – 8, responses were categorized as a behavior-based factor, a person-based factor, or an environment-based factor. Two subject-matter experts performed this classification. Consensus was reached on each item.

Measure	Number of Items Comprising Scale	Alpha
Position (management vs. front-line)	1	-
Impulsivity (Clift, Wilkins, and Davidson 1986)	9	.78
Perception safety issues are included in performance appraisals	1	-
Trust in management intentions (Cook and Wall 1980)	3	.64
Trust in peer intentions (Cool and Wall 1980)	3	.80
Satisfaction with the BBS process	1	-
Amount of positive feedback given related to BBS process	1	-
Amount of positive feedback received related to BBS process	1	-
Amount of negative feedback given related to BBS process	1	-
Amount of negative feedback received related to BBS process	1	-
Trained to give feedback (Yes or No)	1	-
Tenure with the organization	1	-
Perceptions of BBS training	3	.87
Trust in peer abilities (Cook and Wall 1980)	3	.63
Trust in management abilities (Cook and Wall 1980)	3	.75
Involvement in the BBS process	3	.78
Educated in BBS (Yes or No)	1	-

Table 7.6.1 Variables Measured by the Perception Survey

Perception Survey Results

Of the 756 perception surveys returned, 701 were acceptable for analysis. These surveys were returned from 15 of the 20 organizations that participated. The employees at four organizations failed to complete and return the surveys, and the surveys returned from another organization were completed incorrectly. Table 7.6.2 displays the intercorrelation for all variables measured by the perception survey.

Table 7.6.2 Intercorrelations of All Dependent Variables

	Variable	1	2	3	4	5	6	7	8	9	10
1	Position	1.00	-.07	.17*	.15*	-.07	-.16*	-.06	-.05	.24**	-.07
2	Years With Company		1.00	-.02	-.04	-.03	-.04	.06	.11	.09	.09
3	Educated in BB Safety			1.00	.54**	.08	.04	.11	.06	.39**	.07
4	Trained to Observe				1.00	.09	.12	.09	.02	.46**	.07
5	Observations Performed					1.00	.69**	.22**	.09	-.08	.22**
6	Frequency Observed						1.00	.32**	.14	-.08	.35**
7	Frequency Positive Feedback Received							1.00	.65**	.14	.78**
8	Frequency Negative Feedback Received								1.00	.16*	.57**
9	Trained to Give Feedback									1.00	.17*
10	Frequency Positive Feedback Given										1.00

	Variable	11	12	13	14	15	16	17	18	19	20
11	Frequency Negative Feedback Given	1.00	-.02	.04	.10	-.05	-.02	.10	.10	.09	-.11
12	BBS Part of Performance		1.00	-.01	.23**	.17*	.14*	.14*	.19**	.25**	.15
13	Impulsivity			1.00	-.17*	-.20**	-.22**	-.32**	-.17*	-.17*	-.16*
14	Perceptions of BBS Training				1.00	.41**	.50**	.39**	.31**	.59**	.28**
15	Trust in Management Intent					1.00	.67**	.39**	.37**	.42**	.18**
16	Trust in Management Ability						1.00	.30**	.33**	.46**	.26**
17	Trust in Peer Intent							1.00	.65**	.30**	.24**
18	Trust in Peer Ability								1.00	.26**	.28**
19	Involvement in Process									1.00	.28**
20	Process Satisfaction										1.00

	Variable	1	2	3	4	5	6	7	8	9	10
11	Frequency Negative Feedback Given	.07	.02	-.08	-.04	.19**	.29**	.30**	.43**	.08	.60**
12	BBS Part of Performance Evaluations	-.01	-.13	.02	-.01	.07	.03	-.06	-.02	-.03	-.04
13	Impulsivity	-.08	-.15*	-.08	-.07	-.04	-.05	-.08	-.02	-.09	-.07
14	Perceptions of BBS Training	-.01	.10	.32**	.27**	.02	.09	.18*	.13	.25**	.20**
15	Trust in Management Intent	.30**	-.16*	.28**	.32**	-.02	-.10	.08	.03	.38**	.09
16	Trust in Management Ability	.15**	-.10	.31**	.31**	.00	-.03	.10	-.01	.37**	.11
17	Trust in Peer Intent	.03	-.01	.06	.06	-.13	-.05	.09	.00	.05	.14*
18	Trust in Peer Ability	.08	-.02	.09	.10	-.08	-.10	-.04	-.17*	-.04	-.07
19	Involvement in Process	.06	.11	.40**	.33**	.09	.08	.17*	.14*	.31**	.22**
20	Process Satisfaction	-.07	-.06	.16*	.08	-.03	-.02	.01	-.08	-.13	-.01

Predicting involvement in the behavior-based process. A forward entry regression analysis was performed to determine variables most predictive of involvement in the safety process. Employee involvement was determined for each employee by summing their 7-point Likert scale responses to the following three involvement-related items:

- Up to this point I have been very active in this organization's behavior-based safety program,
- I willingly attend and participate in most of the behavior-based safety related activities and meetings that occur in this organization, and,
- It is my intention to remain very active in this organization's behavior-based safety program.

As shown in Table 7.6.3, five variables contributed significantly to predicting self-reported levels of involvement:

- adequacy of the BBS training,
- trust in management abilities,
- extent that safety is used in performance appraisals,
- whether the employee was educated in the BBS process, and
- tenure with the organization.

These variables accounted for 41 percent of the variance in amount of self-reported involvement in the BBS process.

Variable(s) Entered	R	R^2	Adjusted R^2	Std Error of the Estimate
1. Adequacy of BBS training	.54	.29	.29	.54
2. Trust in management abilities	.59	.35	.34	.51
3. Extent that BBS is used in performance appraisals	.61	.37	.36	.50
4. Whether educated in BBS	.63	.39	.38	.50
5. Tenure with organization	.64	.41	.40	.49

Table 7.6.3 Predicting Employee Involvement in a BBS Process

Voluntary vs. mandatory BBS. Table 7.6.4 depicts the results of planned comparison t-tests for the perception survey variables between employees in a mandatory (n = 273) and employees in a voluntary (n = 427) BBS process. As shown in this table, employees in a mandatory BBS process showed higher scores on several questions related to BBS success than did employees in a voluntary BBS process. More specifically, employees in a mandatory process reported significantly higher rates for giving and receiving positive behavior-based feedback and significantly lower rates

Variable	Voluntary BBS Process	Mandatory BBS Process
Average frequency of observation made per employee per month	6.5	7.6
Average frequency of times observed per employee per month	4.0	6.6*
Average frequency positive feedback received per employee per month	1.8	3.6**
Average frequency negative feedback received per employee per month	0.6	0.6
Average frequency positive feedback given per employee per month	3.1	5.3*
Average frequency negative feedback given per employee per month	1.2	1.5
Degree of Involvement	3.4	3.8*
Amount of Satisfaction	3.1	3.3
Trust in Management Abilities	3.8	4.2*
Trust in Management Intentions	3.7	4.1*
Trust in Peer Abilities	3.4	3.7*
Trust in Peer Intentions	3.7	4.0*
Impulsivity	2.5	2.5
BBS is part of performance evaluations	3.3	3.4
Satisfaction with BBS Training	3.5	3.8*

Table 7.6.4 Voluntary versus Mandatory BBS Process
*Notes: *significant at p<.05, **significant at p<.01*

for receiving negative behavior-based feedback. In addition, employees in a mandatory BBS process demonstrated significantly greater levels of trust in management (both abilities and intentions), trust in co-workers (both abilities and intentions), and overall satisfaction with the BBS training received. Obviously, our hypothesis that voluntary participation in BBS would be superior to mandatory participation was not supported.

Discussion

The tremendous improvements in safety and performance of companies who have implemented BBS processes has given this approach to safety management credibility and status (Geller 1999). Unfortunately, little objective research has been performed to elucidate the organizational factors that can facilitate successful implementation of a BBS approach. Instead, many organizations are left to muddle through the unending case studies provided by safety consultants. Although such literature may be enlightening and informative to some degree, it does not adequately inform readers of the underlying behavioral, interpersonal, and organizational processes that lead to successful BBS implementation. The current empirical investigation is a first step in understanding what factors are critical for successful implementation of BBS.

A primary method that has reduced injuries and is used in virtually every organization that employs BBS is observation and feedback. One of the first problems encountered by an organization is deciding if employee participation in the observation and feedback process should be mandatory or voluntary. The obvious concerns associated with making such a process mandatory is that it will lead to negative reactions from the employees due to reduced levels of choice or perceived control (Geller 1998a, b). On the other hand, a voluntary observation and feedback process could lead to minimal employee participation.

The current research enabled us to compare organizations incorporating a voluntary observation and feedback process with organizations opting for mandatory interpersonal observation and feedback. Contrary to our research hypothesis, voluntary observation and feedback processes were not viewed any more favorably than were mandatory observation and feedback processes. In fact, the results for positive regard toward an observation and feedback process seemed to favor the mandatory approach. In addition, findings summarized in Table 7.6.4 indicate that organizations taking the mandatory approach may experience benefits in a number of areas over organizations following an entirely voluntary approach.

At first these results seem to defy logic or common sense. However, upon closer examination of the mandatory approaches, such findings are not too surprising. First, although employees in a mandatory process had no choice in whether or not they participated, they usually had a good deal of choice regarding when and how they conducted observation and feedback sessions. For example, all of the organizations using a mandatory approach gave employees a choice regarding how many times a month they should perform an observation. And all observations occurred with permission from the person being observed. Some of these organizations (37 percent) rotated observers on a monthly basis. Thus, it seems these organizations were able to circumvent reductions in employee perceptions of personal control by building employee choice into the BBS process.

Besides reporting higher levels of employee involvement, mandatory processes also demonstrated significantly higher levels of trust in management and trust in co-workers, a key ingredient for BBS success (as reported in most focus groups). Plus, the use of positive BBS feedback was reported more frequently for mandatory than voluntary programs. It is likely such results are due to the fact that mandatory programs led to more overall involvement. And as involvement in the observation and feedback process increased, more employees increased their trust in the process and realized it was designed to benefit them. Moreover, they probably found out that participation did not lead to negative consequences such as being "ratted-out" by a co-worker or receiving a reprimand from a supervisor.

More experience with interpersonal observation and feedback results in more trust in the intentions and abilities of co-workers and then to more involvement. While these results should not be interpreted as a blanket endorsement for making BBS mandatory in all organizations, it does indicate that mandatory processes are not necessarily detrimental to employee satisfaction, trust, or perceptions of personal control. If instituted properly, mandatory processes can actually facilitate employee perceptions of personal control and increase a number of relationship-based variables beneficial to BBS success. Thus, it seems beneficial for management to set an expectation that everyone participate in the observation and feedback process they have been trained to implement. But management should then give plenty of leeway regarding how the process occurs and offer ongoing support for the procedures customized by the line workers.

Beyond the observation and feedback process, the current investigation offers insight into a number of additional issues relevant to BBS. These results are particularly important because they come from the perspective of line workers or operators. For instance, our data support the notion that employees generally understand that management and supervisors are key determinants of the success or failure of a BBS process. While there was disagreement regarding the extent to which management and supervisors should be involved, everyone acknowledged that these individuals needed to demonstrate visible support of an interpersonal observation and feedback process.

When negative reactions about BBS were given, significant attention was given to perceptions of the BBS steering committee. In such cases, the steering committee was viewed as being out of touch with what is really needed to improve safety in the organization. It was also common to find that a large number of employees perceived safety committee members as simply ingratiating individuals looking to get ahead. This emphasizes the need to select the right people to serve on the BBS steering committee and to rotate membership at regular intervals.

The steering committee is responsible for driving an observation and feedback process, and negative perceptions of its members cannot be beneficial. Further research should investigate the type of involvement required by BBS steering committees from the beginning stages of BBS implementation to eventual institutionalization. For example, active and frequent involvement of a BBS steering committee is clearly needed to get a BBS process started, but it might be advantageous to fade out certain kinds of committee activity as BBS principles and procedures take hold throughout an organization. In other words, the role of the BBS steering committee needs to change as BBS processes mature. It is not clear, however, what roles should remain and how others should change.

Three of the questions asked of employees during focus groups were particularly informative regarding ways to implement and maintain a successful BBS process. We learned useful perspectives from line workers—the employees most directly responsible for BBS and the organization's safety record—by asking them three questions:

- How would they improve their BBS process?
- What were the biggest obstacles to BBS success?
- What are the key ingredients to a BBS process?

Our content analysis of employees' answers to these questions revealed the following noteworthy issues and conclusions:

First, although it may be impressive to collect a lot of behavioral data from behavioral observations, it is probably in an organization's best interest to start "small" with a simple behavioral checklist. Allow employees time to get used to an observation and feedback process before adding complexity. Second, it is imperative that trainers make clear that BBS cannot be a vehicle for blaming individuals or for delivering criticism of any kind. Third, interpersonal trust is essential, from the shop floor to the management boardroom. Trust was mentioned frequently not only as one of the critical factors for success, but also as one of the greatest barriers to success when it is not there.

As with interpersonal trust, employees repeatedly expressed the need for continued support from management. While monetary resources are essential, they are not sufficient. Visible recognition that management appreciates the BBS methods is critical. One way to do this is to provide the necessary time for employees to perform behavioral observations and analyze results. And if the outcome from a BBS process indicates a need for certain changes in environmental conditions or training procedures, management needs to follow through quickly and effectively.

Other critical success factors for BBS success were revealed through the multiple regression analysis. Most important in getting employee involvement was BBS training. These sessions usually represent employees' first exposure to BBS, and if they are not made relevant to the employees' work setting and circumstances, employees could actually leave a training session discouraged, confused, and unlikely to get actively involved.

Research is needed to define the parameters of a BBS training session that will facilitate subsequent employee participation. Individual characteristics of the trainer are probably relevant, as well as the structure and format of the training itself.

Our regression analysis showed that a number of other variables could also help to facilitate employee involvement in a BBS process. As expected from the focus group results, trust in management added significant predictive power to our regression equation. However, we found trust in management abilities and not intentions to be the crucial factor. Perhaps employees believe management is well intentioned when they start the safety process, but if management does not follow through with appropriate support, trust in ability is suspect. So trust in ability means management knows how to facilitate and support BBS. This explanation is consistent with the majority opinion from focus groups that visible management support is essential for a successful BBS process.

In conclusion, we offer a framework for understanding why some BBS processes succeed and others fail. Analogous to the way we analyzed the focus group discussions, we propose the BBS Safety Success Triad illustrated in Figure 7.6.1. The three sides of this triangle are dynamic and interactive. When one side changes the other sides are influenced.

Figure 7.6.1 The Behavior-Based Safety Success Triad

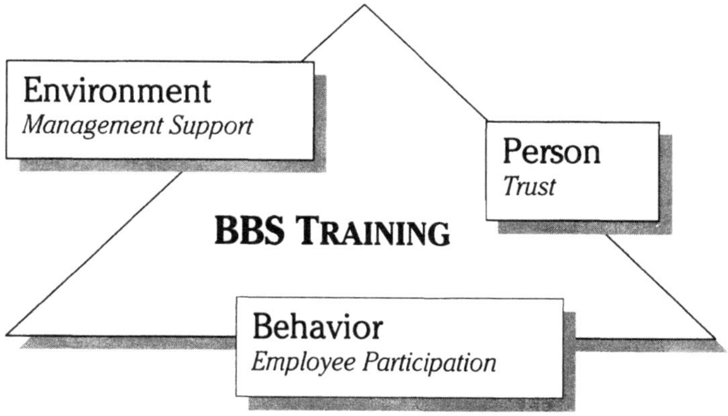

The BBS Safety Success Triad summarizes the discussions from 31 focus groups, as well as the analysis of 701 perception surveys (from 15 different organizations). On the person side of the triangle is interpersonal trust, on the environment side of the triangle is management support, and on the behavior side of the triangle is employee participation/involvement. Training is in the middle of the triangle because of its critical role in facilitating all three sides of the Triad. During training all employees receive the principles, procedures, and tools of a BBS process, and management receives the rationale and the method for supporting BBS. Proper training convinces participants the process works and the people can implement it. This is necessary for self-efficacy (Bandura 1997), a person state deemed necessary for any constructive long-term change among individuals in a work culture.

References

Baer, D. M., M. M. Wolf, and T. Risley. 1968. Some current dimensions of applied behavior analysis. *Journal of Applied Behavior Analysis*, 1:91–97.

Baker, S. P., C. Conroy, and J. J. Johnston. 1992. Occupational injury prevention. *Journal of Safety Research*, 23(2):129–133.

Bandura, A. 1997. *Self-efficacy: The exercise of control.* New York: W. H. Freeman and Company.

Clift, S. M., J. C. Wilkins, and E. A. F. Davidson. 1993. Impulsiveness, venturesomeness and sexual risk-taking among heterosexual gum clinic attenders. *Personality and Individual Differences*, 15(4):403–410.

Cook, J., and T. Wall. 1980. New work attitude measures of trust, organizational commitment and personal need non-fulfillment. *Journal of Occupational Psychology*, 53:39–52.

Daniels, A. C. 1989. *Performance management.* Tucker, GA: Performance Management Publications.

Dose, J. J., and R. J. Klimoski. 1995. Doing the right thing in the workplace: Responsibility in the face of accountability. *Employee Responsibilities and Rights Journal*, 8(1):35–56.

Eysenck, S. B. J., P. R. Pearson, G. Easting, and J. F. Allsopp. 1985. Age norms for impulsiveness, venturesomeness and empathy in adults. *Personality and Individual Differences*, 6:613–619.

Geller, E. S. 1984. Motivating safety belt use with incentives: A critical review of the past and a look to the future. *In Advances in belt restraint systems: design, performance and usage.* Warrendale, PA: Society of Automotive Engineers. 127–142.

———. 1996. *The psychology of safety: How to improve behaviors and attitudes on the job.* Radnor, PA: Chilton Book Company.

———. 1998a. *Beyond safety accountability: How to increase personal responsibility.* Neenah, WI: J. J. Keller & Associates, Inc.

———. 1998b. *Understanding behavior-based safety: Step-by-step methods to improve your workplace.* 2nd ed. Neenah, WI: J. J. Keller & Associates, Inc.

———. 1998c. What's the level of trust in your workplace? *Industrial Safety & Hygiene News*, 32 (3):14–15.

———. 1999. Behavior-based safety: Confusion, controversy, and clarification. *Occupational Health & Safety*, 68(1):40–50.

Geller, E. S., and H. A. Hahn. 1984. Promoting safety belt use at industrial sites: An effective program for blue-collar employees. *Professional Psychology: Research and Practice*, 15(4):553–564.

Geller, E. S., G. R. Lehman, and M. J. Kalsher. 1989 *Behavior analysis training for occupational safety*. Newport, VA: Make-A-Difference, Inc.

Geller, E. S., J. R. Rudd, M. J. Kalsher, F. M. Streff, and G. R. Lehman. 1987. Employee-based programs to motivate safety belt use: A review of short and long-term effects. *Journal of Safety Research*, 18:1–17.

Granovetter, M. 1985. Economic action and social structure: The problem of embeddedness. *American Journal of Sociology*, 91:481–510.

Grant, R. A., C. A. Higgins, and R. H. Irving. 1988. Computerized performance monitors: Are they costing you customers? *Sloan Management Review*, Spring:39–45.

Hackman, R. J., and E. E. Lawler. 1971. Employee reactions to job characteristics. *Journal of Applied Psychology*, 55(3):259–286.

Hart, K. M. 1988. A requisite for employee trust: Leadership. *Psychology: A Journal of Human Behavior*, 25:1–7.

Kello, J. E., E. S. Geller, J. C. Rice, and S. L. Bryant. 1988. Motivating auto safety-belt wearing in industrial settings: From awareness to behavior change. *Journal of Organizational Behavior Management*, 9(2):7–21.

Kelman, H. C. 1958. Compliance, identification, and internalization: Three processes of attitude change. *Conflict Resolution*, 2:51–60.

Leigh, J. 1995. *Causes of death in the workplace*. Westport, CN: Quorum Books.

Ludwig, T. D., and E. S. Geller. 1991. Improving the driving practices of pizza deliverers: Response generalization and moderating effects of driving history. *Journal of Applied Behavior Analysis*, 24:31–34.

———. 1997. Assigned versus participative goal setting and response generalization: Managing injury control among professional pizza deliverers. *Journal of Applied Psychology*, 82(2):253–261.

McAllister, D. J. 1995. Affect- and cognition-based trust as foundations for interpersonal cooperation in organizations. *Academy of Management Journal,* 38(1):24–59.

Miller, T. R. 1997. Estimating the costs of injury to U.S. employers. *Journal of Safety Research*, 28(1):1–13.

National Committee for Injury Prevention and Control. 1989. *Injury prevention: Meeting the challenge*. New York: Oxford University Press.

National Institute for Occupational Safety & Health 1998. *Traumatic occupational injury research needs and priorities: A report by the NORA traumatic injury team*. Special Research Report, Cincinnati, OH: DHHS (NIOSH) Publication No. (PHS):98–134.

National Safety Council. 1998. *Accident facts*. Itasca, IL: National Safety Council.

Petersen, D. 1989. *Safe behavior reinforcement*. New York: Aloray.

Samuelson, C. D., D. M. Messick, C. G. Rutte, and H. Wilke. 1984. Individual and structural solutions to resource dilemmas in two cultures. *Journal of Personality and Social Psychology*, 37:139–155.

Sashkin, M. 1984. Participative management is an ethical imperative. *Organizational Dynamics*, 18(1)5–22.

Seligman, M. E. P. 1975. *Helplessness: On depression development and death*. San Francisco: Freedman.

Streff, F. M., M. J. Kalsher, and E. S. Geller. 1993. Developing efficient workplace safety programs: Observations of response covariation. *Journal of Organizational Behavior Management*, 13(2):3–15.

Sulzer-Azaroff, B., and M. C. De Santamaria. 1980. Industrial safety hazard reduction through performance feedback. *Journal of Applied Behavior Analysis*, 13:287–295.

United States Bureau of Labor Statistics 1997. *Safety and health statistics. Occupational safety and health home page.* url:www.bls.gov.

United States Department of Health and Human Services 1998. Traumatic occupational injury research needs and priorities: A report by the NORA traumatic injury team. *National Institute for Occupational Safety and Health:* Atlanta, GA: Centers for Disease Control and Prevention.

Van Houten, R., and P. A. Nau. 1983. Feedback interventions and driving speed: A parametric and comparative analysis. *Journal of Applied Behavior Analysis*, 16:253–281.

Yamagishi, T., and K. S. Cook 1993. Generalized exchange and social dilemmas. *Social Psychology Quarterly*, 56(4):235–248.

Zand, D. E. 1972. Trust and managerial problem solving. *Administrative Science Quarterly*, 17:229–239.

Notes

[1] Originally appeared in *Industrial Safety & Hygiene News* (April 1999).

[2] Originally appeared in *Industrial Safety & Hygiene News* (January 1999).

[3] Originally appeared in *Industrial Safety & Hygiene News* (June 1996).

[4] Originally appeared in *Industrial Safety & Hygiene News* (August 1996).

[5] Originally appeared in *Professional Safety* (September 2001).

[6] Originally appeared in *Journal of Safety Research* 30:4 (1999):237–249.

Closing Comments from E. Scott Geller

Will BBS Reach its Potential? Cause for Optimism and Alarm

Will Behavior-Based Safety Reach Its Potential: Causes for Optimism and Alarm

E. Scott Geller

Behavior-based safety is founded on the science of the experimental analysis of behavior (Skinner 1953, 1974) and the practical technology of applied behavior analysis (Baer, Wolf, and Risley 1968). Many researchers and practitioners in these disciplines have been devoting their lives to chasing the noble dream—the dream of saving the world with behavior analysis. These "dream chasers" appreciate the value of an objective behavioral approach to problem solving and realize the cost-effectiveness of using the basic three-term contingency (or Activator-Behavior-Consequence Model) to evaluate and improve behavior on a large scale. Indeed, the research literature is replete with impressive demonstrations that this approach works, from improving learning and interpersonal care in educational and residential settings to increasing productivity and work quality in industrial environments (e.g., Goldstein and Krasner 1987; Greene, et al. 1987; Kazdin 1994).

There have been numerous community-based applications of behavior analysis. For example, I have seen firsthand the success of behavior analysis in addressing the human dynamics of such wide-ranging issues as environmental preservation (e.g., Cone and Hayes 1980; Dwyer, Leeming, Cohen, Porter, and Jackson 1993; Geller, Winett, and Everett 1982), transportation safety (Geller 1998; Ludwig and Geller 2000), alcohol abuse and alcohol-impaired driving (Geller and Lehman 1988), and childhood immunization against disease in third-world countries (Lehman and Geller 1987).

The potential for applied behavior analysis to solve societal problems, however, has not been realized. The dream chasers' visions have not been reached. Beyond short-term demonstration projects, there have been very few large-scale implementations of behavior-change strategies for education, transportation safety, environmental protection, alcohol abuse, or public health.

Why Have Our Dreams Fallen Short?

There are a variety of possible reasons for the failure of applied behavior analysis to have far-reaching impact on critical societal problems. The most obvious problem is the audience of these demonstration projects. The research is published in professional journals and books read almost exclusively by other psychologists, mostly behavior analysts. Here, the authors give convincing demonstrations of the efficacy of their behavior change techniques to people who have little interest or influence in large-scale dissemination and application. In other words, the critical social marketing aspects of behavior-change technology have not been addressed (Geller 1989).

Bailey (1991) comments on this dissemination problem as follows, "We have a great science (the experimental analysis of behavior) and a pretty good technology (applied behavior analysis) but no product development or marketing" (p. 39). He explains further that "we do not value marketing" and have "neglected to develop socially acceptable terminology for presenting our concepts to consumers...we have, in our zest for science and technology, taken the human concerns out of behavior analysis" (p. 39).

In addition, long-term maintenance and institutionalization of behavior-change strategies have rarely been studied. Most applications of behavior analysis to address organizational, community, and societal problems have been short-term demonstration projects, conducted to show that a particular intervention procedure has a desired effect. Methods to sustain the beneficial impact of a behavior-change technique are not developed and evaluated.

Boyce and Geller (2001) recently addressed this challenge of maintenance by reviewing the research literature related to applying behavior analysis techniques to improve occupational safety. They found that very few studies included a lengthy evaluation period after an intervention process was withdrawn; therefore, they could not evaluate behavioral maintenance. In addition, Boyce and Geller found no systematic study of factors related to successful institutionalization of behavioral technology. They did, however, identify some variables conducive to sustaining a successful behavior-change process, and these have been verified by behavioral safety practitioners (McSween and Mathews 2001) and discussed throughout this book. For example, the following factors have been shown empirically to contribute to the long-term impact of behavior-change intervention in industrial settings:

- Each level of an organization (from management to line workers) needs education and training to understand the rationale behind an intervention and to realize their specific roles in making the process work (see DePasquale and Geller 1999, and Chapter 7 of this text);
- Indigenous staff need to have substantial input into intervention design and implementation of the intervention procedures (Geller 2001c; McSween 1995);
- A formal accountability system is required, which is best handled by an employee-manned steering committee that monitors intervention results and develops action plans for enhancing intervention impact (Geller 2001c; McSween and Mathews 2001);
- A formal procedure for collecting, reviewing, and using behavioral data is needed to support the accountability system and enable continuous improvement; and
- Group and individual rewards are needed to support ongoing participation in the process, as well as to recognize exemplary achievements (Geller 1996a, 1997, and Chapter 4 of this book).

Behavior-Based Safety: A Cause for Optimism

For almost a decade, behavior-based safety (BBS) has been flourishing in industrial settings nationwide and, more recently, throughout the world. In fact, BBS training and implementation has become a priority for most industries that want to improve their safety performance. Consultants and consulting firms claiming to have expertise in BBS have been steadily increasing, and insurance companies are adding special divisions or areas of expertise to teach their clients BBS principles and help them implement behavior-change techniques for injury prevention.

In my more than 30 years of attempting to address large-scale quality-of-life problems with applied behavior analysis, I have never witnessed more extensive marketing, application, and long-term success than in BBS. The papers in this book defined the principles and procedures of BBS and offered guidelines for sustaining these techniques. It should be clear, however, that long-term maintenance of BBS tools and methods require an understanding and appreciation of non-behavioral concepts and principles, as discussed in this book (see especially Chapter 7). In other words, it is necessary to go beyond the basic theory and technology of applied behavior analysis to get the most extensive and sustained use out of behavioral technology. Many safety scholars, trainers and consultants do not realize this and have made BBS seem narrow and elementary. This has resulted in substantial confusion, conflict, and misunderstanding about BBS.

Behavior-Based Safety: A Cause for Alarm

The following quotations from *Professional Safety* illustrate bewilderment and discord with regard to the potential of BBS. Some reflect a basic misinterpretation of the tools and methods of BBS; others presume BBS is narrow and only about the behavior of line workers or wage employees. All of these quotes indicate we have a lot of beliefs to change if BBS is ever to reach its potential in reducing workplace injuries. Having read this book, you know how off-the-mark and actually ridiculous each of the following statements are. But the authors obviously believed them.

- ※ "Behavior-based safety is not an upstream venture. It is as far downstream as one can go" (Manuele 1998, p. 34).
- ※ "...as firms purchase packages...behavior-based safety moves toward becoming nothing more than the next program of the month" (Petersen 1998, p. 34).
- ※ "...too many behavior-based approaches seem to encourage management to abandon its responsibility" (Manuele 1998, p. 35).
- ※ "...the focus is improving worker behavior rather than an organization's culture or the work system" (Manuele 1998, p. 33).
- ※ "...'behavior-based' concepts...in many settings, have little practical use in productivity, quality and other areas" (Petersen 1998, p. 35).
- ※ "...not enough attention is given to the reasons for the at-risk behavior...Behavior-based safety is but a small element of a successful initiative" (Zeller 1998, p. 41).
- ※ "It's fascinating to me that no convincing evidence of the efficacy of

the behavioral approach has been produced by its proponents" (English 1998, p. 42, 57).

More critical of BBS, however, have been certain labor unions—the United Automobile Workers and the Oil, Chemical and Atomic Workers International Union, in particular. Consider, for example, the following quotations from union documents. All of these are inaccurate, of course, but they reflect the severe misunderstanding of the behavioral approach to occupational safety and health.

From the UAW Health and Safety Department (1999):

- "Behavior-based safety is one of the oldest and most outdated approaches to health and safety" (p. 1).
- "Behavior-based safety programs will drive problems underground, inject fear into the workplace, and discourage workers from reporting injuries and illnesses" (p. 7).
- "Behavior-based safety programs sentence workers to a work life of exposure to serious health and safety hazards" (p. 7).

From the Assistant Director, UAW Health and Safety Department (Howe 1998):

- "Behavior-based safety programs are just a retread of old outdated ideas and strategies that have never been proven effective" (p. 5).
- "Behavior-based safety advocates...blaming workers for almost all health and safety problems" (p. 9).
- "Behavior-based programs emphasize working downstream...(and) are completely reactive recommending the use of the most outdated, ineffective methods of control" (p. 10).
- "Many behavior-based safety proponents say that regulations and standards are not needed" (p. 12).

From a representative of the Oil, Chemical and Atomic Workers International Union (Hoyle 1998):

- "Behavior modification programs are the least effective way to prevent accidents because they focus on the narrow issue of correcting worker mistakes" (p. 4).

- "Changing worker behavior is one of the least effective methods for accident prevention" (p. 7).
- "In all behavior modification programs the central thing that is looked for is the use of personal protective equipment" (p. 10).
- "The research shows that positive reinforcement is as bad as negative reinforcement. It's just a different side of the same coin" (p. 13).
- "Behavior modification safety programs are a perfect fit for management's avoidance of responsibility for health, safety and accidents" (p. 15).

Moreover, the chairman of the Labor Dividion, U. S. Occupational Safety and Health Administration (Foster, 2000) urged the National Safety Council to develop an official policy statement for "Behavior-Based Safety Training" with the following introduction:

> The National Safety Council urges caution in the implementation and administration of Behavior Modification Programs, Behavior-Based Safety Training (BBS), incentive and/or performance-based safety, peer-group safety or other similar behavior modification programs, hereinafter referred to as BBS or Behavior-Based Safety Training programs. If such programs are improperly implemented and administered, they may place blame on employees, pit worker against worker for reward competition, cause underreporting of accidents and the ability to learn from accidents and close calls. Key components of a good safety program must include adequately identifying hazards, promoting training for hazard recognition, promoting employee buy-in for safe practices and hazard reduction.

What's the Problem?

This confusion, disregard, and lack of appreciation for BBS is reminiscent of the nationwide controversy regarding the application of contingency management programs in prisons 30 years ago (Geller, Johnson, Hamlin, and Kennedy 1977; Milan and McKee 1974; Reppucci and Saunders 1974). The basic issues are quite analogous. Then and now, the fact that a behavioral approach works is disturbing to some people.

In the 1970s, lawyers for the American Civil Liberties Union (ACLU) were concerned that program planners would use applied behavior analysis to benefit the prison administrators rather than individual inmates. Today, certain union officials claim BBS allows management to shirk their responsibilities and put inappropriate pressure on hourly workers to reduce injuries by changing their behaviors. While the ACLU lawyers were upset that prison inmates could feel coerced to participate because their deprived environment gave unfair influence to even small rewards, certain union representatives fear BBS forces employees to continue working in hazardous work settings. In other words, the concern is that a technology that is effective at improving behavior will be abused.

Backstabbing by Safety Consultants

An additional barrier to BBS reaching its potential is the meddling and interference by various consultants. The popularity of BBS has caused several safety consultants to criticize and demean behavior analysis as a means to boost their own approach. For example, consultant Steve Simon promotes his culture change business by claiming "A reduction in unsafe behaviors is certainly well and good, but it doesn't produce a fundamental cultural shift" (Hans 1996, p. 45). Likewise, Don Ostrander (Director of the National Safety Council's Consulting Division) reportedly "regards the behavior-based approach as a potentially useful tool but cautions that it is not a magic bullet.... [T]he all-consuming focus on employee behaviors can mask management inadequacies that otherwise might come to light" (Hans 1996, pp. 45-46).

Another safety consultant, T.A. Smith (1995), claims that a focus on behavior "destroys creativity" (p. 40) and takes important attention away from system influences and management responsibility. Lessin (1997) supports Smith's perspective with statements like, "Behavioral safety programs are another non-solution.... [They] ignore the system of work that underlies the 'choices' workers make on their jobs" (p. 42). Both Smith and Lessin view BBS as victim blaming or holding employees responsible for aspects of safety only management can control.

Consultant Michael Topf advocates "changing attitudes first" (Manuele 1998, p. 35) and applying a holistic approach to addressing the human dynamics of safety (Topf 1997, 1998). While Topf and his associates assert that BBS is narrow and limiting, they use the same interpersonal observation and feedback methods developed and applied by behavior analysts. In other words, this successful consulting firm follows essentially the same intervention plan advocated by BBS consultants, but uses non-behavioral language to describe their procedures and offer rationale.

What Should We Do?

The confusion, controversy, and discord regarding BBS is certainly disconcerting. Is there reason for alarm? Well, we can't sit back and let others demean this successful approach to empowering workers to reduce the frequency and severity of personal injury. Consider that stifling BBS from reaching its potential will mean that many more people will suffer occupational injuries. Consider that without a behavior-based approach to the human dynamics of safety, organizations have no objective and fair assessment procedures to define the people aspects of safety performance, and they have no research-tested intervention strategies to improve work practices.

We can't allow the confusion and discord about BBS to continue. We each need to play a role in helping BBS reach its potential. For example, we can all talk with others about the principles and procedures covered in this book. We can all teach the essence of BBS. We can emphasize that BBS addresses more than behaviors. It is about developing ownership, empowerment, systems thinking, interdependency, and a sense of belonging. It is about "actively caring."

Actively caring for safety means going beyond the call of duty to remove environmental hazards and improve safety-related behaviors. How do we improve safety-related behavior? We do this, of course, by applying behavior analysis techniques to support safe behavior and correct at-risk behavior. However, we must be careful not to imply we are targeting only a select group of individuals or behaviors in organizational settings. Obviously, the methods and tools of BBS can be customized for all levels of an organization and for every position within a given level. We need to demonstrate this by our actions—from teaching principles to planning interventions.

Realize also that the common retort "Let the results speak for themselves" is not sufficient. Sure, BBS technology does produce cost-effective reductions in workplace injuries. That is the reason for its remarkable popularity. But others use the same procedures and outcome data to advocate their "special" non-behavioral or culture-based approaches. We need to continually teach people the basic principles behind their observations of performance improvement, and we need to do this with language they understand and appreciate.

Bottom line: We need to speak for our safety-improvement results by explaining the principles and procedures behind objective success, as reviewed in this book. Then BBS might get the credit it deserves, and eventually reach its potential to help produce injury-free workplaces worldwide.

References

Baer, D. M., M. M. Wolf, and T. Risley. 1968. Some current dimensions of applied behavior analysis. *Journal of Applied Behavior Analysis*, 1:91–97.

Bailey, J. S. 1991. Marketing behavior analysis requires different talk. In E. S. Geller ed. *Science, theory, and technology: Varied perspectives*, Monograph Number 6, Lawrence, KS: Society for the Experimental Analysis of Behavior.

Boyce, T. E., and E. S. Geller. 2001. Applied behavior analysis and occupational safety: The challenge of response maintenance. *Journal of Organizational Behavior Management*.

Cone, J. D., and S. C. Hayes 1980. *Environmental problems/Behavioral solutions*. Monterey, CA: Brooks/Cole.

DePasquale, J. P., and E. S. Geller. 1999. Critical success factors for behavior-based safety: A study of 20 industry-wide applications. *Journal of Safety Research,* 30(4):237–249.

Dwyer, W. O., F. C. Leeming, M. K. Cobern, B. E. Porter, and J. M. Jackson. 1993. Critical review of behavioral interventions to preserve the environment: Research since 1980. *Environment and Behavior*, 25:275–321.

English, W. 1998. How to measure success. *Professional Safety*, 43(10):57.

Foster, R. D. May 2000. *National Safety Council position/policy request form*. Dallas, TX: US/DOL OSHA.

Geller, E. S. 1989. Applied behavior analysis and social marketing: An integration to preserve the environment. *Journal of Social Issues*, 45:17–36.

———. 1996. The truth about safety incentives. *Professional Safety*, 41(10):34–39.

———. 1997. Key processes for continuous safety improvement: Behavior-based recognition and celebration. *Professional Safety*, 42(10):40–44.

———. 1998. *Applications of behavior analysis to prevent injury from vehicle crashes*. 2nd ed. Monograph published by the Cambridge Center for Behavioral Studies, Cambridge, MA.

———. 2001a. *Beyond safety accountability*. Rockville, MD: Government Institutes.

———. 2001b. From conversation to commitment. *Occupational Health & Safety*, 70(1):58–63.

———. 2001c. *The psychology of safety handbook*. Boca Raton, FL: CRC Press.

Geller, E. S., D. F. Johnson, P. H. Hamlin, and T. D. Kennedy. 1977. Behavior modification in a prison. *Criminal Justice and Behavior*, 4(1):11–43.

Geller, E. S., and G. R. Lehman. 1988. Drinking-driving intervention strategies: A person-situation-behavior framework. In *The social control of drinking and driving*. M. D. Laurence, J. R. Snortum, and F. E. Zimring, eds. Chicago, IL: University of Chicago Press.

Geller, E. S., R. A. Winett, and P. B. Everett. 1982. *Preserving the environment: New strategies for behavior change*. New York: Pergamon Press.

Goldstein, A. P., and L. Krasner. 1987. *Modern applied psychology*. New York: Pergamon Press.

Greene, B. F., R. A. Winett, R. Van Houten, E. S. Geller, and B. A. Iwata, eds. 1987. *Behavior analysis in the community: Readings from the Journal of Applied Behavior Analysis*. Lawerence, KS: University of Kansas.

Hans, M. 1996. Does behavior-based safety work? *Safety and Health*, National Safety Council. June:44–49.

Howe, J. Jan. 1998. *A union critique of behavioral safety*. Paper presented at the ASSE Behavioral Safety Symposium, Orlando, FL.

Hoyle, B. 1998. *Fixing the workplace, not the worker: A workers' guide to accident prevention*. Lakewood, CO: Oil, Chemical and Atomic Workers International Union.

Kazdin, A. E. 1994. *Behavior modification in applied settings*. 5th ed. Pacific Grove, CA: Brooks/Cole Publishing Company.

Lehman, G. R., and E. S. Geller. 1987. *Educational technology for health workers in Nigeria*. Washington, D.C.: Academy for Educational Development.

Lessin, N. 1997. Workers need real rights. *Industrial Safety & Hygiene News,* 31(10):42.

Ludwig, T. D., and E. S. Geller. 2000. Intervening to improve the safety of occupational driving: A behavior-change model and review of empirical evidence. *Journal of Organizational Behavior Management,* 19(4).

Manuele, F. A. 1998. Perspectives on behavioral safety: Observations of ASSE's behavior safety symposium. *Professional Safety*, 43(8):32–37.

McSween, T. E. 1995. *The value-based safety process: Improving your safety culture with a behavioral approach*. New York: Van Nostrand Reinhold.

McSween, T. E., and G. A. Mathews. 2001. Maintenance in organizational behavior management. *Journal of Organizational Behavior Management*.

Milan, M. A., and J. M. McKee. 1974. Behavior modification: Principles and applications in corrections. In *Handbook of criminology*. D. Glaser, ed. New York: Rand McNally.

Petersen, D. 1998. Behavior-based safety: Magic pill, fad, or what? *Professional Safety,* 43(8):34.

Reppucci, N. D., and J. T. Saunders. 1974. Social psychology of behavior modification: Problems of implementation in natural settings. *American Psychologist,* 29:649–660.

Skinner, B. F. 1953. *Science and human behavior*. New York: Macmillan.

———. 1974. *About behaviorism*. New York: Alfred A. Knopf.

Smith, T. A. 1995. Viewpoint: Rebutting behaviorism. *Industrial Safety & Hygiene News*, 40(3):40.

Topf, M. 1997. Take the holistic approach. *Industrial Safety & Hygiene News,* 31(10):38.

———. 1998. Behavioral safety: A multifaceted approach. *Professional Safety*, 43(8):34–35.

UAW health and Safety Department. 1999. *Warning: Behavior-based safety can be hazardous to your health & safety program*. Available online: http://www.uaw.org/publications/h&s/0998/hs_03.html.

Zeller, D. 1998. Still searching. *Professional Safety*, 43(11):41.

Index

D

E

F

G

P

Q

R

S

Government Institutes Mini-Catalog

PC #	ENVIRONMENTAL TITLES	Pub Date	Price*
629	ABCs of Environmental Regulation	1998	$65
672	Book of Lists for Regulated Hazardous Substances, 9th Edition	1999	$95
4100	CFR Chemical Lists on CD ROM, 1999-2000 Edition (single user)	1999	$150
512	Clean Water Handbook, Second Edition	1996	$115
581	EH&S Auditing Made Easy	1997	$95
673	EH&S CFR Training Requirements, Fourth Edition	1999	$99
825	Environmental, Health and Safety Audits, 8th Edition	2001	$115
548	Environmental Engineering and Science	1997	$95
643	Environmental Guide to the Internet, Fourth Edition	1998	$75
820	Environmental Law Handbook, Sixteenth Edition	2001	$99
688	Environmental Health & Safety Dictionary, Seventh Edition	2000	$95
821	Environmental Statutes, 2001 Edition	2001	$115
4099	Environmental Statutes on CD ROM for Windows-Single User, 1999 Ed.	1999	$169
707	Federal Facility Environmental Compliance and Enforcement Guide	1999	$115
708	Federal Facility Environmental Management Systems	2000	$99
689	Fundamentals of Site Remediation	2000	$85
515	Industrial Environmental Management: A Practical Approach	1996	$95
510	ISO 14000: Understanding Environmental Standards	1996	$85
551	ISO 14001: An Executive Report	1996	$75
588	International Environmental Auditing	1998	$179
518	Lead Regulation Handbook	1996	$95
582	Recycling & Waste Mgmt Guide to the Internet	1997	$65
615	Risk Management Planning Handbook	1998	$105
603	Superfund Manual, 6th Edition	1997	$129
566	TSCA Handbook, Third Edition	1997	$115
534	Wetland Mitigation: Mitigation Banking and Other Strategies	1997	$95

PC #	SAFETY and HEALTH TITLES	Pub Date	Price*
697	Applied Statistics in Occupational Safety and Health	2000	$105
547	Construction Safety Handbook	1996	$95
553	Cumulative Trauma Disorders	1997	$75
663	Forklift Safety, Second Edition	1999	$85
709	Fundamentals of Occupational Safety & Health, Second Edition	2001	$69
612	HAZWOPER Incident Command	1998	$75
662	Machine Guarding Handbook	1999	$75
535	Making Sense of OSHA Compliance	1997	$75
718	OSHA's New Ergonomic Standard	2001	$95
558	PPE Made Easy	1998	$95
683	Product Safety Handbook	2001	$95
598	Project Management for EH&S Professionals	1997	$85
658	Root Cause Analysis	1999	$105
552	Safety & Health in Agriculture, Forestry and Fisheries	1997	$155
669	Safety & Health on the Internet, Third Edition	1999	$75
668	Safety Made Easy, Second Edition	1999	$75
590	Your Company Safety and Health Manual	1997	$95

Government Institutes

4 Research Place, Suite 200 • Rockville, MD 20850-3226
Tel. (301) 921-2323 • FAX (301) 921-0264
Email: giinfo@govinst.com • Internet: http://www.govinst.com

Please call our customer service department at (301) 921-2323 for a free publications catalog.

CFRs now available online. Call (301) 921-2355 for info.

* All prices are subject to change; please call for current prices.

Government Institutes Order Form

4 Research Place, Suite 200 • Rockville, MD 20850-3226
Tel (301) 921-2323 • Fax (301) 921-0264
Internet: http://www.govinst.com • E-mail: giinfo@govinst.com

4 EASY WAYS TO ORDER

1. Tel: **(301) 921-2323**
Have your credit card ready when you call.

2. Fax: **(301) 921-0264**
Fax this completed order form with your company purchase order or credit card information.

3. Mail: **Government Institutes Division**
ABS Group Inc.
P.O. Box 846304
Dallas, TX 75284-6304 USA
Mail this completed order form with a check, company purchase order, or credit card information.

4. Online: **Visit http://www.govinst.com**

PAYMENT OPTIONS

❑ **Check** *(payable in US dollars to **ABS Group Inc. Government Institutes Division**)*

❑ **Purchase Order** *(This order form must be attached to your company P.O. Note: All International orders must be prepaid.)*

❑ **Credit Card** ☐ VISA ☐ MasterCard ☐ American Express

Exp. ___ /___

Credit Card No. ____________________

Signature ____________________

(Government Institutes' Federal I.D.# is 13-2695912)

CUSTOMER INFORMATION

Ship To: (Please attach your purchase order)

Name: ____________________

GI Account # *(7 digits on mailing label)*: ____________________

Company/Institution: ____________________

Address: ____________________
(Please supply street address for UPS shipping)

City: __________ State/Province: __________

Zip/Postal Code: __________ Country: __________

Tel: (   ) ____________________

Fax: (   ) ____________________

E-mail Address: ____________________

Bill To: (if different from ship-to address)

Name: ____________________

Title/Position: ____________________

Company/Institution: ____________________

Address: ____________________
(Please supply street address for UPS shipping)

City: __________ State/Province: __________

Zip/Postal Code: __________ Country: __________

Tel: (   ) ____________________

Fax: (   ) ____________________

E-mail Address: ____________________

Qty.	Product Code	Title	Price

Subtotal __________

MD Residents add 5% Sales Tax __________

Shipping and Handling (see box below) __________

Total Payment Enclosed __________

30 DAY MONEY-BACK GUARANTEE

If you're not completely satisfied with any product, return it undamaged within 30 days for a full and immediate refund on the price of the product.

SOURCE CODE: BP03

Shipping and Handling	Sales Tax
Within U.S:	Maryland 5%
1-4 products: $6/product	Texas 8.25%
5 or more: $4/product	Virginia 4.5%
Outside U.S:	
Add $15 for each item (Global)	

www.ingramcontent.com/pod-product-compliance
Ingram Content Group UK Ltd.
Pitfield, Milton Keynes, MK11 3LW, UK
UKHW021036150926
475561UK00003B/131